AF843180

Ice-Houses

Ice-Houses

ENERGY, ARCHITECTURE, AND SUSTAINABILITY

ALIREZA DEHGHANI-SANIJ
Senior Research Fellow, Waterloo Institute for Sustainable Energy (WISE),
University of Waterloo, Waterloo, ON, Canada

MEHDI N. BAHADORI
Professor (Emeritus) of Mechanical Engineering,
Sharif University of Technology, Tehran, Iran

ELSEVIER

ACADEMIC PRESS
An imprint of Elsevier

Academic Press is an imprint of Elsevier
125 London Wall, London EC2Y 5AS, United Kingdom
525 B Street, Suite 1650, San Diego, CA 92101, United States
50 Hampshire Street, 5th Floor, Cambridge, MA 02139, United States
The Boulevard, Langford Lane, Kidlington, Oxford OX5 1GB, United Kingdom

Copyright © 2021 Elsevier Inc. All rights reserved.

No part of this publication may be reproduced or transmitted in any form or by any means, electronic or mechanical, including photocopying, recording, or any information storage and retrieval system, without permission in writing from the publisher. Details on how to seek permission, further information about the Publisher's permissions policies and our arrangements with organizations such as the Copyright Clearance Center and the Copyright Licensing Agency, can be found at our website: www.elsevier.com/permissions.

This book and the individual contributions contained in it are protected under copyright by the Publisher (other than as may be noted herein).

Notices
Knowledge and best practice in this field are constantly changing. As new research and experience broaden our understanding, changes in research methods, professional practices, or medical treatment may become necessary.

Practitioners and researchers must always rely on their own experience and knowledge in evaluating and using any information, methods, compounds, or experiments described herein. In using such information or methods they should be mindful of their own safety and the safety of others, including parties for whom they have a professional responsibility.

To the fullest extent of the law, neither the Publisher nor the authors, contributors, or editors, assume any liability for any injury and/or damage to persons or property as a matter of products liability, negligence or otherwise, or from any use or operation of any methods, products, instructions, or ideas contained in the material herein.

Library of Congress Cataloging-in-Publication Data
A catalog record for this book is available from the Library of Congress

British Library Cataloguing-in-Publication Data
A catalogue record for this book is available from the British Library

ISBN 978-0-12-822275-1

For information on all Academic Press publications
visit our website at https://www.elsevier.com/books-and-journals

Publisher: Brian Romer
Acquisitions Editor: Lisa Reading
Editorial Project Manager: Rachel Pomery
Production Project Manager: Prem Kumar Kaliamoorthi
Cover Designer: Christian Bilbow

Typeset by SPi Global, India

ELSEVIER Book Aid International
Working together to grow libraries in developing countries
www.elsevier.com • www.bookaid.org

This book is dedicated to my dear father, Amrollah Dehghani-Sanij, who wisely taught me how to overcome difficulties in life's experiences. May God rest his soul in peace.

Alireza Dehghani-Sanij

Contents

Preface

Today's world faces a number of ongoing and emerging challenges, such as the energy crisis, water scarcity, and social and environmental threats. Additionally, these challenges are intensifying day by day. For example, the global energy demand, which is mainly supplied by fossil fuel resources (oil, coal, and natural gas), is dramatically growing for a variety of reasons, the primary reason being the rapidly increasing population worldwide. Therefore, finding, developing and using effective, economical, and practical solutions to overcome or at least reduce ongoing and emerging crises and threats is essential.

Buildings (including residential, commercial, and office) are responsible for a significant share of the world's total energy use and escalating environmental problems (e.g., greenhouse gas (GHG) emissions, global warming, climate change, and resource depletion). On the other hand, meeting the energy demands of buildings (for power and heat) and providing good indoor environmental quality (IEQ) for the inhabitants are very important because many people spend most of their time inside buildings. Thus, optimizing the energy consumption in buildings and improving the performance of their support systems, for example, heating, ventilation, and air-conditioning (HVAC) systems, are crucial. Passive cooling and natural ventilation technologies can play a substantial role in meeting the energy needs of buildings in an ecologically friendly way and thus reducing their contribution to environmental degradation.

This book introduces a novel technology/approach that relies on various methods of daily/seasonal ice storage (e.g., heat pumps and ice-houses) to reduce the use of electricity generated from fossil-fuel power plants, and subsequently, energy consumption for the cooling, ventilating, and air-conditioning of buildings, especially at peak times. It thus helps meet inhabitants' thermal comfort requirements at the same time reducing air pollution and GHG emissions (particularly that of CO_2). This new technology/approach actually originates from natural ice-makers or traditional ice-houses, called *Yakhchals*[a] in Persian, that flourished in Iran centuries ago. In the essence ofYakhchals, which once produced ice to cool water or other beverages during extremely hot summers, lies a mystery that shows

[a] The word *Yakhchal* is a compound noun in Persian. *Yakh* means *ice* and *chal* means *pit/underground reservoir.*

the miracle of adobe and the technical know-how of the Iranian architects and engineers who established these masterpieces.

The present book is based on our earlier book in Farsi entitled *Natural and Traditional Ice Making in Iran*, published by Yazda Publications in Tehran, Iran, in April 2010. However, the authors have revised, updated, and extended that version and added almost as much new content as the existing one. This book consists of nine chapters covering different aspects of ice-houses and cold energy production, storage, and utilization. Chapter 1 discusses global energy resources (renewable and nonrenewable) and consumption, the role of energy in social, economic, and sustainable development, energy storage technologies, and the environmental impact of fossil fuel use. Chapter 2 studies and describes the natural production of ice during winter nights and the operation of traditional ice-houses used in the past centuries. Chapter 3 discusses and analyzes the methods of producing artificial ice via refrigeration systems, focusing on compression and absorption refrigeration. In Chapters 4 and 5, the history and architecture of traditional Iranian ice-houses, respectively, are explained in detail. Chapter 6 introduces and describes a new approach to producing ice using a low-energy technology in areas with cold winters and storing it in underground reservoirs for summer use (e.g., space cooling). Chapter 7 discusses and assesses the methods of cold production and storage in the form of ice and their role in cooling buildings during peak times while reducing electricity consumption. Chapter 8 describes the use of technologies such as seasonal ice storage, heat pumps, renewable energies (e.g., solar energy) to meet the cooling/heating demands of buildings. Finally, Chapter 9's display of pictures taken in different cities of Iran and also in Merv, Turkmenistan illustrate the grandeur, beauty, and variety of traditional ice houses.

The authors would like to thank the following people, who in different ways were of great help in making this book a reality: Dr. A. Sayigh, Dr. F. Fakhar Tehrani, Dr. M. Rasekhi, Dr. A. Sebt Hosseini, Dr. M.R. Khani, Dr. V. Enjilela, Dr. M. Soltani, Dr. H. Jørgensen, Dr. M.B. Dusseault, Dr. S.B. Mahbaz, Dr. A.A. Semsar Yazdi, Dr. J.S. Nathwani, Dr. Y. Leonenko, H. Massarat, M.A. Mokhlesi, N. Nazarieh, F. Moradi Kashkooli, M. Roostaie, and A. Mehrtash, all of whom studied the initial draft of the book and made valuable comments and suggestions. In addition, the authors would like to thank Dr. R. Rashidi Meybodi, M.H. Dehghan, H.R. Dehghanisanij, M.R. Dehghanisanij, S. Vesali Barazandeh, S. Mohammadrezakhani, M.R. Ameri Sefideh, H. Karrabi, S.A. Hejri, M. Goodarzi, M. Hatami, A. Shahriyar-Panah, H. Zohour Al-Ain, T. Norouz Zadeh, M. Mohammadnia, F. Rezanejad,

M. Anisi, R. Bagheri, and P. Zohour Al-Ain for supplying photographs/drawings of different Iranian ice-houses (Yakhchals). The assistance of the managers and employees of the Central Library under the auspices of the Cultural Heritage, Handicrafts and Tourism Organization of Iran, particularly S. Eshaghi, is greatly appreciated. Special thanks go to S. Tajalizadeh and J.M. McPherson who kindly helped us with the final edit of the book in terms of language editing.

No piece of writing is perfect. The authors would be very thankful if the readers of this book could send us their comments and/or pictures of other ice-houses, which still exist all over the world, and so help us to improve and enrich later editions of the book.

Alireza Dehghani-Sanij
Mehdi N. Bahadori

CHAPTER 1

Energy consumption and environmental consequences

1.1 Introduction

Energy, one of the most important and fundamental requirements of today's world, has also played a vital role in human civilization. The most basic type of energy for primitive humans was their physical strength, provided through nutrition. Later, they learned how to tame and utilize animals to perform various tasks, and to use dry wood, leaves, or plants to make fire for cooking, heating, and lighting. Until the mid-1800s, wood and agricultural residues were the world's primary sources of energy [1]. However, for the past few centuries, humans have also been able to access fossil fuel resources—primarily coal, later oil, and then natural gas. These energy sources have caused a great change in people's living standards [2]. Nowadays, the range of human uses of different energy sources (mainly fossil fuels) has expanded to the point where without them modern life would be impossible, or at least very difficult. In other words, human life today is highly dependent on energy consumption. In general, energy can be analogized as a leading dynamic engine or force driving the achievement of numerous desired objectives, such as human comfort and well-being; social welfare and development; economic growth and prosperity; and technological, industrial, and sustainable development [3–6].

Energy is conceptualized differently by different sciences, and even by different individuals such as consumers and traders [7, 8]. For example, in physical science and engineering terminology, *energy* is defined as the capability of doing work [2, 7, 9], with *work* meaning to apply a force to move, or displace, an object. In simple terms, energy is force times distance, or in more sophisticated terms, energy is "the potential effectiveness of a system on an environment [10]." Historical evidence suggests that the concept of energy was first described in the works of Aristotle (384 BC–322 BC). However, due to many changes in and rewritings of ancient texts, it is not possible to confirm exactly who first introduced the basic idea [3, 11]. Experts in history, language, and lexicography have traced the root of the

Ice-Houses
https://doi.org/10.1016/B978-0-12-822275-1.00012-6
Copyright © 2021 Elsevier Inc.
All rights reserved.

modern English word *energy* to the French word *énergie*, which is itself from the Latin root *energia*. This Latin word was derived from the Ancient Greek words *ἐνέργεια* (*enérgeia*), which means *activity* or *operation*, *ἐνεργός* (*energós*), meaning *active* or *working*, or from a combination of the two Greek words *ἐν* (*en*) and *ἔργον* (*ergon*), meaning *in work* [12–16].

Energy exists in different forms, as indicated in Fig. 1.1, and may be divided into two principal but broad groups: kinetic and potential. The former means the energy of moving objects or particles, including electrical energy, thermal energy, radiant or electromagnetic energy, motion energy, and sound energy. The latter is defined as energy that is stored in an object or substance and consists of chemical energy, nuclear energy, stored mechanical energy (e.g., compressed springs and stretched rubber bands), magnetic potential energy, and gravitational potential energy. According to the first law of thermodynamics, which describes only the quantity of energy [17], energy is never created or destroyed, but it can be transferred from one place to another or transformed into different forms. The first law is also known as the Law of Conservation of Energy.

In addition to the quantity of energy, the quality of energy is also important. In practice, this quality is related to the capability of an energy unit to be used in making products and services for individuals [18]. The second law of thermodynamics, which concerns the quality of energy, states that whenever energy is transferred or transformed, its quality and usefulness are degraded. Energy quality varies from one form to another. In other words, different forms of energy have different qualities and degrees of efficiency [19, 20] and may be classified as high-quality, medium-quality, or low-quality energy [21–23]. For instance, electrical energy [24, 25], nuclear energy [3], and some forms of chemical stored energy [25] exemplify high-quality energy. Fossil fuels fall into the category of medium-quality energy [23, 24]. Finally, biomass energy [24], ocean thermal energy, and molecular energy

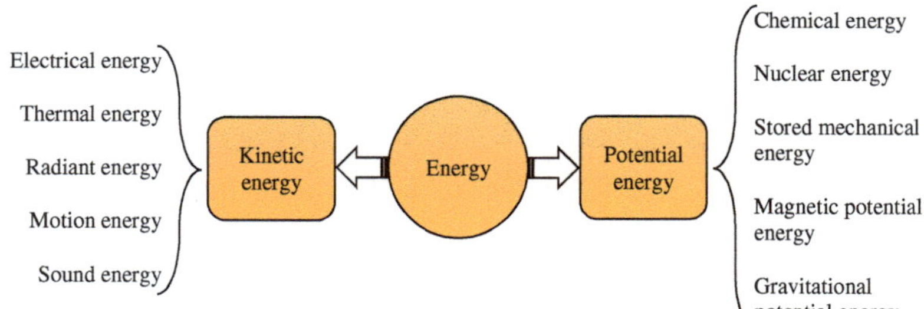

Fig. 1.1 Categorization of energy into different forms.

are examples of low-quality energy [26]. In 1851, William Thomson (1824–1907), later Lord Kelvin, analyzed and classified the quality of energy for the first time [27] and suggested the concept of the *availability* of energy, meaning that we can actually apply energy to carry out work. Thereafter, Zoran Rant (1904–1972) continued working on this concept and, in 1956, invented the term *exergy* (*exergie* in German), coined from a combination of the two Ancient Greek words, ἐξ (*ex*) and ἔργον (*ergon*), that means *from work* [28, 29]. In 1873, Josiah Willard Gibbs (1839–1903) developed the related concept of exergy [30].

The basic units employed to measure energy vary among the various systems and purposes for measurement. For instance, the units of energy in the International System of Units (SI) and the British system are, respectively, the Joule (J) and the British Thermal Unit (BTU), with 1 BTU equaling 1055.06 J. The Calorie (Cal), another unit of energy, is broadly utilized in nutrition (1 Cal = 4.184 kJ). The units of energy often used in issues related to energy production and consumption consist of the barrel of oil equivalent (BOE) and the ton of oil equivalent (TOE or toe). In the SI system, 1 BOE and 1 TOE are equal to 6119 MJ and 41,868 MJ, respectively. Additionally, there is the kilowatt-hour (kWh), an energy unit commonly employed in electrical power engineering [31]—for example, as a billing unit for electricity delivered to consumers by electric utilities [32–34]. One kWh is equal to 3600 kJ.

Energy demand and consumption in different sectors—such as industry, transport, and building—have dramatically risen globally and will continue to increase for various reasons, the most important ones being the world's swiftly growing population, urbanization, industrialization, and economic development [4, 35–39]. Currently, nonrenewable carbon-based fuels are the main sources of energy in the world, as detailed in the next section. In 1970 and 2000, for example, the world's populations were, respectively, ~3.7 and ~6.1 billion people; by 2020, the number had reached roughly 7.7 billion people [40]. It is anticipated that by 2050, the world's population will exceed 9 billion people [41, 42]. Moreover, Earth's population is not only growing in number, but also redistributing, with a general trend toward greater urbanization. Fig. 1.2 illustrates the changes in the world population between 1970 and 2020. In 1970, 37% of the global total were classified as urban dwellers, but by 2020, that group had risen to 56% [40], an increasing trend shown in Fig. 1.2. Simultaneous with population growth and urbanization, the rate of industrialization has also increased significantly over the past decades. Therefore, most countries have experienced swift economic

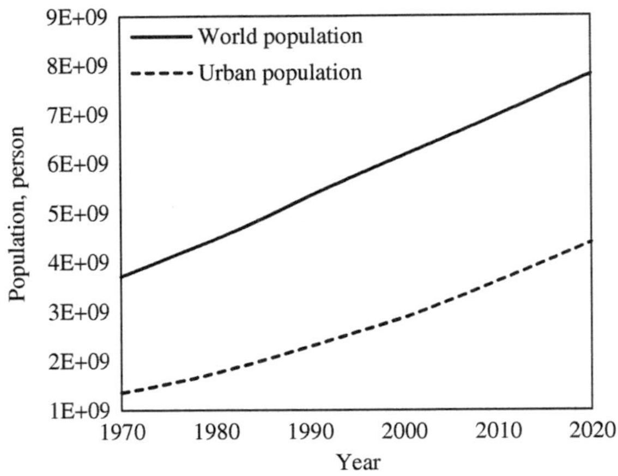

Fig. 1.2 World's population from 1970 to 2020. *(Data from Worldometer, World Population by Year, 2020. <http://www.worldometers.info/world-population> [Accessed 17 October 2020]).*

development, particularly China, which has had the world's fastest economic development since the 1980s [43–45]. The swift development of industrial economies has promoted the increasing consumption of energy, primarily generated by high-carbon fuels.

Global primary energy[a] use was $\sim 3.9 \times 10^{14}$ MJ in 2000 [47], $\sim 5.06 \times 10^{14}$ MJ in 2010, and it reached $\sim 5.84 \times 10^{14}$ MJ in 2019 [48], showing that the consumption of primary energy worldwide grew by nearly 50% between 2000 and 2019. In addition, the world's primary energy consumption per capita[b] was ~70.2 GJ per capita in 2009 and ~75.7 GJ per capita in 2019 [48], indicating a rise of around 8% over ten years. It is predicted that energy use worldwide will increase by about 50% from 2019 to 2050 [49]. With this escalation in global energy consumption comes the corresponding depletion of fossil fuel resources, with the attendant promise of future energy resource shortages and significant environmental problems that will be detailed in Section 1.4. Therefore, reductions in the consumption of carbon-based fuels and the incorporation of more green, renewable, and sustainable energy sources, along with energy storage, saving, management, and optimization, are highly desirable goals.

[a]Primary energy means a form of energy that is found in nature and has not been subjected to human-engineered transformation processes [46], such as crude oil extracted from oil fields [3].

[b]Per capita is a Latin phrase that translates to "by head," meaning the average per person.

1.2 Global energy resources and consumption

The energy needed by the world can be provided from two main but broad categories of resources [3, 9, 50]: nonrenewable (or conventional) and renewable (or alternative). The former consists of all types of fossil fuels—such as oil (also called petroleum), coal, and natural gas—and radioactive elements, including uranium, radium, and thorium [3, 51]. The former category is termed nonrenewable because the resources are extracted from nature, whose supplies are not infinite, and their consumption rate is much higher than the time needed for their replenishment or recovery [52, 53]. The latter category will be detailed in Section 1.5. Fig. 1.3 displays worldwide energy production from various resources (both renewable and nonrenewable) from 1970 to 2018, and a projection to 2040, indicating that nonrenewables—mainly fossil fuels—were the primary sources of energy supply in the world up to 2018, and most likely will continue to be for at least a few decades more. However, renewable energy resources are starting to make a substantial contribution to meeting the world's future energy needs (Fig. 1.3).

Based on a report published by the BP Statistical Review of World Energy (SRWE) in 2020 [48], oil[c] is still the main fuel for primary energy

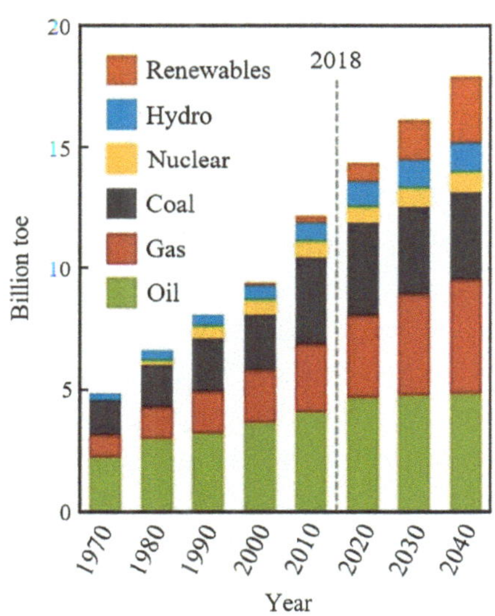

Fig. 1.3 World energy production from renewable and nonrenewable sources. *(Modified from B. Dudley, BP Energy Outlook: 2019 edition, p. 141, BP p.l.c., London, UK; 2019).*

[c]In the 20th century, oil played the most important part in meeting global energy needs [26].

Table 1.1 Share of the world's fossil fuels in primary energy production.

Resources	Production			Share of primary energy (%)
	2000	2010	2019	2019
Oil[a] (thousands of barrels per day)	74,893	83,409	95,192	33.1%
Coal[b] (exajoules)	98.49	150.82	167.58	27.0%
Natural gas[c] (billion cubic meters)	2413.4	3146.2	3989.3	24.2%

[a] Contains all types of oil.
[b] Commercial solid fuels only.
[c] Excludes gas flared or recycled.
Data from B. Dudley, BP Statistical Review of World Energy June 2011, p. 45, BP p.l.c., London, UK; June 2011; B. Looney, Statistical Review of World Energy 2020, 69th edition, p. 65, BP p.l.c., London, UK; June 2020.

production worldwide, as shown in Table 1.1. A number of Middle Eastern countries—Saudi Arabia, Iran,[d] Iraq, the United Arab Emirates (UAE), Kuwait, and Qatar—are the leading oil producers in the world, responsible for ~47.5% of global total production at the end of 2019 [48]. Currently, the United States (19.7%), China (14.3%), and India (5.4%) are the largest consumers of oil worldwide. Coal is now the second major fuel for the generation of primary energy globally, accounting for about a quarter of total primary energy production (Table 1.1). In 2019, the United States (23.3%) and Russian Federation (15.2%) held the majority of the world's coal reserves, but China was the main producer and consumer of this source of primary energy, with 47.6% and 51.7% of the total share, respectively [48].

According to the BP SRWE [48], in 2019, the Russian Federation, Iran, and Qatar had the highest proven gas reserves, with 19.1%, 16.1%, and 12.4% of the global share, respectively, totaling a combined share of ~47.6, but the United States was the world's largest producer of natural gas, with a ~23.1% share. The United States (21.5%), Russian Federation (11.3%), and China (7.8%) are currently the world's greatest consumers of natural gas. The contribution of nuclear energy in producing global primary energy was 4.3% in 2019, and the United States (30.5%), France (14.3%), and China (12.5%) were the main consumers of this type of energy [48]. The data provided by the BP SRWE in 2019 show that nonrenewable resources were used to produce ~88.6% of the total share of primary energy worldwide.

[d]Currently, Iran's oil production has fallen sharply because of US sanction.

A notable point is that energy-supply security concerns are associated with fossil fuels for a variety of reasons, such as their uneven international distribution; high costs for exploration, exploitation, transportation, and storage [36, 54–56]; political and environmental issues [57]; price fluctuations; and even weather impacts (for instance, icing problems on marine vessels and offshore structures during the cold months of the year) on supply and transportation [58–61]. Most developed and developing economies cannot meet their domestic energy demand without massive fossil fuel imports, using a process that is complex and has the potential for many problems. Transitioning to inherently local renewable energy sources would avoid or alleviate the risks related to complex international supply chains and political issues.

1.3 The role of energy in social, economic, and sustainable development

No one can deny the significance of the energy sector in each country's prosperity. For the last two centuries, nonrenewable energy resources—largely oil, coal, and natural gas, and to a much lesser extent other nonrenewables like uranium—have been governing modern economic development. Crude oil could replace coal as the main source of energy due to its exceptional qualities, such as liquidity, easy transportation and storage, as well as high energy intensity [62, 63]. In fact, crude oil—also called "black gold"—has long been considered the invisible engine of growth, especially since the middle of the 20th century. This raw material plays an important role in the lives of many people, not only in terms of the energy that it provides, but also because of its contributions to many human endeavors, such as manufacturing, food production, and even medicine. Oil has been an important factor in mankind's history and development, plus in the Earth's geological and biological evolution and history [64]. The International Energy Agency (IEA) has estimated that fossil fuels will most likely remain the dominant energy source used to satisfy human demands until 2035 [65].

The rapid growth of the modern economy started with developments and advancements in the field of transportation, initiated by the invention of the combustion engine [64, 66]. These engines enabled the automotive and airplane industries to progress, creating many new and fast means of transport for both goods and people. In fact, the swift and constant development of inventions and innovations in the energy sector has caused rapid progress in developed countries, driven by their quest to control natural forces, increase mobility, and boost productivity [64, 67, 68]. For example, early harnessing of wind energy (e.g., for sailing and windmills), animal power, and hydro energy (e.g., in the watermills used by monasteries and villages) gave way to the superior technologies of water wheels and turbines, which for many years attracted more attention than steam engines [67, 68].

More recently, the rise in energy demand and the scarcity of energy resources have induced the countries with rich underground oil and gas reservoirs to begin extraction and exploitation, and thus accelerate their economic development. As an example, considerable oil deposits discovered off the coast of Equatorial Guinea in 1996 made the country the third largest exporter of oil in Sub-Saharan Africa—producing 360,000 barrels per day in 2004 [64]. After nearly three years, the average per capita income of this country passed USD 20000, placing Equatorial Guinea among high-income economies. Despite this rapid growth in the economy, the country has the average life expectancy of ~50 years, a relatively high infant mortality rate for Africa, and a poverty line of USD 2 a day, below which many of the population live [64] because the wealth from oil is held by a powerful few. At the national level, oil exports have offered Equatorial Guinea an opportunity to relax its balance-of-payments constraints.

1.3.1 The role of energy in social development

Energy directly affects economic transformation and social well-being. For instance, heat, light, and power are considered crucial for building or running factories and plants, irrigating crops, and supporting education and health services. Secure, available, reliable, and cost-effective energy services have benefited industrialized economies and underpinned their development and prosperity [69, 70]. The absence of modern energy services, which is a serious barrier to social and economic development, must be overcome in order for all countries to improve their human development, including their productivity, gender equality, health and safety, and education [71, 72].

According to Table 1.2, in 2009, globally, ~1.4 billion individuals did not have access to electrical energy, and ~2.7 billion individuals employed traditional biomass methods for cooking [65, 73]. Lack of access to electrical energy is projected to remain a problem at the global level well into the future, although with a slight reduction in the number of people affected. For instance, it is estimated that ~1.2 billion individuals will lack access to electrical energy worldwide in 2030, with about 87% of this number living in rural regions [73, 74]. Those affected will mainly reside in Sub-Saharan Africa, India, and other developing Asian countries except China. Similarly,

Table 1.2 Number of individuals (in millions) with no access to electrical energy, and relying on traditional biomass-fueled cooking in 2009 [73].

Region	Number of people with no access to electrical energy	Number of people who use traditional biomass fuels for cooking
Africa	587	657
Sub-Saharan Africa	585	653
Developing Asia	799	1937
China	8	423
India	404	855
Other countries in Asia	387	659
Latin America	31	85
Developing countries[a]	1438	2679
Worldwide[b]	1441	2679

[a]Includes Middle East countries.
[b]Contains Organization for Economic Cooperation and Development (OECD) and transition economies.

Table 1.3 Individuals (in millions) without access to electrical energy by world region in 2009 and 2020 [73].

	2009			2020
Region	**Rural**	**Urban**	**Total**	**Total**
Africa	466	121	587	644
Sub-Saharan Africa	465	120	585	640
Developing Asia	716	82	799	650
China	8	0	8	2
India	380	23	404	342
Other countries in Asia	328	59	387	307
Latin America	27	4	31	16
Developing countries[a]	1229	210	1438	1350
Worldwide[b]	1232	210	1441	1352

[a]Includes Middle East countries.
[b]Contains OECD and transition economies.

the number of individuals who rely on biomass as a traditional heat source for cooking will increase to ~2.8 billion people in 2030, with 82% of them living in rural regions [73–76]. As illustrated in Table 1.3, over one billion individuals (~1.35 billion) currently lack access to electrical energy globally, a slight decrease from 2009. Many of them (~85%) reside in rural regions, mostly in Sub-Saharan Africa and India [73].

Achieving worldwide access to modern energy conflicts with the need to bring down the emission of greenhouse gases (GHGs), which is caused largely by high-consumption lifestyles and has become worse with today's access to relatively cheap fossil fuels and the associated technologies for utilizing them. Unfairly, the numerous negative impacts of high energy use and affluent lifestyles are mainly felt by the societies in which most energy demand is satisfied by mining and processing hydrocarbon resources, and their populations are principally rural, poor, and powerless [77]. The Niger Delta of Nigeria, for example, has experienced oil spills and flares that have caused massive local contamination, loss of life and livelihoods, plus spiraling poverty and corruption [78]. Sub-Saharan Africa is considered the greatest challenge as only 31% of its population have access to electrical energy, the lowest level worldwide. If we exclude South Africa, the share drops to 28%, which is approximately equivalent to the residential electricity consumption of New York with its ~19.5 million inhabitants [73, 74].

Current demographic, economic, social, and technological trends are seriously challenging the global energy system's long-term sustainability [74, 76, 79]. If appropriate policies are not implemented by governments and those already in place are not followed, by 2030, it is anticipated that [65, 73].

- Energy use will rise by more than half (53%);
- Fossil fuels will dominate the energy mix (80% share);
- Global energy-related CO_2 emissions will rise by more than half (55%); and
- The number of individuals with no access to electrical energy or modern cooking and heating methods due to poverty will grow.

The main goal of national energy-development activities is to improve the quality and quantity of modern energy and to determine how improvements will affect the different framings of energy access. In September 2015, world leaders at the United Nations (UN) adopted targets known as the "17 Sustainable Development Goals (SDGs)," for integrating related economic, social, and environmental aspects and recognizing their interlinkages in achieving sustainable development. In view of the need for better energy access and the SDGs, there is agreement that multiple sustainability objectives must be adopted [80]. In line with this thinking, Rao and Pachauri [81] have emphasized the provision of access to electrical energy, clean cooking energy, and improved water and sanitation as the vital factors in enhancing people's living standards. Their assessments stress that the countries of Sub-Saharan Africa would have to significantly improve their populations' energy access to attain the goal of universal access by 2030. Additionally, the quality and quantity of energy services in these regions would need to be far higher than at present, as energy is the key factor in achieving several other goals, such as those associated with the health, education, and production sectors.

In addition, another ongoing challenge, mainly in developing countries, is cooking systems. More than three billion people around the world use polluting and inefficient cooking systems that are associated with health hazards, gender inequality, and harm to local environments and the climate system, mostly from deforestation and black carbon emissions [71]. A major health challenge arises from the inefficient combustion of cooking fuels in low-income countries. The resulting smoke and fumes and indoor air pollution in fact lead to suffering and death [82], and women and children face injury and personal security threats through physically looking for, gathering, and carrying fuels. Furthermore, unsustainable biomass harvesting for combustion and charcoal production directly connects to deforestation, land use problems, and net GHG emissions [83].

The SDGs of the 2030 Agenda identify electrification as a major parameter that significantly influences a country's Human Development Index (HDI). For instance, in 2003, Brazil launched a program to electrify ten million households in rural areas within a five-year period, thereby raising its Municipal Human Development Index (MHDI), which is based on income, education, and health [84].

1.3.2 The role of energy in economic development

Without thermal energy, light, and power, we could not build or run the factories and cities that provide goods, jobs, and homes, and we would not enjoy the facilities and services that make life more comfortable and pleasant. The energy industry affects economic growth in two ways [85–87]: (1) as an important sector of the economy by extracting, transforming, and distributing energy goods and services throughout the economy, thus creating new jobs and value in the long run, and (2) as an input for nearly all goods and services underpinning the rest of the economy. In fact, energy can be considered as the oxygen and lifeblood of the economy and its growth, particularly at the mass industrialization level that causes the increase in per capita gross domestic product (GDP) in emerging economic giants [85, 88]. For instance, the American Petroleum Institute (API) estimates that the industry directly and indirectly supports more than nine million jobs in the United States, that is, over 5% of the country's total employment [85, 89]. Furthermore, the energy industry in 2009 supported a total value of more than 1 trillion USD added to the national economy, representing 7.7% of the US GDP, of which 4% was accounted for by the energy industry. The share is even higher in countries that are heavily dependent on energy exports, such as Nigeria (30%), Venezuela (35%), and Kuwait (57%) [85, 90].

Beyond the direct connection of energy to the economy, it is also associated with other sectors in indirect ways that are not immediately obvious. For instance, the preparation of each Calorie's worth of food requires an average of five Calories of fossil fuel. This situation may rise to an average of 80 Cal for high-end products such as beef. The biggest industrial user of freshwater (40% of all freshwater withdrawals in the United States) is the energy sector [85, 91]. The energy sector, labor, and the economy are directly interconnected. That is, the industry employs labor and capital to produce goods and services that drive the economy, and capital and labor are utilized to produce energy, which also drives the economy. High priorities of economic growth and job creation around the world make this connection more important [85, 92–94]. Fig. 1.4 indicates the share of the energy sector in the business GDP, accompanied by other industries, in a number of OECD countries.

Fig. 1.5 displays the share of employment created by the energy sector in comparison with the shares of other sectors in a number of OECD countries. In Norway, for example, 20% of the business sector GDP and only

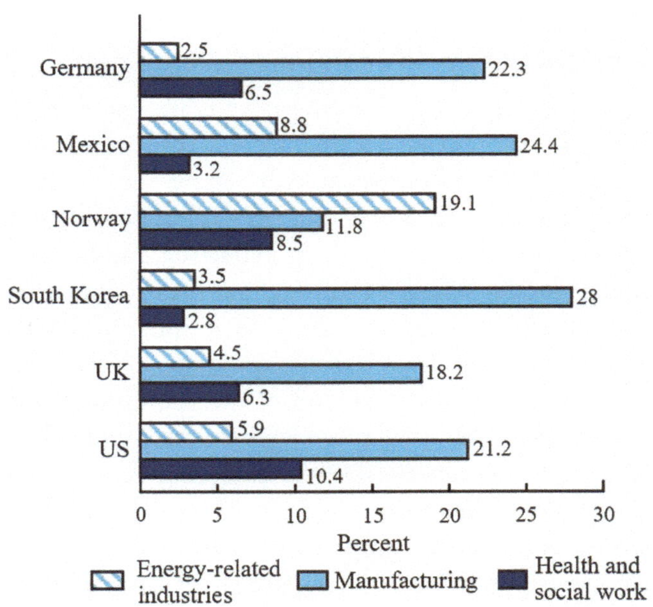

Fig. 1.4 Share of the energy sector in business GDP in comparison with other industries in a number of OECD countries [85, 94].

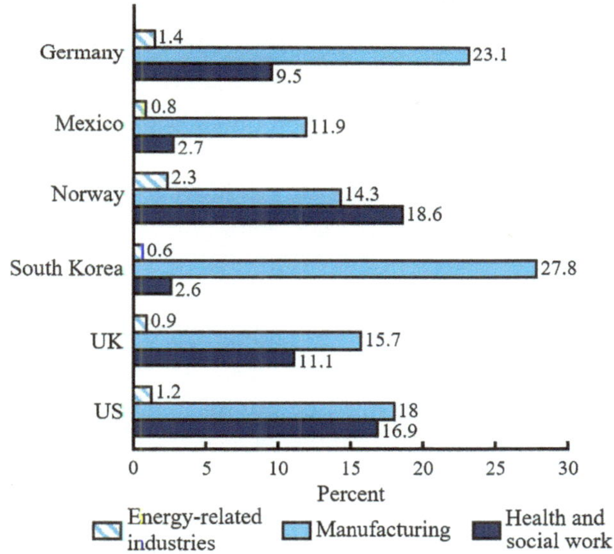

Fig. 1.5 Share of the energy sector in business employment in comparison with other industries in a number of OECD countries [85, 94].

2.3% of business sector employment are accounted for by energy-related industries. Oil may constitute a great share of Norway's wealth, but it should be noted that other sectors, particularly service industries, are supported by this wealth. For instance, the number of individuals who work in health care is about eight times the number who work in energy extraction [85, 94]. However, several studies have demonstrated that the energy industry in the United States supports many more jobs than it directly creates because the country's or the industry's long supply chains require many employees and suppliers [85, 95].

In addition to the role of the energy sector in national economies, relatively lower and more stable energy prices can help stimulate other contributors to the economy. Lower energy prices not only decrease costs for consumers and businesses, but also reduce the expenses for nearly all goods and services. This reduction will in turn increase the disposable income that can be spent in other ways and make the goods and services more affordable.

1.3.3 The role of energy in sustainable development

The whole structure of society as we know it today would collapse without energy. For instance, imagine how a 24-h cut in electrical energy supplies affects a city. It would be very destructive, as the lights would go out, computers and lifts stop operation, and hospitals fall to a lower level of care and maintenance. Just these few examples represent how dependent we are on this particular form of energy. As mentioned earlier, energy demand increases with the growth of population, much faster than the average 2% [96]. As a result, high-consumption lifestyles and enhanced energy demands will rise together, along with the wealthy industrialized economies that are a relatively big consumer of the world's energy supply (~75%), constituting about 25% of the population of the world [97].

Energy supply and use are connected not only to global warming, but also to other environmental challenges and threats, which will be described in the next section. These issues need to be considered in order to attain a sound energy future accompanied by minimal environmental effects, although the global demand for energy is projected to rise continuously until 2050 as both populations and industries grow [96]. Unless changes are made to how we meet the rising worldwide energy demand, current environmental degradation and climate change problems will be exacerbated. One solution that can avoid the impending energy scarcity and benefit the environment is to employ renewable and sustainable energy sources and technologies, as detailed in Section 1.5.

To paraphrase the 1987 definition from the World Commission on Environment and Development (WCED), sustainable energy development can be defined as satisfying the energy requirements of the current generation while preserving global energy resources for future generations [98, 99]. Various levels of economic development, politics, religion, culture, and customs all around the world are based on different opinions about energy sustainability. People in the most developed countries tend to use environmentally friendly technologies and products due to high levels of economic development, modern education systems, and political maturity in their societies. On the other hand, countries with very low or medium levels of economic development have many problems and do not have the appropriate infrastructure for energy-sustainability initiatives. Although some citizens in these countries are aware of environmental problems thanks to their education, and thus support sustainability development, the poor conditions in their regions' hamper further advancements in energy sustainability [64].

Many factors contribute to achieving sustainable development. The main objective is to meet the demand for sustainable energy supplies that are accessible at reasonable cost, readily available for all required tasks, and that do not have negative impacts [100–102]. Clearly, there is an intimate connection between green, renewable, energy sources and sustainable development. Fossil fuels, uranium, and other nonrenewable sources are normally acknowledged to be finite. However, sunlight, wind, falling water, and other renewables are commonly considered to be eco-friendly and sustainable over a relatively long time. In addition, waste that is convertible to useful energy forms and biomass fuels can also be viewed as sustainable energy sources depending on how sustainable is defined [102].

Of course, sustainability itself must be pursued with caution. For instance, environmental challenges and threats can considerably affect sustainable development. For example, activities that cause health, ecological, and other problems over time continually degrade the environment and are not sustainable. Moreover, because a large portion of environmental impacts are connected to the use of energy resources, reaching sustainable development in a society requires that only energy resources that cause no environmental problems should be applied [103]. Overall though, the shift to employing renewable energy sources is in line with sustainable development and can provide a cleaner system than is possible by tightening controls on conventional energy.

To sum up, a number of activities will affect a society's move toward sustainable development. The following are among the most important [104–109]:

- Increasing public awareness through the media and public and professional organizations;
- Employing public and government channels to inform the public about energy utilization; environmental effects; and green, renewable energy resources;
- Providing environmental education and training to the public through specialized agencies and training facilities;
- Developing innovative energy strategies to achieve an effective sustainable energy plan based on the latest information and methods;
- Utilizing renewable energy resources to attain eco-friendly sustainable energy programs;
- Improving financing as a very important tool for reaching the main goal and accelerating the acceptance and utilization of renewable energy systems and technologies for sustainable energy development;
- Monitoring and evaluation on an ongoing basis to assess program success.

1.4 Environmental impacts of fossil fuel use

Today, one of the most serious challenges and threats the world faces is dealing with the environmental impacts of using nonrenewable, carbon-based fuels. Extensive and prolonged reliance on these fuels has contributed significantly to widespread environmental degradation. Among the consequences are pollution of the air, water, and land; diminishing biodiversity among both plants and animals; and a range of human health problems [4, 110–112]. Since the beginning of the industrial revolution in the 18th century, the extraction and burning of fossil fuels for energy production and other uses have substantially raised atmospheric concentrations of GHGs—particularly the primary one, carbon dioxide (CO_2)—and air pollutants (e.g., particulate matter) [113–115]. The vast majority of these emissions and pollutants, however, have been released since the end of World War II (1945) due to increasing materialism driving industrialization and fuel consumption. Fig. 1.6 schematically represents the processes of energy production, distribution, and consumption, as well as the energy waste and environmental pollution associated with these processes.

Yearly anthropogenic CO_2 emissions have grown around twofold in the last four decades [116–118]. For example, global fossil fuel CO_2 emissions were about 3.1 gigatonnes carbon per year (Gt C yr^{-1}) in the 1960s [119] and ~6.4 Gt C yr^{-1} in 1995, reaching ~9.8 Gt C yr^{-1} in 2013 [120, 121].

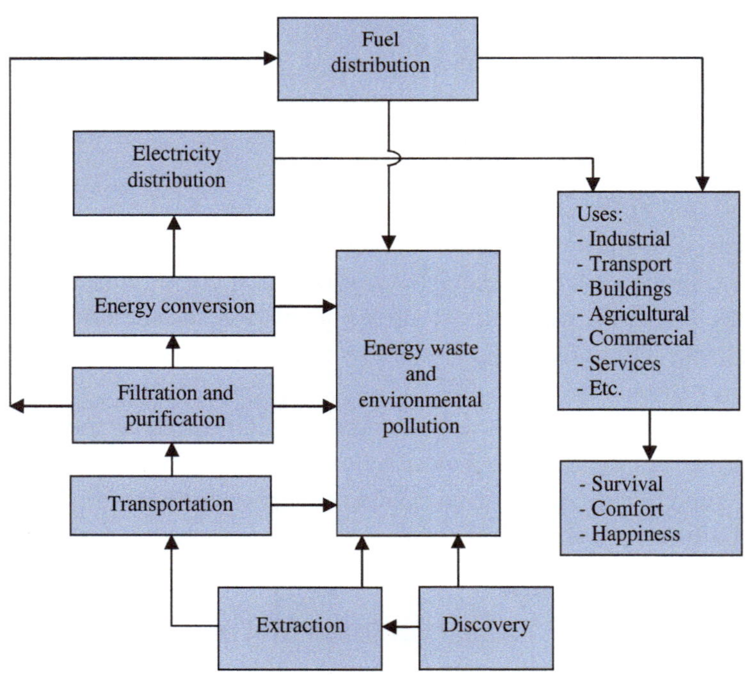

Fig. 1.6 Schematic diagram of processes of production, distribution, and consumption of energy, plus their associated energy waste and environmental pollution.

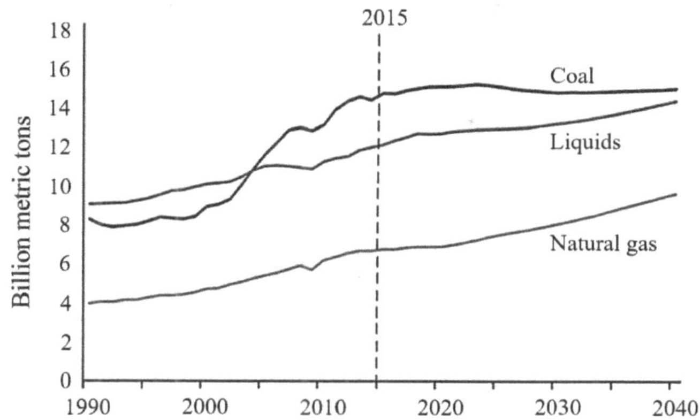

Fig. 1.7 History of CO_2 emissions by type of fossil fuel from 1990 to 2015, and predicted future trends up to 2040 [113, 126].

Additionally, the atmospheric concentration of CO_2 was about 277 parts per million (ppm) in 1750 [122], reaching ~280 ppm in 1850 [123]. Thereafter, it rose to ~360 ppm in 1995 and above 400 ppm in 2015 [120, 121], before reaching ~414.7 ppm in 2019 [124]. Atmospheric CO_2 concentrations will most likely continue to increase over the coming decades despite scientists' warnings and political commitments to address the problem [123, 125]. Fig. 1.7 shows the history and anticipated future of CO_2 emissions by different types of fossil fuels. In this figure, the term "liquids" refers to conventional petroleum liquids—such as conventional crude oil, natural gas-plant liquids, and refinery gain—as well as unconventional liquids (including biofuels, bitumen, coal-to-liquids, coal- and biomass-to-liquids, gas-to-liquids, extra-heavy oils, and oil shale). One point of interest is that reduced industry and travel during the COVID-19 pandemic caused CO_2 emissions to drop significantly in 2020. However, they will certainly rise again as the global economy recovers [127]. China and the United States of America are currently the world's major emitters of CO_2 [48, 125], in 2019, respectively, accounting for 28.8% and 14.5% of total CO_2 emissions [48].

Increasing GHGs emissions have led to many catastrophic environmental consequences such as melting polar ice; rising sea levels; the expansion of seas and oceans; acid rain; atmospheric disturbances; global warming; and climate change [113, 114, 128–130]. Table 1.4 lists a variety of pollution types caused by energy conversion and consumption at the local, regional, and global levels. The harmful effects of different types of pollutions on human health and the environment are illustrated in Table 1.5. It is predicted that, by 2050, global warming and climate change will have forced about

Table 1.4 Various pollutions resulting from conversion and consumption of energy.

Local level	Regional level	Global level
– Air, water, and soil pollution – Destruction of pastures, forests, and natural landscapes – Loss of peace and silence – Various forms of radiation	– Acid rain, waterbody pollution – Earthquakes due to large dam construction – Oil spills into waterbodies – Dispersion of radioactive materials in nuclear accidents	– Greenhouse effect due to emissions of carbon dioxide, methane, and other gases, leading to rising temperatures, climate change, and adverse environmental consequences

Table 1.5 Adverse impacts of various forms of pollution on human health and the environment [131–137].

Air pollution in cities	Eye irritation, shortness of breath, respiratory diseases, lung cancer, damage to plant life
Acid rain	Pollution of rivers, lakes, and groundwater, as well as loss of aquatic life Damage to forests and agricultural products Damage to buildings and other structures (e.g., bridges), equipment, clothing, etc.
Greenhouse gases (GHGs) emissions	Increasing global temperatures, rising sea levels, greater probability of hurricanes and tornadoes, changes in precipitation trends, changes in agricultural production, harmful effects on human health

150–200 million people to leave their homes—either to move elsewhere in their own countries or to other countries entirely. This migration of so-called environmental or climate refugees will bring widespread political and social turmoil as well as other consequences [138–141]. Therefore, swift and basic changes are essential in all sectors of carbon emitters worldwide, especially the energy sector [142], to achieve a cleaner environment and sustainability. These changes will require new, government-enforced policy frameworks at the global level and the deployment of technology [6].

Effective and practical solutions exist for overcoming or minimizing the global environmental challenges and threats posed by the use of fossil fuels. Among the solutions are to improve energy efficiency in all sectors; manage and optimize energy in all systems; save energy in all its forms; capture CO_2 by various methods and store it underground (e.g., in deep, porous geological formations); utilize renewable, carbon-free, or low-carbon energies; and store energy in various ways and forms. The latter two are explained next.

1.5 Renewable energy resources: Current status and future prospects

In this section, different aspects of renewable energy sources are discussed. Renewable energy can be defined as a type of energy derived from natural resources that can recover at almost the same or a faster rate than their consumption rate. Scientists have already introduced different types of renewable energies, directly or indirectly derived either from the sun or from the heat produced in regions deep inside the Earth. In other words, clean, green, and renewable energy can be directly or indirectly produced from solar, wind, geothermal (shallow, deep, or a combination of both [36, 143]), hydro, hydrogen, ocean, or marine resources (e.g., current, tidal, ocean thermal energy [144]), solid biomass (e.g., wood, agricultural and forestry residues, and municipal solid waste [145]), biogas, and liquid biofuels. Note that biomass can only be considered a renewable energy if its consumption rate is not higher than its regeneration rate [146]. Many technologies and equipment related to energy generation have been created and expanded over time to produce usable energy from these natural and eco-friendly resources, in the form of electrical energy, thermal energy, and industrial heat for various purposes such as space and water conditioning, and fuels for transportation [147, 148].

Despite the fact that both renewable and conventional energy sources can help meet future energy demand, the use of these sources will expand unequally. It is expected that renewable energy sources use will more than quadruple by 2040 [38, 149]. In other words, the share of clean, renewable, and sustainable energy sources will be larger than that of nonrenewable ones in the world's future energy supply, as technological advancements will increase efficiency, expand technical capabilities, and reduce the unit energy costs of these sources, thereby enhancing their competitiveness against conventional energy sources such as high-carbon fuels [150].

The main renewable and sustainable energy sources are currently hydro, wind, solar, geothermal, and biomass [3, 4, 118]. For example, hydropower provided the largest share of electricity production (~71%) among other renewables in 2016 [151], and other sources of renewable and sustainable energy contributed a lower share of global energy. For instance, all forms of bioenergy in 2016 met ~10% of the world's energy supply [151]. In addition, by 2016, only about 1% of all electricity consumed worldwide was provided by solar energy, but this particular source of energy will have excellent prospects as the cost of solar systems and their installation is continually dropping, accompanied by improvements in their efficiency [151, 152]. At the end of 2015, wind energy produced ~435 GW of power, and its use is expected to grow 5.5% annually, reaching ~977 GW by 2030 from both onshore and offshore wind farms [151, 153]. Note that power generated by the sun and wind has been largely nonuniform around the world because of those sources' natures, which are highly dependent on geographical and climate conditions. Only ~0.5 GW of commercial energy in 2016 was produced by marine and ocean energy, but the amount is anticipated to rise up to ~1.7 GW in the near future [151, 154]. Geothermal energy was responsible for less than 1% of worldwide electricity production in 2016 and could provide a small portion of the primary energy used globally [151]. For example, in 2015, the geothermal sources generated output equal to ~75 TWh for thermal energy and the same amount for electrical energy at the global level [151, 155].

Worldwide, the clean and sustainable energy sectors are increasingly attracting investors, as well as government support in the form of policies introduced by nations to secure their energy supply and diversify their energy mix, plus stimuli such as tax credits for renewable energy supplies, discounts for installing renewable energy generators, renewable energy portfolio measures, and markets for renewable energy certificates [156–158]. The correlation between renewable energy demand and economic development confirms the importance of designing and introducing appropriate national environmental and energy policies [159, 160]. Ideally, national policymakers need to develop harmony among energy, the environment, and the economy driven by suitable strategies.

Many developing countries, much like developed ones, have decided to reduce their high dependency on fossil fuels by investing more in renewable and environmentally friendly energy sources. Global investments in clean and sustainable energy by developing countries have increased from 18% in 2004 to 42% in 2018 [159]. China, Brazil, and India have the highest investment in green and renewable energy among these nations, as well as ranking, respectively, first, fifth, and eighth among all the countries of the world, and representing approximately 37% of global clean energy investments [159]. Currently, several countries in the world use a certain type of renewable energy to meet a substantial portion of their energy needs.

Table 1.6 Use of various renewable energy sources in EU-28 countries in 2015, and predictions for 2020 [162].

Source of technology	Renewable energy source use by Eurostat in 2015 (kilotonnes)	Renewable energy source use by NREAPs in 2020
Hydropower	29,858	31,786
Onshore wind	20,843	30,303
Solid biomass	85,396	94,346
Solar photovoltaic	8669	7062
Biogas	8279	10,601
Offshore wind	3784	11,740
Geothermal	1244	3589
Concentrated solar power	469	1633
Bioliquids	740	5512
Tidal, wave, and ocean energy	41	559
Solar thermal	2004	6455
Renewable energy from heat pump	9697	12,289
Biodiesels	11,427	20,920
Biogasoline	2622	7324
Other biofuels	168	746
Compliant biofuels	13,239	28,989
All biofuels	14,217	28,990

For example, in 2019, Denmark produced about 47% of its electricity from wind [161]. Table 1.6 indicates the rapid growth of renewable energy implementation in the EU[e]-28 countries based on the data provided by Eurostat and the National Renewable Energy Action Plan (NREAP) for 2015 and 2020, respectively. Table 1.7 shows the share of the various renewable energy sources each EU-28 country employed in 2015 and those predicted for 2020.

In some countries (for instance, Canada and the United States), government policies, such as on the construction of nuclear reactors, have already reduced the carbon footprint of global energy systems [163]. However, this type of energy is very costly and comes with high security hazards [164]. Recent policies in Europe, the United States, China, and India support and have expanded the use of wind and solar photovoltaic (PV) systems [165, 166]. Wind power, whose development and expansion started in the

[e]European Union.

Table 1.7 Share of EU-28 member states' use of total global renewable energy sources in 2015, and predictions for 2020 [162].

	Share of total renewable energy		
EU-28 member states	**2015 (%)**	**2020 (%)**	**Point(s) off/on target**
Sweden	53.9	49	−4.9
Latvia	37.6	40	2.4
Finland	39.3	38	−1.3
Austria	33	34	1
Denmark	30.8	30	−0.8
Croatia	29	20	−9
Portugal	28	31	3
Estonia	28.6	25	−3.6
Romania	24.8	24	−0.8
Lithuania	25.8	23	−2.8
Slovenia	22	25	3
Bulgaria	18.2	16	−2.2
Italy	17.5	17	−0.5
Spain	16.2	20	3.8
Greece	15.4	18	2.6
France	15.2	23	7.8
Germany	14.6	18	3.4
Czech Republic	15.1	13	−2.1
Slovakia	12.9	14	1.1
Poland	11.8	15	3.2
Hungary	14.5	13	−1.5
Cyprus	9.4	13	3.6
Ireland	9.2	16	6.8
Belgium	7.9	13	5.1
United Kingdom[a]	8.2	15	6.8
Netherlands	5.8	14	8.2
Malta	5	10	5
Luxemburg	5	11	6

[a]The United Kingdom formally left the EU on January 31, 2020.

late 1990s, now supplies over 5% of global power [167]. For example, it currently accounts for about 4% of total electricity production in Canada [168]. Biofuel energies have developed continuously to now, meeting 3% of today's global transport energy demands, driven energy policies, and production targets in Brazil, the United States, and the EU [169]. According to the IEA, in 2019 [170], the energy produced from the burning of biofuel and urban waste increased by 47% from 1990 (902,367 ktoe) to 2017 (1,329,064 ktoe).

By the end of 2018, China had installed around 729 GW of renewable energy capacity out of approximately 1900 GW of total installed capacity [171]. During 2018, 26.7% (1868 TWh) of total electrical energy produced (6994 TWh) in China was supplied from various types of renewable energies [171]. To highlight China's transition to renewables, in 2007, coal accounted for 81% of its electricity generation, whereas coal produced only 59% of its total energy in 2018, despite an 8.5% rise in electric energy production [171]. The advancement of China in responding to its energy demand by means of renewable energy sources can be extended to all countries in the world, as the new energy technologies that helped China in achieving its energy goals can also be developed and adopted in other countries in the upcoming years. Three important results associated with the use of ongoing renewable energy in China are as follows:

- Coal-fired thermoelectric plants will be removed from the market as their operation hours drop. Government prioritizing of renewable electricity to supply its power grid resulted in financial losses of about 47% for China's coal-power companies from January to August 2018, with restricted (1.1%) profitability recorded for others [171, 172].
- New solar PV energy generation will gradually make all thermal plants involved in energy production obsolete. In May 2019, the first batch of 250 unsubsidized grid-parity wind and solar projects was approved for 16 provinces in China, with the goal of producing about 21 GW of energy. Large-scale government initiatives like the above have not only been observed in China but also regularly reported in other countries and have led to significant development of wind power and solar PV generation in, for instance, Italy [173], Germany [174], and Spain [175].
- Continuous development of renewable power generation, particularly of solar PV, in most parts of the world, has reduced the price of electricity. This effect will eventually accelerate the expansion of electric vehicles (EVs), which will significantly reduce oil use, as transportation accounts for about 50% of yearly oil use worldwide [176, 177]. Recent reports reveal that, due to the adoption of EVs, the world's cumulative yearly oil demand has dropped since 2011, by around 352,000 barrels per day (b/d), roughly equal to Portugal's yearly oil use [178]. It is estimated that EVs could displace 7.3 million to 13.7 million b/d of oil demand by 2040, as the production of lithium-ion (Li-ion) batteries is predicted to more than quadruple (e.g., ranging from 294 GWh in 2018 to 1235 GWh in 2023) [179, 180]. China is increasingly investing in the production of both EVs and their charging stations.

India, the second most populous country in the world, produced 33.8 GW of electrical energy from installed PV solar power between 2015 and 2019 [181]. In addition, around 77.64 GW (~21.8%) of the country's total installed power capacity of 356.07 GW was being generated from renewable energy sources by March 2019 [182]. India now produces about 36.62 GW of wind power annually and is ranked as the country with fourth greatest wind power production worldwide. Moreover, the country is constructing the world's largest solar power plant, with ~2.25 GW of final capacity, in the Bhadla Solar Park of Rajasthan [171].

The dominance of renewable energy sources, plus their considerable potential for exploitation and development in the energy industries of both developed and developing countries herald the forthcoming new energy era [183]. Europeans welcome possible reductions in their need to import energy from nonrenewable energy sources, mainly supplied by Russia. Switching to renewable energy sources, preferably local, will strengthen their energy and economic security [183, 184]. In EU countries, the primary energy production from renewables has risen over a long time; for example, it grew by 184% from 1990 to 2015. In addition, for the same period, overall electricity production from different sources of renewable energy has swelled by 203% [185], mainly supplied by hydroelectric plants. The share of wind energy in power generation has increased significantly in recent years as well, to become the second largest source of renewable power by 2000 [185, 186]. Moreover, 18.6% of the overall energy utilization for heating/cooling purposes was provided by different sources of renewables in 2015. The EU aims to rapidly develop renewable sources to meet its energy requirements and sustainability goals, with a plan for them to produce at least 27% of its total energy by 2030 [185, 186].

A similar trend can also be extended for the other countries of the world. In 1978, United Nations General Assembly Resolution 33/148 first legitimated the notion of and need for new and renewable energy sources internationally. Since then, renewable energy sources have rapidly come to play an important role in the economic policies of states and also in the global economy. From 2004 to 2014, the number of countries developing programs to take advantage of renewable energy sources rose from 48 to 164 [187], causing total power generation from renewable energy sources to reach significant levels in the global energy balance. The swift growth in renewable energy sources' capacities and shares in both the production and consumption of energy is a logical consequence of the even faster growth of investments in this field, which globally multiplied more than five times between 2004 and 2015 [188]. It is anticipated that the share of renewable energy sources in the use of primary energy and electricity generation worldwide will, respectively, reach 25% and 46% by 2040 [183], which about 30% of the latter one will be from solar energy, whereas its share together with wind energy was only ~5% in 2014 [189, 190].

Even though renewable energies have many considerable advantages, such as being eco-friendly, producing no or only slight GHG emissions, distributing energy equally around the world, solving the energy poverty problem, and having low operation expenses, they do have a number of limitations [191]. The primary disadvantages of these types of energies are their higher initial costs and also expensive storage systems [191]. Some renewable energies (such as hydro, solar, and wind) are greatly dependent on weather conditions, which may result in energy deficiencies if unpredictable and unfavorable climate conditions prevail for a relatively long time [192]. In addition, some renewable energy technology installations require large land areas; for instance, Brook and Bradshaw [193] reported that about 50% of the total energy demand in the United States can be met through renewable energy sources if more than 17% of country's land is used [194]. The following paragraph discusses some specific constraints of each renewable energy source.

Among the limitations on utilizing wind energy are noise pollution; visual unattractiveness; interference with telecommunication signals [195]; and its direct or indirect negative effects on wildlife, particularly migratory birds [195]. Geothermal energy may release small amounts of GHGs into the atmosphere, but in significantly lower amounts than the emissions produced by fossil fuels. Even though geothermal energy is accepted as a renewable source, various factors affect its thermal recharge [191]. Moreover, this source of energy may discourage many investors due to the high initial installation costs [196]. One of the main risks of using hydropower is its potential damage to aquatic habitats and consequent reduction in their populations [197]. Furthermore, this energy source is highly dependent on precipitation and the availability of adequate water, which cannot be controlled. Like geothermal energy, hydropower also entails a high initial financial outlay [195]. For ocean/marine energy, the main disadvantage is the high costs of the required technology. Generally, the existing technology for energy production from ocean or marine sources requires further research as it is not fully developed yet [198]. Additionally, the environmental and socioeconomic impacts of this energy source need to be investigated thoroughly [199].

Even briefly considering the constraints of renewable energy sources in general or individually makes it clear that a single renewable energy source often cannot meet continuous energy demand during a year. However, the integration of several renewable energy sources can help overcome or minimize the limitations of one working alone [200–202]. Consequently, the concept of hybrid renewable energy (HRE) has been introduced, mainly for use in remote and isolated areas [201]. In 1981, Castle et al. [203] examined the feasibility of integrating PV systems and wind-powered generators. Losie et al. [204] considered a diesel generator as a backup source in HRE systems in order to cover the intermittence of renewable energy. These studies have attracted much attention to hybrid PV and wind power systems [205–209]. Similarly, geothermal and biomass energies have shown great potential for integration with solar PV systems, particularly for regions that are either heavily forested or agriculturally developed [210, 211] or have the necessary features for geothermal energy production [212–214]. Fig. 1.8 provides a

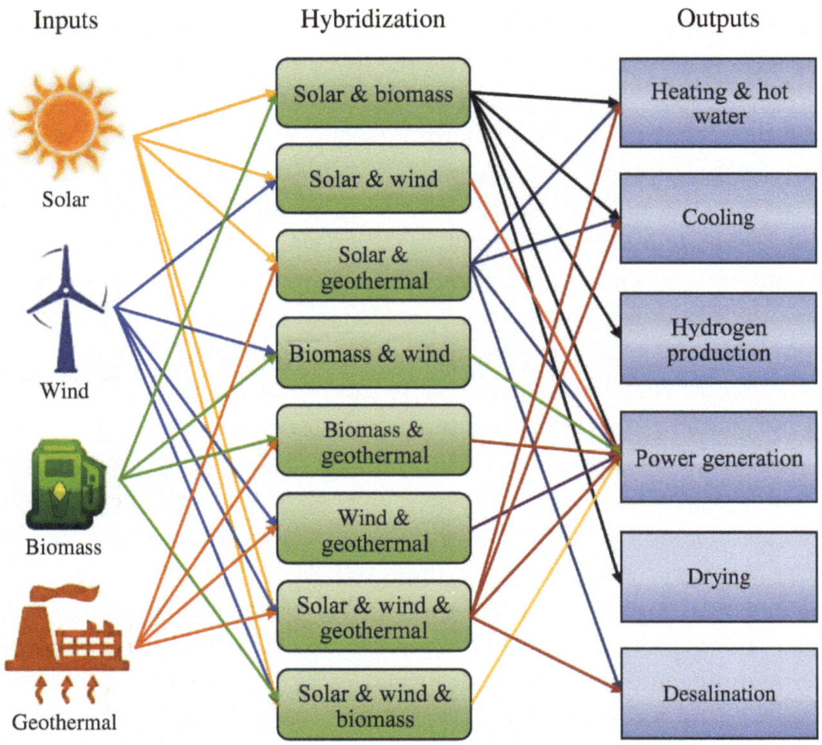

Fig. 1.8 Graphical illustration of the use of hybrid renewable energy (HRE) systems with a number of inputs and outputs [200].

graphical representation of the use of HRE systems with more than two inputs and several outputs. It generally seems that hybrid energy systems (both renewables and conventional) accompanied by energy storage capabilities are efficient and practical solutions for meeting and securing the world's future energy requirements and for reducing environmental problems.

1.6 Energy storage technologies

Energy storage can generally be defined as a technology/approach that is utilized for capturing and storing (mainly temporarily) the energy generated during low load (or demand) periods, in various forms, for later use during high load (or demand) periods [215, 216]. The storage of energy has become an essential component of energy generation and distribution systems for a variety of reasons, the primary ones being the need to balance or match energy loads (or demands) between production and utilization over time, and the desire to make energy systems more flexible and reliable in providing service [217, 218]. For example, the intermittent nature and significant fluctuations of wind and solar energy, along with variable electricity demands on grids, highlight the importance of employing energy storage systems (ESSs) in power generation plants to efficiently use the electricity they produce [219, 220]. Thus, large-scale ESSs need to be developed and used for better energy management [221]. In addition to storing excess energy, which is the main advantage of ESSs, they also

- Facilitate the improvement of renewable energy plants, which cannot be stand-alone without an ESS. Consequently, nonrenewable carbon-based fuel use and the associated emissions of CO_2 can be phased out [222].
- Create an equilibrium between energy generation and requirements, especially in renewable energy systems. Doing so will also reduce problems in electrical power generation systems [223].
- Decrease load shedding risk by shaving peak energy loads, particularly with large capacity ESSs.
- Improve power plant efficiency, thereby diminishing operating expenses in the long run [224].
- Due to their flexibility, provide an effective and suitable means of delivering power to remote and isolated regions that are typically under-serviced [225].

According to the storage technologies employed, ESSs can be categorized into five broad groups [3, 4, 226–228]: (1) mechanical, (2) chemical, (3) electrochemical, (4) electrical, and (5) thermal systems. The specifications of each group of ESSs will briefly be discussed in the following paragraphs.

- **Mechanical energy storage systems (MESSs)**
  These storage systems have a number of advantages not shared by other types, particularly when considering factors such as environmental effect, expense, and sustainability [164]. MESSs can be divided into three major categories: flywheel energy storage (FES); pumped hydro energy storage (PHES); and compressed air energy storage (CAES) [4, 229, 230]. These types of storage systems can also be found in other shapes, for example, in liquid-piston, gravity, and mechanical springs [164, 231, 232]. The most important factors in selecting a suitable MESS include the source of energy, nature of loads, and available space [164, 233]. MESSs can substantially contribute to enhancing the quality of air via producing less pollutants at the operational and construction levels, thereby reducing health risks [164].
- **Chemical energy storage systems (CESSs)**
  The concept of chemical energy storage is based on the absorption of surplus heat in an endothermic reaction. For energy to discharge, an exothermal process must occur by which the reaction products are combined; after that, the resultant heat (energy) can be utilized [234]. Batteries (inclusive secondary and flow ones) and renewable generated chemicals, such as hydrogen and fuel cells, are the main forms of CESSs [235]. Batteries can also be considered as electrochemical energy storage systems (EESSs) applicable in small-scale to large-scale ranges. Some examples of battery-based energy storage systems are portable consumer electronics (for instance, mobile phones); EVs; small-scale grid-based systems and the integration of renewable energy sources in dwellings; and emergency electrical energy in power plants at large-scale grid-based ranges for stabilization and load balancing of fluctuating renewable sources [235, 236].

  Current and future energy generation systems will essentially rely on CESSs, in the form of biomass, coal, and gas, plus renewable energies [235, 237]. The CESSs will have the largest storage capacity, each facility accounting for amounts in the terawatt-hour (TWh) range [235, 237]. This energy storage system is also the only option applicable in seasonal mode, because it alone can use power-to-gas charging technology in combination with the existing gas infrastructure to store and convert gas into electricity [237, 238].
- **Electrochemical energy storage systems (EESSs)**
  There is a clear connection between CESSs and EESSs: batteries. For example, Li-ion types constitute a part of EESSs. Other types of EESSs

are fuel cells and flow batteries [239]. EESSs have shown great potential to contribute significantly to the implementation of sustainable energy [239]. In situations when there is a delay in electricity supply in a power network, batteries can help the system through achieving demand-side management, reducing the peak-to-valley difference at night, and strengthening the self-regulation of peak and valley loads in local and regional power networks [240–243]. Furthermore, they are able to provide safety and consistency in power networks, improve the quality of power, enhance transmission and transformation capabilities, increase electricity supply reliability, and boost large-scale access to renewable energy sources with the features of swift response, robust regulation, and flexible configuration [240, 244–247]. The battery types in ESSs are mainly based on lead-acid (Pb-A), nickel-cadmium (Ni–Cd), Li-ion, sodium-sulfur (Na–S), and vanadium redox flow [248–253].

- **Electrical energy storage systems (EESSs*)**
  Some conventional examples of EESSs* are batteries, supercapacitors, dielectric capacitors, and superconducting magnetic energy storage (SMES) [4, 254]. The key parameters affecting the efficiency and specifications of these types of storage systems include energy density and power density [254, 255]. Batteries benefit from high energy density, but have only modest power density due to the relatively slow kinetics of the redox processes involved [254, 256–258]. Although dielectric capacitors provide only limited energy density, the electron transportation during charging/discharging means that these capacitors actually offer high power density [254, 259]. The energy and power densities in supercapacitors stand between those of batteries and dielectric capacitors, with values ranging from about 3–10 W/kg to 3 kW/kg [254, 256].
- **Thermal energy storage systems (TESSs)**
  The mechanism driving TESSs is based on heating or cooling a storage medium, which can conserve the thermal energy for later use for various purposes [260, 261]. The main types of TESSs are demonstrated in Fig. 1.9. The primary applications of TESSs are for space heating and cooling in buildings and industrial processes to increase overall efficiency and boost reliability [227]. They can also improve national economies by reducing the need to import power, decreasing the financial outlay and running costs of projects, and lowering environmental pollution [227, 260]. Unlike high-efficiency PV systems, which are mainly used in residential buildings, TESSs, which work mainly with solar energy, have industrial applications, utilizing a main portion of the sun's thermal

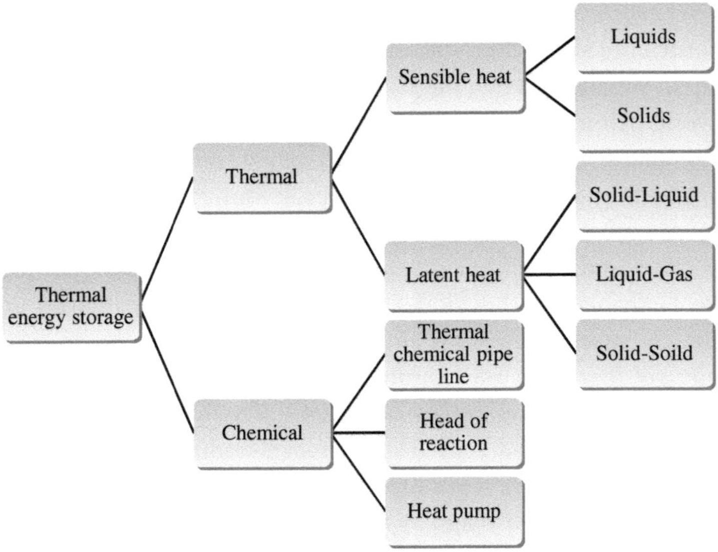

Fig. 1.9 Categorization of thermal energy storage systems (TESSs). *(Modified from I. Sarbu, C. Sebarchievici, A comprehensive review of thermal energy storage, Sustainability, vol. 10, No. 1, 191, 2018).*

energy in daytime [260]. However, it should be mentioned that these systems usually cannot operate in low or no sunlight times as they have no thermal backup. Combining TESSs with concentrating solar power (CSP) plants to make these storage systems store solar heat is a sound solution for producing electrical energy when sunlight is not accessible. Recently, researchers have focused on selecting and characterizing new materials with specific thermophysical properties capable of serving 24-h operations in an efficient TESS [262].

1.6.1 Comparison of different energy storage systems (ESSs)

Due to the wide range of ESS properties, different ESSs may be employed in various applications, time frames, and magnitude scales [4, 226]. For instance, supercapacitors and CAESs are, respectively, suitable for specific narrow uses and wider applications [4]. A number of criteria are used in comparing the various ESSs, such as [4, 226, 263, 264] capacity reserve; short- or long-time storage; response time; stationary or portable setup; energy density rating; conversion rate; storage costs; security; end use (e.g., stand-alone or grid-connected); environmental effects; and storage time limits. Table 1.8 compares several important specifications of various ESSs, including lifetime, cycling times, cycle efficiency, energy density, and power density.

Table 1.8 Comparison of most important factors among several energy storage systems (ESSs) [4, 228, 265].

System	Lifetime (years)	Cycling times (cycles)	Cycle efficiency (%)	Energy density (Wh/L)	Power density (W/L)
PHES	40–60 [266], 40+ [267], 30+ [268]	10,000–30,000 [269]	70–85 [266], 70–80 [268], 87 [270], 75–85 [271]	0.5–1.5 [266], 1–2 [272]	0.5–1.5 [266], ~1 [272]
Large-scale CAES	20–40 [266], 30 [273], 20+ [267, 271]	8000–12,000 [269]	42–54 [266, 274], AA-CAES 70 [271, 275]	3–6 [266], 2–6 [272]	0.5–2 [266], ~1 [272]
Over-ground small CAES	23+ [276]	Test 30,000 stop/starts [276]	–	Higher than large-scale CAES	Higher than large-scale CAES
FES	~15 [266], 15+ [267], 20 [277]	20,000+ [266], 21,000+ [267]	~ 90–95 [266], 90 & 95 [273]	20–80 [266, 272, 278]	1000–2000 [266], ~5000 [272]
TES	10–20 [266], 5–15 [266], 30 [271]	–	~30–60 [266]	80–120, 120–200, 200–500 [266]	–
SMES	20+ [266], 30 [277]	100,000+ [266], 20,000+ [269]	~95–97 [266], 95–98 [279], 95 [273]	0.2–2.5 [266], ~6 [272]	1000–4000 [266], ~2500 [272]
Capacitor	~5 [266], ~1–10 [280]	50,000+ [266], 5000 (100% DoD) [281]	~60–70 [266], 70+ [281]	2–10 [266], ~0.05 [282]	100,000+ [266]

Continued

Table 1.8 Comparison of most important factors among several energy storage systems (ESSs) [4, 228, 265]—cont'd

System		Lifetime (years)	Cycling times (cycles)	Cycle efficiency (%)	Energy density (Wh/L)	Power density (W/L)
Supercapacitor		10–30 [266], 10–12 [279]	100,000+ [266], 50,000+ [267]	~ 90–97 [266], 84–95 [279]	10–30 [266], ~10–30 [278]	100,000+ [266]
Hydrogen fuel cell		5–15 [266], 20 [283], 20+ [279]	1000+ [266], 20,000+ [279]	~20–50 [266], 32 [284], 45–66 [285]	500–3000 [266]	500+ [266]
Battery energy storage (BES)	Pb-A	5–15 [266, 286], 13 [267]	500–1000 [266], 200–1800 [287]	70–80 [266], 63–90 [269], 75–80 [288]	50–80 [266], 50–90 [273]	10–400 [266]
	Li-ion	5–15 [266], 14–16 [289]	1000–10,000 [266], up to 20,000 [290]	~90–97 [266], 75–90 [291]	200–500 [266], 200–400 [272], 150 [273]	1500–10,000 [272]
	Na-S	10–15 [266], 15 [267], 12–20 [292]	2500 [266], 3000 [293], 2500–4500 [269]	~75–90 [266], 75 [293], 75–85 [288]	150–250 [266], 150–300 [272]	~140–180 [272]
	Ni-Cd	10–20 [266], 3–20 [287], 15–20 [286]	2000–2500 [266], 3500 [294]	~60–70 [266], 60–83 [269]	60–150 [266], 15–80 [272], 80 [273]	80–600 [272]
	Vanadium redox	5–10 [266], 20 [295]	12,000+ [266], 13,342 [267]	75–85 [266, 296], 65–75 [291]	16–33 [266], 25–35 [297]	~<2 [272]
	Zn-Br	5–10 [266], 10 [267], 8–10 [289]	2000+ [266], 1500 [267]	~65–75 [266], 66–80 [269], 66 [277]	30–60 [266], ~55–65 [272]	~<25 [272]

Table 1.9 Comparison of different kinds of energy storage systems (ESSs) based on different expenses [4, 228, 265].

System	Power capital expense ($/kW)	Energy capital expense ($/kW h)	Operating and maintenance expense
PHES	2500–4300 [291], 2000–4000 [268]	5–100 [296], 10–12 [277]	0.004 $/kWh [273], ~3 $/kW/year [298]
Large-scale CAES	400–800 [266], 800–1000 [268]	2–50 [266], 2–120 [299], 2 [273]	0.003 $/kWh [273], 19–25 $/kW/year [298]
Over-ground small CAES	517 [277], 1300–1550 [300]	1MVA from £296 k [276], 200–250 [300]	Very low [276]
FES	250–350 [266]	1000–5000 [266], 1000–14,000 [299]	~0.004 $/kWh [273], ~20 $/kW/year [298]
TES	200–300 [266], 250 [271], 100–400 [271]	20–50 [266], 30–60 [266], 3–30 [266]	–
SMES	200–300 [266], 300 [277], 380–489 [300]	1000–10,000 [266], 500–72,000 [277]	0.001 $/kWh [273], 18.5 $/kW/year [298]
Capacitor	200–400 [266]	500–1000 [266]	13 $/kW/year [298], < 0.05 $/kWh [281]
Supercapacitor	100–300 [266], 250–450 [300]	300–2000 [266]	0.005 $/kWh [273], ~6 $/kW-year [277]
Hydrogen fuel cell	500 [277], 1500–3000 [301]	15 [277], 2–15 €/kWh [288]	0.0019–0.0153 $/kW [301]

Continued

Table 1.9 Comparison of different kinds of energy storage systems (ESSs) based on different expenses [4, 228, 265]—cont'd

System		Power capital expense ($/kW)	Energy capital expense ($/kW h)	Operating and maintenance expense
Battery energy storage (BES)	Pb-A	300–600 [266], 200–300 [277], 400 [293]	200–400 [266], 50–100 [286], 330 [293]	~50 $/kW/year [298]
	Li-ion	1200–4000 [266], 900–1300 [286], 1590 [291]	600–2500 [266], 2770–3800 [291]	–
	Na-S	1000–3000 [266], 350–3000 [299]	300–500 [266], 350 [293], 450 [302]	~80 $/kW/year [298]
	Ni-Cd	500–1500 [266]	800–1500 [296], 400–2400 [286]	~20 $/kW/year [298]
	Vanadium redox	600–1500 [266]	150–1000 [266], 600 [302]	~70 $/kW/year [298]
	Zn-Br	700–2500 [266], 400 [303], 200 [277]	150–1000 [266], 500 [304]	–

Table 1.9 illustrates a comparison of different kinds of ESSs according to their expenses, such as for power capital, energy capital, and operating and maintenance. Tables 1.8 and 1.9 present data derived from scientific research conducted by both academic centers and industry [4, 228]. However, the data provided in these tables will change over time. Table 1.10 shows an analysis of the strengths, weaknesses, opportunities, and threats, known as SWOT analysis, of a number of ESSs. In addition, Fig. 1.10 compares the specific energy and specific power of various types of ESSs.

As illustrated in Tables 1.8–1.10 and Fig. 1.10, ESSs have a wide variety of characteristics, uses, benefits, qualities, and limitations. For instance, high efficiency, power density, and stability, plus rapid response times have been reported for FES systems, but some of their disadvantages are high self-discharge rates, low overall magnitudes, certain safety issues, and high expense [226, 245, 265, 307]. The advantages of CAESs include having grid-scale potential, flexibility, long life, relatively low operation and maintenance expenses, and low self-discharge rates [4, 308]. On the other hand, in comparison with others, this type of storage is only moderately efficient [265], with key restriction on the geological suitability of the storage site [4, 308]. Batteries are currently the most common storage systems, being efficient, convenient, reliable, and easy to use, and needing little maintenance [4, 309]. However, their limitations include environmental problems, high expense (in comparison with utility power), the need for critical and even rare materials, low energy density, and a restricted shelf life [4, 309].

Electricity supplies must be sufficient to meet variable requirements, as electrical energy production and use occur at the same [4, 226, 310]. The short- and long-term storage characteristics of ESSs can help not only in balancing this supply and requirement, but also in controlling the frequency and voltage fluctuations at local and large grid scales [311]. Therefore, to store the electric energy generated in ESSs, it must be converted into another form of energy [312]. As mentioned above, batteries are one of the best choices available for storing electricity among other ESSs, despite their problems and limitations. Additionally, they aid in smoothing output and improving renewable energy versatility in microgeneration systems through their integration with alternative renewable energy sources (e.g., wind and solar), allowing these sources of energy to provide and distribute electrical power steadily [4, 313–315].

In TESSs, storage density is a key factor in optimizing the solar ratio—meaning the amount of solar radiation available for heating/cooling purposes—efficiency of appliances, for example, in solar thermal collectors and

Table 1.10 SWOT analysis carried out on a number of grid-scale energy storage systems (ESSs) [4, 226, 263, 305, 306].

System	Strengths	Weaknesses	Opportunities	Threats
CAES	High capacity; Low cost per kWh; Minor needs for power electronic converters; Negligible storage losses; Storing energy for more than one year	Need for underground cavities; Need for fuel (e.g., H_2 and CH_4) if gas turbines used	Can prospectively be adapted for distributed storage	Popularity related to thermal power plants; Probably, increasing fuel costs over time
PHES	High capacity; Low cost per kWh; Minor needs for power electronic converters; Long lifetime; Reliable	Centralized storage; Geographical restrictions; High investment cost of installation; Environmental concerns	Can be used for offshore wind parks and with a lower reservoir under sea level	Can become obsolete when distributed storage preferred; Increasing public opposition due to environmental damage
BES	Distributed storage; Good configurability; Fast response time; High energy efficiency and density	High investment costs; Short life span; Temperature issues in cold climates	Emerging technologies, most likely BES will be a distributed system (many cars)	Constant development phase complicates selection; Raw materials' limits; Environmental impacts
Hydrogen	Distributed storage; Other uses for produced hydrogen; Minor environmental issues	Low efficiency; High investment costs; Need for power electronics and control; Need for stable load	Market penetration; Perspective nanotube storage media; Dedicated converters	Maturing battery technologies; EMI issues related to the use of power electronic converters

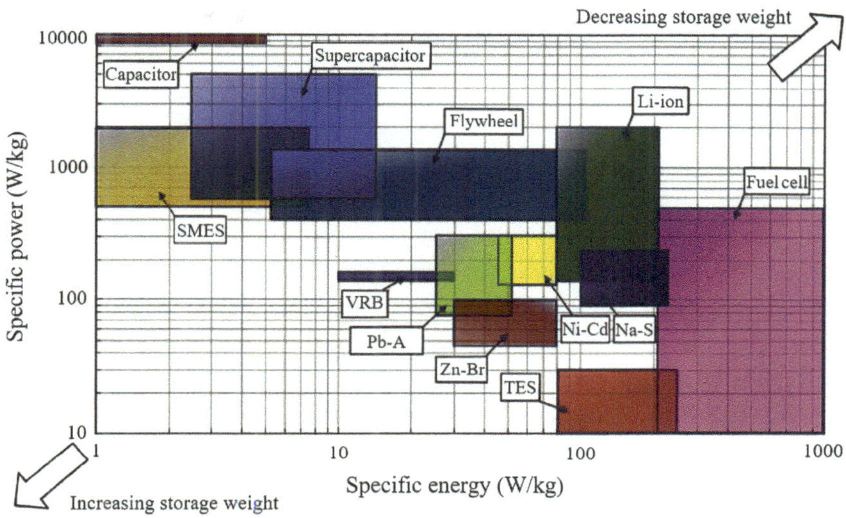

Fig. 1.10 Comparing energy storage systems (ESSs) based on specific energy and specific power [4, 228].

absorption chillers, and the energy consumed to heat/cool spaces. In this regard, phase change materials (PCMs), as storage media, have shown potential to help improve solar system applications [260]. In small-sized water storage tanks, energy density may increase by means of PCMs, eventually will either reduce the solar storage volume for a defined solar fraction, or increase the solar fraction for a defined available volume [316]. Thermal storage can be utilized on both the hot side and cold side of a plant. For example, the hot water from collectors and auxiliary heaters can be provided for cooling and heating modes either in the generator of the absorption chiller or directly to the users [260]. In fact, the cold water made via an absorption chiller can be stored to supply the cooling demands within buildings. In three situations—hot (80–90°C), warm (40–50°C), and cold (7–15°C)—storage are usually identified according to the different temperature ranges [317].

Solar plants always have heat storage on their hot side, present to produce heating and/or domestic hot water. Cold storage implemented in larger plants would be more cost-effective (e.g., lower electricity expenses in electric compression chillers), lower installed cooling power, and longer continuous operation of chillers [318]. Systems with thermal storage do not benefit from having effective backup, but it does help them achieve thermal stabilization. Therefore, thermal storage is most helpful in solar-assisted thermal systems [319]. A number of studies have investigated the applications and effects of TES technologies and both sensible and latent heat storage in various systems [320–324]. It should be noted that the focus of these studies is only on the cold or hot side, or on components of the system, or on one of its integral mechanisms. This type of energy storage will be detailed in Chapter 7.

TESSs can improve the balance between supply and demand through energy conservation. In addition, they enhance system performance and thermal reliability, which emphasizes the importance of developing efficient and economical TESSs, especially at large scales [260]. Currently, the commercially available thermal storage systems are based on sensible/latent heat, but systems based on PCMs are under development and demonstration. Research and development groups focusing on new storage materials should be supported by policy measures and investment incentives in order to improve the implementation of TESSs in different types of buildings, for industrial applications, and for variable renewable power generation. Another application of TESSs will be in future greenhouses for either heating, cooling, or dehumidification purposes, providing poly-generation possibilities. To make this option widely appealing and cost-effective, further research is required on appropriate PCMs and the possibility of thermochemical energy storage.

1.7 Conclusion

Energy is one of the most disputed and important subjects in the modern world since it contributes significantly to many aspects of each country, including economic, industrial, social, and sustainable development. In other words, energy plays a basic and unavoidable role in sustaining and improving people's quality of life. Due to environmental and health problems caused by the use of nonrenewable high-carbon energy sources, we need to phase them out and transition to clean, carbon-free (or at least low-carbon), and renewable energy sources coupled with environmentally friendly technologies and sustainable practices. This chapter, which has introduced energy and its associated issues, highlights the importance of revisiting the sustainable technology used by earlier generations (for instance, traditional ice-houses), cold energy production, storage and usage (specifically in the form of ice), and of learning from it to develop hybrid ecologically friendly technologies for meeting buildings' thermal energy needs. These and other related topics will be discussed in detail in Chapters 2–9.

References

[1] U.S. Energy Information Administration (U.S. EIA), Biomass Explained: Wood and Wood Waste, July 27, 2020. https://www.eia.gov/energyexplained/biomass/wood-and-wood-waste.php. (Accessed 12 October 2020).

[2] What do you know about new energies? Biomass energy, Fourth report, Iran Renewable Energy Organization (SANA), Ministry of Energy, Tehran, Iran; 2007 (in Farsi).

[3] A.R. Dehghani-Sanij, in: A. Sayigh (Ed.), Cisterns: Sustainable Development, Architecture and Energy, River Publishers, Denmark, 2016.

[4] A.R. Dehghani-Sanij, E. Tharumalingam, M.B. Dusseault, R. Fraser, Study of energy storage systems and environmental challenges of batteries, Renew. Sustain. Energy Rev. 104 (2019) 192–208.

[5] R. Parameshwaran, S. Kalaiselvam, S. Harikrishnan, A. Elayaperumal, Sustainable thermal energy storage technologies for buildings: a review, Renew. Sustain. Energy Rev. 16 (2012) 2394–2433.

[6] IEA, Perspectives for the Clean Energy Transition: The Critical Role of Buildings, International Energy Agency (IEA), Paris, France, April 2019.

[7] R.A. Lancor, Using student-generated analogies to investigate conceptions of energy: a multidisciplinary study, Int. J. Sci. Educ. 36 (1) (2014) 1–23.

[8] J. Kester, The Politics of Energy Security: Critical Security Studies, New Materialism and Governmentality, Routledge, New York, NY, USA, 2018.

[9] M.O. Abdullah, Applied Energy: An Introduction, CRC Press, 2019.

[10] S.A. Ashrafizadeh, Z. Tan, Energy Balance, Chapter 5, in: Mass and Energy Balances: Basic Principles for Calculation, Design, and Optimization of Macro/Nano Systems, Springer, Cham, 2018, pp. 127–158.

[11] I. Asimov, Life and Energy, Doubleday, New York, NY, 1962.

[12] A.C. Baugh, T. Cable, A History of the English Language, sixth ed., Routledge, London, UK, 2012.

[13] P. Hjertholm, An Inquiry into the Meaning of Energy, PhD Thesis, Faculty of the Humanities, University of Copenhagen, Denmark, Nov. 2019.

[14] H.G. Liddell, R. Scott, An Intermediate Greek-English Lexicon, Abridged, Simon Wallenberg Press, UK, 2007.

[15] N.M. Azari, Study the Foreign Trade Effects on Economic Development in some Developing Countries Including Iran and Turkey, Master's Thesis, Çankaya University, Ankara, Turkey, Feb. 2016.

[16] C.P. Fernandes, A contribution to the analysis of energy security, Part I, in: C.P. Fernandes, T.F. Rodrigues (Eds.), Geopolitics of Energy and Energy Security, Instituto da Defesa Nacional, 2017, pp. 61–78.

[17] A.C. Baytas, A.F. Bayyas, Entropy generation in porous media, Chapter 8, in: D.B. Ingham, I. Pop (Eds.), Transport Phenomena in Porous Media III, Elsevier Science, 2005, pp. 201–226.

[18] C.J. Cleveland, C. Morris, Part II: foundations, introduction, in: Handbook of Energy, vol. II: Chronologies, Top Ten Lists, and Word Clouds, Elsevier, Amsterdam, The Netherlands, 2014, pp. 339–341.

[19] M.G. Patterson, Approaches to energy quality in energy analysis, Int. J. Global Energy Issues 5 (1) (1993) 19–28.

[20] M.T. Brown, S. Ulgiati, Emergy indices of biodiversity and ecosystem dynamics, Chapter 4, in: S.E. Jørgensen, F.-L. Xu, R. Costanza (Eds.), Handbook of Ecological Indicators for Assessment of Ecosystem Health, CRC Press, 2010, pp. 89–112.

[21] C. Arndt, The concept of entropy in physics, Chapter 4, in: Information Measures: Information and its Description in Science and Engineering, Springer-Verlag, Berlin Heidelberg, 2001, pp. 85–132.

[22] H. Imura, Resources, Energy, and Environmental Load, in: Environmental Systems Studies: A Macroscope for Understanding and Operating Spaceship Earth, Springer, Tokyo, 2013, pp. 115–138.
[23] S. Zhu, Z. Zhu, Smart energy management and comprehensive utilization of nuclear energy, in: Proceedings of the 2012 20th International Conference on Nuclear Engineering, ASME 2012 Power Conference, ICONE20-POWER2012, Anaheim, CA, USA, July 30–August 3, 2012.
[24] W.P.S. Dias, S.P. Pooliyadda, Quality based energy contents and carbon coefficients for building materials: a systems approach, Energy 29 (4) (March 2004) 561–580.
[25] J. Jensen, Energy quality, Chapter 2, in: Energy Storage, Elsevier, 1980, pp. 5–9.
[26] M.N. Bahadori, M. Yaghoubi, Ventilation and Natural Cooling System in Traditional Buildings of Iran, Markaz-e Nashr-e Daneshgahi, Tehran, Iran, 2006 (in Farsi).
[27] T. Ohta, Energy Technology: Sources, Systems and Frontier Conversion, Elsevier, 1994. Imprint by Pergamon.
[28] R. Zoran, Exergie, Ein neues Wort für Technische Arbeitsfähigkeit, Forsch. Auf dem Gebiete des Ingenieurw. 22 (1956) 36–37.
[29] Y.A. Çengel, M.A. Boles, Thermodynamics an Engineering Approach, sixth ed., McGraw-Hill Education, New York, NY, 2008.
[30] W. van Gool, J.J.C. Bruggink (Eds.), Energy and Time in the Economic and Physical Sciences: Papers and Comments, North-Holland, Amsterdam, The Netherlands, 1985, pp. 61–78.
[31] ASTM, SI10-10, IEEE/ASTM SI 10 American National Standard for Metric Practice, ASTM International, West Conshohocken, PA, 2010.
[32] L. Wu, G. Kaiser, D. Solomon, R. Winter, A. Boulanger, R. Anderson, Improving efficiency and reliability of building systems using machine learning and automated online evaluation, in: 2012 IEEE Long Island Systems, Applications and Technology Conference (LISAT), IEEE, Farmingdale, NY, USA, May 2012, pp. 1–6.
[33] S. Harrison, Sustainability and environmental awareness in ceramics: the solar-fired electric kiln with lP gas reduction, J. Aust. Ceram. Soc. 59 (1) (April 2020) 89–93.
[34] R. Hadizadeh, A.A. Shojaie, A. Shojaie, A. Shahabi, Strategies for energy pricing to modify energy consumption using reward process functions, Energy Policy Res. 4 (1) (July 2017) 65–70.
[35] M.N. Bahadori, A.R. Dehghani-Sanij, in: A. Sayigh (Ed.), Wind Towers: Architecture, Climate and Sustainability, Springer International Publishing, New York, NY, 2014.
[36] C. Kinney, A.R. Dehghani-Sanij, S.B. Mahbaz, M.B. Dusseault, J.S. Nathwani, R.A. Fraser, Geothermal energy for sustainable food production in Canada's remote northern communities, Energies 12 (2019) 1–25.
[37] M. Soltani, A.R. Dehghani-Sanij, A. Sayadnia, F.M. Kashkooli, K. Gharali, S.B. Mahbaz, M.B. Dusseault, Investigation of airflow patterns in a new design of wind tower with a wetted surface, Energies 11 (2018) 1–23.
[38] M. Soltani, F.M. Kashkooli, A.R. Dehghani-Sanij, A. Nokhosteen, A. Ahmadi-Joughi, K. Gharali, S.B. Mahbaz, M.B. Dusseault, A comprehensive review of geothermal energy evolution and development, Int. J. Green Energy 16 (13) (2019) 971–1009.
[39] U.K. Pata, The effect of urbanization and industrialization on carbon emissions in Turkey: evidence from ARDL bounds testing procedure, Environ. Sci. Pollut. Res. 25 (8) (2018) 7740–7747.
[40] Worldometer, World Population by Year, 2020. http://www.worldometers.info/world-population. (Accessed 17 October 2020).
[41] F. Kreith, S. Krumdieck, Principles of Sustainable Energy Systems, second ed., CRC Press, Taylor & Francis Group, Boca Raton, FL, USA, 2013.
[42] A.R. Dehghani-Sanij, M.R. Khani, B. Zhang, S. Narimannejad, M. Mohammadnia, Water quality analysis of underground reservoirs in hot and arid regions, J. Appl. Environ. Biol. Sci. 6 (7) (2016) 149–161.

[43] X.Q. Lao, W.J. Ma, T. Sobko, Y.H. Zhang, Y.J. Xu, X.J. Xu, D.M. Yu, S.P. Nie, Q.M. Cai, X.L. Wei, L. Xia, Dramatic escalation in metabolic syndrome and cardiovascular risk in a Chinese population experiencing rapid economic development, BMC Public Health 14 (2014) 983.
[44] C.D. Gomersall, Critical care in the developing world—a challenge for us all, Crit. Care 14 (2010) 131.
[45] X.Q. Lao, Y.H. Zhang, M.C.S. Wong, Y.J. Xu, H.F. Xu, S.P. Nie, W.J. Ma, G.N. Thomas, I.T. Sun Yu, The prevalence of metabolic syndrome and cardiovascular risk factors in adults in southern China, BMC Public Health 12 (2012) 64.
[46] M. Gustafsson, R. Thygesen, B. Karlsson, L. Ödlund, Rev-changes in primary energy use and CO_2 emissions—an impact assessment for a building with focus on the Swedish proposal for nearly zero energy buildings, Energies 10 (2017) 978.
[47] B. Dudley, BP Statistical Review of World Energy June 2011, BP p.l.c, London, UK, June 2011, p. 45.
[48] B. Looney, Statistical Review of World Energy 2020, sixty ninth edition, BP p.l.c, London, UK, June 2020, p. 65.
[49] L. Capuano, International Energy Outlook 2020 (IEO2020), Center for Strategic and International Studies, U.S. Energy Information Administration (U.S. EIA), Washington, DC, USA, October 14, 2020.
[50] A. Demirbaş, Global renewable energy resources, Energy Sources, Part A 28 (8) (2006) 779–792.
[51] B.M. Berkovsky, Energy for Tomorrow's World, The United Nations, Educational, Scientific and Cultural Organization (UNESCO), Paris, France, 1978, pp. 5–7.
[52] M. Schilling, L. Chiang, The effect of natural resources on a sustainable development policy: the approach of non-sustainable externalities, Energy Policy 39 (2) (2011) 990–998.
[53] C.O. Osueke, C.A.K. Ezugwu, Study of Nigeria energy resources and its consumption, Int. J. Sci. Eng. Res. 2 (12) (2011) 1–8.
[54] J. Chow, R.J. Kopp, P.R. Portney, Energy resources and global development, Science 302 (2003) 1528–1531.
[55] B. Yilmaz, B. Yüksel, G. Orhan, D. Aydin, Z. Utlu, Synthesis and characterization of salt-impregnated anodic aluminum oxide composites for low-grade heat storage, Int. J. Miner. Metall. Mater. 27 (2020) 1–7.
[56] A.R. Dehghani-Sanij, Theoretical and Experimental Study of Heat Loss and Ice Accretion from Large Structures on the Marine Vessels and Offshore Structures, PhD Thesis, Memorial University of Newfoundland (MUN), St. John's, NL, Canada, 2017.
[57] M.J. Burkea, J.C. Stephens, Political power and renewable energy futures: a critical review, Energy Res. Soc. Sci. 35 (2018) 78–93.
[58] A.R. Dehghani-Sanij, S.R. Dehghani, G.F. Naterer, Y.S. Muzychka, Sea spray icing phenomena on marine vessels and offshore structures: review and formulation, Ocean Eng. 132 (2017) 25–39.
[59] A.R. Dehghani-Sanij, M. Mahmoodi, S.R. Dehghani, Y.S. Muzychka, G.F. Naterer, Experimental investigation of vertical marine surface icing in periodic spray and cold conditions, ASME J. Offshore Mech. Arct. Eng. 141 (2019) 021502-1–021502-16.
[60] A.R. Dehghani-Sanij, Y.S. Muzychka, G.F. Naterer, Analysis of ice accretion on vertical surfaces of marine vessels and structures in arctic conditions, in: The ASME 2015 34th International Conference on Ocean, Offshore and Arctic Engineering (OMAE2015), OMAE2015-41306, St. John's, NL, Canada, May 31–June 5, 2015.
[61] A.R. Dehghani-Sanij, Y.S. Muzychka, G.F. Naterer, Predicted ice accretion on horizontal surfaces of marine vessels and offshore structures in Arctic Regions, in: The ASME 2016 35th International Conference on Ocean, Offshore and Arctic Engineering (OMAE2016), OMAE2016-54054, Busan, South Korea, June 19–24, 2016.
[62] Y.H. Lee, H.N. Hu, J.S. Chiou, Jump dynamics with structural breaks for crude oil prices, Energy Econ. 32 (2) (2010) 343–350.

[63] D.M. Coutinho, D. França, G. Vanini, L.A.N. Mendes, A.O. Gomes, V.B. Pereira, B.M. Ávila, D.A. Azevedo, Rapid hydrocarbon group-type semi-quantification in crude oils by comprehensive two-dimensional gas chromatography, Fuel 220 (2018) 379–388.
[64] G. Carbonnier, J. Grinevald, Energy and development, in: G. Carbonnier (Ed.), International Development Policy: Energy and Development, Palgrave Macmillan, London, UK, 2011, pp. 3–20.
[65] F. Birol, World Energy Outlook 2010, International Energy Agency (IEA), Paris, France, Nov. 2010.
[66] G. Kalghatgi, Is it really the end of internal combustion engines and petroleum in transport? Appl. Energy 225 (2018) 965–974.
[67] N. Von Tunzelmann, G.N. Von Tunzelmann, T. Von, Steam Power and British Industrialization to 1860, Oxford University Press, Oxford, UK, 1978.
[68] V. Smil, Energy in World History, Routledge, New York, NY, 2018.
[69] Y. Mulugetta, E.B. Hagan, D. Kammen, Energy access for sustainable development, Environ. Res. Lett. 14 (2) (2019) 020201.
[70] R. Fouquet, Path dependence in energy systems and economic development, Nat. Energy 1 (8) (2016) 1–5.
[71] P. Alstone, D. Gershenson, D.M. Kammen, Decentralized energy systems for clean electricity access, Nat. Clim. Chang. 5 (4) (2015) 305–314.
[72] B.K. Sovacool, I.M. Drupady, Energy Access, Poverty, and Development: The Governance of Small-Scale Renewable Energy in Developing Asia, Routledge, London, UK, 2016.
[73] K. Kaygusuz, Energy for sustainable development: a case of developing countries, Renew. Sustain. Energy Rev. 16 (2) (2012) 1116–1126.
[74] N. Tanaka, O. Kjorven, K. Yumkella, Energy Poverty: How to Make Modern Energy Access Universal? Special Early Excerpt of the World Energy Outlook 2010 for the United Nations General Assembly on the Millennium Development Goals, International Energy Agency (IEA), The United Nations Development Programme (UNDP), and United Nations Industrial Development Organization (UNIDO), Paris, France, Sept. 2010.
[75] T. Abbasi, S.A. Abbasi, Renewable Energy Sources: Their Impact on Global Warming and Pollution, PHI Learning Pvt. Ltd., New Delhi, Delhi, India, 2010.
[76] B.K. Hodge, Alternative Energy Systems and Applications, John Wiley & Sons, Inc., New York, NY, 2017.
[77] P. Siakwah, Are natural resource windfalls a blessing or a curse in democratic settings? Globalised assemblages and the problematic impacts of oil on Ghana's development, Resour. Policy 52 (2017) 122–133.
[78] C.C.N. Ugochukwu, J. Ertel, Negative impacts of oil exploration on biodiversity management in the Niger De area of Nigeria, Impact Assess. Proj. Appraisal 26 (2) (2008) 139–147.
[79] OECD, Integrating Science & Technology into Development Policies: An International Perspective, Organization for Economic Cooperation and Development (OECD), Paris, France, May 2007.
[80] F.F. Nerini, J. Tomei, L.S. To, I. Bisaga, P. Parikh, M. Black, A. Borrion, C. Spataru, V.C. Broto, G. Anandarajah, B. Milligan, Mapping synergies and trade-offs between energy and the sustainable development goals, Nat. Energy 3 (1) (2018) 10–15.
[81] N.D. Rao, S. Pachauri, Energy access and living standards: some observations on recent trends, Environ. Res. Lett. 12 (2) (2017) 025011.
[82] WHO, World Health Statistics 2016: Monitoring Health for the SDGs, Sustainable Development Goals, World Health Organization (WHO) Press, Geneva, Switzerland, 2016.

[83] T. Bruckner, I.A. Bashmakov, Y. Mulugetta, H. Chum, A. de la Vega Navarro, J. Edmonds, A. Faaij, B. Fungtammasan, A. Garg, E. Hertwich, D. Honnery, D. Infield, M. Kainuma, S. Khennas, S. Kim, H.B. Nimir, K. Riahi, N. Strachan, R. Wiser, X. Zhang, E. Systems, Climate change 2014, in: O. Edenhofer, R. Pichs-Madruga, Y. Sokona, E. Farahani, S. Kadner, K. Seyboth, J.C. Minx (Eds.), Mitigation of Climate Change, Contribution of Working Group III to the Fifth Assessment Report of the Intergovernmental Panel on Climate Change, Cambridge University Press, Cambridge, UK, New York, NY, USA, 2014.

[84] P.B. da Silveira Bezerra, C.L. Callegari, A. Ribas, A.F. Lucena, J. Portugal-Pereira, A. Koberle, A. Szklo, R. Schaeffer, The power of light: socio-economic and environmental implications of a rural electrification program in Brazil, Environ. Res. Lett. 12 (9) (Aug. 2017) 095004.

[85] WEF, Energy for Economic Growth: Energy Vision Update 2012, Industry Agenda, World Economic Forum (WEF) and IHS CERA, Geneva, Switzerland, 2012. http://www3.weforum.org/docs/WEF_EN_EnergyEconomicGrowth_IndustryAgenda_2012.pdf. (Accessed 7 December 2020).

[86] T. Moss, Job Creation and Energy in Africa, Energy for Growth Hub, Sept. 2018. https://www.energyforgrowth.org/wp-content/uploads/2018/09/Job-Creation-and-Energy-in-Africa.pdf. (Accessed 19 November 2020).

[87] M. Ram, A. Aghahosseini, C. Breyer, Job creation during the global energy transition towards 100% renewable power system by 2050, Technol. Forecast. Soc. Chang. 151 (2020) 119682.

[88] M.A. Destek, A. Aslan, Renewable and non-renewable energy consumption and economic growth in emerging economies: evidence from bootstrap panel causality, Renew. Energy 111 (2017) 757–763.

[89] C. Kreger, D. Moran, S. Therson, A transdisciplinary approach to unemployment in the United States' petroleum industry, Transdiscip. J. Eng. Sci. 8 (2017) 83–102.

[90] I. Ozturk, A literature survey on energy–growth nexus, Energy Policy 38 (1) (2010) 340–349.

[91] O. Fricko, S.C. Parkinson, N. Johnson, M. Strubegger, M.T. van Vliet, K. Riahi, Energy sector water use implications of a 2 C climate policy, Environ. Res. Lett. 11 (2016) 034011.

[92] UN, Sustainable Development Goals (SDGs), Department of Economic and Social Affairs, United Nations (UN), New York, NY, 2015. https://sustainabledevelopment.un.org/sdgs. (Accessed 19 November 2020).

[93] UN, Transforming Our World: the 2030 Agenda for Sustainable Development, Knowledge Platform, United Nations (UN), New York, NY, 2015. https://sustainabledevelopment.un.org/post2015. (Accessed 19 November 2020).

[94] IHS CERA and OECD Structural Analysis Database, Data are 10-year averages of the most recent data available—2000–2009 for the United States, 1993–2002 for Norway and 1994–2003 for all other countries, 11211–7&8.IHS CERA and OECD Structural Analysis Database, Data are 10-year averages of the most recent data available—2000–2009 for the United States, 1993–2002 for Norway and 1994–2003 for all other countries, 11211–7&8.

[95] J. Meckling, L. Hughes, Globalizing solar: global supply chains and trade preferences, Int. Stud. Q. 61 (2) (2017) 225–235.

[96] I. Dincer, Renewable energy and sustainable development: a crucial review, Renew. Sustain. Energy Rev. 4 (2) (2000) 157–175.

[97] J. Dunderdale (Ed.), Energy and the Environment, Royal Society of Chemistry, Cambridge, UK, 1990.

[98] G.H. Brundtland, Brundtland report, Our common future, Comissão Mundial 4 (1) (1987) 17–25.

[99] O. Langhelle, Sustainable development: exploring the ethics of our common future, Int. Polit. Sci. Rev. 20 (2) (1999) 129–149.
[100] R. Norton, An Overview of a Sustainable City Strategy, Report Prepared for the Global Energy Assessment Planning for Cities and Municipalities, Montreal, QC, Canada, Oct. 1991.
[101] M. MacRae, Realizing the Benefits of Community Integrated Energy Systems, Study No. 45, Canadian Energy Research Institute, Calgary, AB, Canada, 1992.
[102] I. Dincer, M.A. Rosen, A worldwide perspective on energy, environment and sustainable development, Int. J. Energy Res. 22 (15) (Dec. 1998) 1305–1321.
[103] J.O. Jaber, O.O. Badran, N. Abu-Shikhah, Sustainable energy and environmental impact: role of renewables as clean and secure source of energy for the 21st century in Jordan, Clean Techn. Environ. Policy 6 (3) (2004) 174–186.
[104] I. Dincer, Impact of science and technology on energy and environmental research and development, Int. J. Global Energy Issues 15 (3) (2001) 164–185.
[105] E.N. Manolis, E.N. Manoli, Raising awareness of the sustainable development goals through ecological projects in higher education, J. Clean. Prod. 279 (2020) 123614.
[106] A. Kowalska-Pyzalska, What makes consumers adopt to innovative energy services in the energy market? A review of incentives and barriers, Renew. Sustain. Energy Rev. 82 (2018) 3570–3581.
[107] P.A. Østergaard, N. Duic, Y. Noorollahi, H. Mikulcic, S. Kalogirou, Sustainable development using renewable energy technology, Renew. Energy 146 (2020) 2430–2437.
[108] A. Plastun, Y. Yelnikova, A. Shelyuk, A. Vorontsova, A. Artemenko, The role of public investment policy and responsible investment in sustainable development financing, Agric. Res. Econ.: Int. Sci. E-J. 6 (2) (2020) 108–125.
[109] I. Delponte, I. Pittaluga, C. Schenone, Monitoring and evaluation of sustainable energy action plan: practice and perspective, Energy Policy 100 (2017) 9–17.
[110] J.B. Jimoh, C.A. Ajagbe, The perspective and Association of Geography with environment and society, Int. J. Recent Res. Social Sci. Humanities 2 (4) (2015) 248–256.
[111] L. Horrigan, R.S. Lawrence, P. Walker, How sustainable agriculture can address the environmental and human health harms of industrial agriculture, Environ. Health Perspect. 110 (5) (2002) 445–456.
[112] S.D. Sadhu, M. Garg, A. Kumar, Major environmental issues and new materials, Chapter 4, in: C.M. Hussain, A.K. Mishra (Eds.), New Polymer Nanocomposites for Environmental Remediation, Elsevier, 2018, pp. 77–97.
[113] A.R. Kazemi, S.B. Mahbaz, A.R. Dehghani-Sanij, M.B. Dusseault, R. Fraser, Performance evaluation of an enhanced geothermal system in the Western Canada Sedimentary Basin, Renew. Sustain. Energy Rev. 113 (2019) 109278.
[114] A. Mardiana, S.B. Riffat, Building energy consumption and carbon dioxide emissions: threat to climate change, J. Earth Sci. Clim. Change S3 (2015) 001.
[115] R.H. Moss, The next generation of scenarios for climate change research and assessment, Nature 463 (2010) 747–756.
[116] IEA, CO_2 Emissions from Fuel Combustion (Highlights), 2013 edition, International Energy Agency (IEA), Paris, France, 2013.
[117] P. Nejat, F. Jomehzadeh, M.M. Taheri, M. Gohari, M.Z.A. Majid, A global review of energy consumption, CO_2 emissions and policy in the residential sector (with an overview of the top ten CO_2 emitting countries), Renew. Sustain. Energy Rev. 43 (2015) 843–862.
[118] M. Soltani, F.M. Kashkooli, A.R. Dehghani-Sanij, A.R. Kazemi, N. Bordbar, M.J. Farshchi, M. Elmi, K. Gharali, M.B. Dusseault, A comprehensive study of geothermal heating and cooling systems, Sustain. Cities Soc. 44 (2019) 793–818.
[119] C. Le Quéré, R.M. Andrew, P. Friedlingstein, et al., Global carbon budget 2018, Earth Syse. Sci. Data 10 (4) (2018) 2141–2194.

[120] T. Boden, B. Andres, Global CO_2 Emissions from Fossil-Fuel Burning, Cement Manufacture, and Gas Flaring: 1751–2013 Oak Ridge, Tennesse Carbon Dioxide Information Analysis Centre, Oak Ridge National Laboratory, Oak Ridge, TN, USA, 2016.
[121] B.P. Heard, B.W. Brook, T.M.L. Wigley, C.J.A. Bradshaw, Burden of proof: a comprehensive review of the feasibility of 100% renewable-electricity systems, Renew. Sustain. Energy Rev. 76 (2017) 1122–1133.
[122] F. Joos, R. Spahni, Rates of change in natural and anthropogenic radiative forcing over the past 20000 years, Proc. Natl. Acad. Sci. U. S. A. 105 (5) (2008) 1425–1430.
[123] A. Fernández-Cirelli, Environment—water interactions, in: M.C. Donoso (Ed.), Water Interactions with Energy, Environment, Food and Agriculture, vol. I, Eolss Publishers Co. Ltd., Oxford, UK, 2009, pp. 31–59.
[124] U.S. National Oceanic and Atmospheric Administration (NOAA), Carbon dioxide levels in atmosphere hit record high in May, Department of Commerce, 2019. https://www.noaa.gov/news/carbon-dioxide-levels-in-atmosphere-hit-record-high-in-may. (Accessed 21 October 2020).
[125] UNEP, Emissions Gap Report 2019, The United Nations Environment Programme (UNEP), Nairobi, Kenya, 2019, p. 81.
[126] EIA, International Energy Outlook 2017 (IEO2017), U.S. Energy Information Administration (U.S. EIA), Washington, DC, USA, 2017.
[127] B. Dudley, BP Energy Outlook, 2019 edition, BP p.l.c, London, UK, 2019, p. 141.
[128] L. Pérez-Lombard, J. Ortiz, C. Pout, A review on buildings energy consumption information, Energ. Buildings 40 (3) (2008) 394–398.
[129] A. Khajuria, N.H. Ravindranath, Climate change vulnerability assessment: approaches DPSIR framework and vulnerability index, J. Earth Sci. Clim. Change 3 (2012) 109.
[130] A.Y. Htut, S. Shrestha, V. Nitivattananon, A. Kawasaki, Forecasting climate change scenarios in the Bago river basin, Myanmar, J. Earth Sci. Clim. Change 5 (2014) 228.
[131] A. Ghorani-Azam, B. Riahi-Zanjani, M. Balali-Mood, Effects of air pollution on human health and practical measures for prevention in Iran, J. Res. Med. Sci. 21 (2016) 65.
[132] N.A.B. Mabahwi, O.L.H. Leh, D. Omar, Human health and wellbeing: human health effect of air pollution, Procedia Soc. Behav. Sci. 153 (2014) 221–229.
[133] L. Leena, Impact of pollution on human health: a mini review article, Environ. Conserv. J. 19 (1&2) (2018) 11–15.
[134] R.P. Singh, Z. Islam, Environmental Studies, Concept Publishing Company Pvt. Ltd., New Delhi, Delhi, India, 2012.
[135] S.M. Khopkar, Environmental Pollution: Monitoring and Control, New Age International (P) Ltd., New Delhi, Delhi, India, 2007.
[136] T. Husain, J.R. Chaudhary, Human health risk assessment due to global warming—a case study of the Gulf countries, Int. J. Environ. Res. Public Health 5 (4) (2008) 204–212.
[137] R.S. Ahima, Global warming threatens human thermoregulation and survival, J. Clin. Invest. 130 (2) (2020) 559–561.
[138] J. Scheffran, M. Brzoska, H.G. Brauch, P.M. Link, J. Schilling (Eds.), Climate Change and Human Security and Violent Conflict: Challenges for Societal Stability, vol. 8, Springer-Verlag, Berlin Heidelberg, 2012.
[139] L. Merone, P. Tait, 'Climate refugees': is it time to legally acknowledge those displaced by climate disruption? Aust. N. Z. J. Public Health 42 (6) (2018) 508–509.
[140] N. Myers, Environmental refugees: a growing phenomenon of the 21st century, Philos. Trans. R. Soc. Lond. B Biol. Sci. 357 (1420) (2002) 609–613.
[141] D. Bhowmik, Problems of refugee and the climate change, in: S.K. Das, N. Chowdhary (Eds.), Refugee Crises and Third-World Economies, Emerald Publishing Ltd., Bingley, UK, 2020, pp. 45–72.
[142] M. Ram, D. Bogdanov, A. Aghahosseini, A. Gulagi, A.S. Oyewo, M. Child, U. Caldera, K. Sadovskaia, J. Farfan, L.S.N.S. Barbosa, M. Fasihi, S. Khalili, B. Dalheimer,

G. Gruber, T. Traber, F. De Caluwe, H.-J. Fell, C. Breyer, Global Energy System Based on 100% Renewable Energy—Power, Heat, Transport and Desalination Sectors, Study by Lappeenranta University of Technology and Energy Watch Group, Lappeenranta, Berlin, Germany, March 2019.
[143] S.B. Mahbaz, A.R. Dehghani-Sanij, M.B. Dusseault, J.S. Nathwani, Enhanced and integrated geothermal systems for sustainable development of Canada's northern communities, Sustainable Energy Technol. Assess. 37 (2020) 100565.
[144] M.Z. Malik, F. Musharavati, S. Khanmohammadi, M.M. Baseri, P. Ahmadi, D.D. Nguyen, Ocean thermal energy conversion (OTEC) system boosted with solar energy and TEG based on exergy and exergo-environment analysis and multi-objective optimization, Sol. Energy 208 (2020) 559–572.
[145] Z. Wang, Q. Bui, B. Zhang, T.L.H. Pham, Biomass energy production and its impacts on the ecological footprint: an investigation of the G7 countries, Sci. Total Environ. 743 (2020) 140741.
[146] H. AlMohamadi, S. Gunukula, W.J. DeSisto, M.C. Wheeler, Formate-assisted pyrolysis of biomass: an economic and modeling analysis, Biofuels Bioprod. Biorefin. 12 (1) (2018) 45–55.
[147] J.K. Kazak, J.A. Kamińska, R. Madej, M. Bochenkiewicz, Where renewable energy sources funds are invested? Spatial analysis of energy production potential and public support, Energies 13 (21) (2020) 5551.
[148] H. Dagdougui, A. Ouammi, R. Sacile, A regional decision support system for onsite renewable hydrogen production from solar and wind energy sources, Int. J. Hydrogen Energy 36 (22) (2011) 14324–14334.
[149] B. Dudley, BP Energy Outlook: 2018 edition, BP p.l.c, London, UK, 2018, p. 124.
[150] N. Abas, A. Kalair, N. Khan, Review of fossil fuels and future energy technologies, Futures 69 (2015) 31–49.
[151] WEC, World Energy Resources, The World Energy Council (WEC), London, UK, 2016. https://www.worldenergy.org/assets/images/imported/2016/10/World-Energy-Resources-Full-report-2016.10.03.pdf. (Accessed 27 November 2020).
[152] J.C.A.Y. Rizk, Y. Chaiko, Solar tracking system: more efficient use of solar panels, World Acad. Sci. Eng. Technol. 41 (2008) 313–315.
[153] M.Z. Jacobson, C.L. Archer, Saturation wind power potential and its implications for wind energy, Proc. Natl. Acad. Sci. 109 (39) (Sept. 2012) 15679–15684.
[154] T. Wilberforce, Z. El Hassan, A. Durrant, J. Thompson, B. Soudan, A.G. Olabi, Overview of ocean power technology, Energy 175 (2019) 165–181.
[155] M. Melikoglu, Geothermal energy in Turkey and around the world: a review of the literature and an analysis based on Turkey's vision 2023 energy targets, Renew. Sustain. Energy Rev. 76 (2017) 485–492.
[156] J. Yang, X. Li, H. Hu, X. Zhang, Y. Yu, Y. Chen, Growth and lipid accumulation properties of a freshwater microalga, *Chlorella ellipsoidea* YJ1, in domestic secondary effluents, Appl. Energy 88 (10) (2011) 3295–3299.
[157] N. Bowden, J.E. Payne, Sectoral analysis of the causal relationship between renewable and non-renewable energy consumption and real output in the US, Energy Sources Part B 5 (4) (2010) 400–408.
[158] R.A. Salim, K. Hassan, S. Shafiei, Renewable and non-renewable energy consumption and economic activities: further evidence from OECD countries, Energy Econ. 44 (2014) 350–360.
[159] B. Ozcan, I. Ozturk, Renewable energy consumption-economic growth nexus in emerging countries: a bootstrap panel causality test, Renew. Sustain. Energy Rev. 104 (2019) 30–37.
[160] N. Wohlgemuth, Evaluation of renewable energy policies in an integrated economic-energy-environment model, Forest Policy Econ. 10 (3) (2008) 128–139.

[161] S. Gerst, C. Hettig, M. Matheis, Denmark Government Measures for Exporters in Times of COVID-19 Crisis, IfTI Working Paper 5, Institute for Trade and Innovation (IfTI), Offenburg University, Germany, 2020.
[162] S. Saint Akadiri, A.A. Alola, A.C. Akadiri, U.V. Alola, Renewable energy consumption in EU-28 countries: policy toward pollution mitigation and economic sustainability, Energy Policy 132 (2019) 803–810.
[163] B. Pan, Y. Zhang, Impact of affluence, nuclear and alternative energy on US carbon emissions from 1960 to 2014, Energ. Strat. Rev. 32 (2020) 100581.
[164] M. Mahmoud, M. Ramadan, A.G. Olabi, K. Pullen, S. Naher, A review of mechanical energy storage systems combined with wind and solar applications, Energ. Conver. Manage. 210 (2020) 112670.
[165] A. Ali, W. Li, R. Hussain, X. He, B.W. Williams, A.H. Memon, Overview of current microgrid policies, incentives and barriers in the European Union, United States and China, Sustainability 9 (7) (2017) 1146.
[166] M. DeCastro, S. Salvador, M. Gómez-Gesteira, X. Costoya, D. Carvalho, F.J. Sanz-Larruga, L. Gimeno, Europe, China and the United States: three different approaches to the development of offshore wind energy, Renew. Sustain. Energy Rev. 109 (2019) 55–70.
[167] V. Smil, The long slow rise of solar and wind, Sci. Am. 310 (1) (2014) 52–57.
[168] Government of Canada, National Energy Board (NEB), Provincial and Territorial Energy Profiles—Canada, 2020. https://www.neb-one.gc.ca/nrg/ntgrtd/mrkt/nrgsstmprfls/cda-eng.html. (Accessed 27 November 2020).
[169] G.R. Timilsina, Biofuels in the long-run global energy supply mix for transportation, Philos. Trans. R. Soc. A Math. Phys. Eng. Sci. 372 (2006) (2014) 20120323.
[170] IEA, World Energy Balances 2019: Data and Statistics, International Energy Agency (IEA), Paris, France, 2019.
[171] M. Pagliaro, F. Meneguzzo, Distributed generation from renewable energy sources: ending energy poverty across the world, Energ. Technol. 8 (7) (2020) 2000126.
[172] L. Myllyvirta, Guest post: Why China's CO_2 emissions grew 4% during first half of 2019, CarbonBrief: Clear on Climate, 5 Sept. 2019. www.carbonbrief.org/guest-post-why-chinas-co2-emissions-grew-4-duringfirst-half-of-2019. (Accessed 1 April 2020).
[173] M. Pagliaro, F. Meneguzzo, F. Zabini, R. Ciriminna, Assessment of the minimum value of photovoltaic electricity in Italy, Energy Sci. Eng. 2 (2) (2014) 94–105.
[174] J. Cludius, H. Hermann, F.C. Matthes, V. Graichen, The merit order effect of wind and photovoltaic electricity generation in Germany 2008–2016: estimation and distributional implications, Energy Econ. 44 (2014) 302–313.
[175] A. Ciarreta, M.P. Espinosa, C. Pizarro-Irizar, Is green energy expensive? Empirical evidence from the Spanish electricity market, Energy Policy 69 (2014) 205–215.
[176] L. Albanese, R. Ciriminna, F. Meneguzzo, M. Pagliaro, The impact of electric vehicles on the power market, Energy Sci. Eng. 3 (4) (2015) 300–309.
[177] D. Maugouber, D. Doherty, 2019 Road Fuel Outlook, Bloomberg New Energy Finance (BNEF), 2019.
[178] BMI, The Three Tiers of Battery Megafactories, Benchmark Mineral Intelligence (BMI), 13 Sept. 2019. www.benchmarkminerals.com/the-three-tiers-of-battery-megafactories. (Accessed 27 November 2020).
[179] BNEF, Long-Term Electric Vehicle Outlook 2018, Bloomberg New Energy Finance (BNEF), 2018.
[180] BNEF, Long-Term Electric Vehicle Outlook 2019, Bloomberg New Energy Finance (BNEF), 2019.
[181] P. Sanjay, India Added 5.4 GW of Solar in 9M 2019, with 2.2 GW Installed in Q3 2019, Mercom India, 21 Nov. 2019. https://mercomindia.com/india-solar-installed-q3-2019. (Accessed 27 November 2020).

[182] ITA, India—Energy, The International Trade Administration (ITA), U.S. Department of Commerce Manages, Washington, DC, USA, May 2019. https://www.export.gov/apex/article2?id=India-Energy. (Accessed 27 November 2020).
[183] B.N. Porfiriev, S.A. Roginko, Energy on renewable sources: prospects for the world and for Russia, Her. Russ. Acad. Sci. 86 (6) (2016) 433–440.
[184] B.N. Porfiriev, New global trends in energy power development—challenges and risks for the Russia integration into the world economy, Stud. Russ. Econ. Dev. 26 (1) (2015) 32–36.
[185] Eurostat: Statistics Explained, Archive: Energy from Renewable Sources, Jan. 2018. http://ec.europa.eu/eurostat/statistics-explained/index.php/Energy_from_renewable_sources#Further_Eurostat_information. (Accessed 27 November 2020).
[186] R. Priddle, World Energy Outlook 2014, International Energy Agency (IEA), Paris, France, Nov. 2014.
[187] J.L. Sawin, K.M. Seyboth, F. Sverrisson, Renewables 2016: Global Status Report, Renewable Energy Policy Network for the 21st Century (REN21) Secretariat, 2016, p. 271. Paris, France.
[188] L. Mills, J. Byrne, Clean Energy Investment: Q4 2015 Factpack, Bloomberg New Energy Finance (BNEF), Jan. 2016.
[189] BNEF, New Energy Outlook, Long-Term Projections of the Global Energy Sector, Executive Summary, Bloomberg New Energy Finance (BNEF), 2015.
[190] IEA, Energy and Climate Change, World Energy Outlook Special Report, International Energy Agency (IEA), Paris, France, June 2015.
[191] G.E. Halkos, E.C. Gkampoura, Reviewing usage, potentials, and limitations of renewable energy sources, Energies 13 (11) (2020) 2906.
[192] P.A. Owusu, S. Asumadu-Sarkodie, A review of renewable energy sources, sustainability issues and climate change mitigation, Cogent Eng. 3 (1) (2016) 1167990.
[193] B.W. Brook, C.J. Bradshaw, Key role for nuclear energy in global biodiversity conservation, Conserv. Biol. 29 (3) (2015) 702–712.
[194] J.J. Klemeš, P.S. Varbanov, P. Ocłoń, H.H. Chin, Towards efficient and clean process integration: utilisation of renewable resources and energy-saving technologies, Energies 12 (21) (2019) 4092.
[195] J. Mohtasham, Review article-renewable energies, Energy Procedia 74 (2015) 1289–1297.
[196] R. Shortall, B. Davidsdottir, G. Axelsson, Geothermal energy for sustainable development: a review of sustainability impacts and assessment frameworks, Renew. Sustain. Energy Rev. 44 (2015) 391–406.
[197] D.K. Okot, Review of small hydropower technology, Renew. Sustain. Energy Rev. 26 (2013) 515–520.
[198] M. Melikoglu, Current status and future of ocean energy sources: a global review, Ocean Eng. 148 (2018) 563–573.
[199] G. Wright, A.M. O'Hagan, J. de Groot, Y. Leroy, N. Soininen, R. Salcido, M.A. Castelos, S. Jude, J. Rochette, S. Kerr, Establishing a legal research agenda for ocean energy, Mar. Policy 63 (2016) 126–134.
[200] S. Guo, Q. Liu, J. Sun, H. Jin, A review on the utilization of hybrid renewable energy, Renew. Sustain. Energy Rev. 91 (2018) 1121–1147.
[201] C.V. Nayar, W.B. Lawrance, S.J. Phillips, Solar/wind/diesel hybrid energy systems for remote areas, in: Proceedings of the 24th Intersociety Energy Conversion Engineering Conference, Aug. 6–11, 1989, IEEE, Washington, DC, 1989, pp. 2029–2034.
[202] T. Markvart, Sizing of hybrid photovoltaic-wind energy systems, Sol. Energy 57 (4) (1996) 277–281.
[203] J.A. Castle, J.M. Kallis, S.M. Moite, N.A. Marshall, Analysis of merits of hybrid wind/photovoltaic concept for stand-alone systems, in: 15th Photovoltaic Specialists Conference, Kissimmee, FL, May 12–15, 1981, pp. 738–744.

[204] B.J. Losie, J.G. Steeves, Case study of a hybrid wind/photovoltaic/diesel power system at a radio repeater site, in: INTELEC '84—International Telecommunications Energy Conference, Nov. 4–7, 1984, IEEE, New Orleans, LA, USA, 1984, pp. 159–165.
[205] J.G. McGowan, J.F. Manwell, C. Avelar, C.L. Warner, Hybrid wind/PV/diesel hybrid power systems modeling and South American applications, Renew. Energy 9 (1–4) (1996) 836–847.
[206] P.G. Arul, V.K. Ramachandaramurthy, R.K. Rajkumar, Control strategies for a hybrid renewable energy system: a review, Renew. Sustain. Energy Rev. 42 (2015) 597–608.
[207] M.D. Al-Falahi, S.D.G. Jayasinghe, H.J.E.C. Enshaei, A review on recent size optimization methodologies for standalone solar and wind hybrid renewable energy system, Energ. Conver. Manage. 143 (2017) 252–274.
[208] A. Balleh, A. Faydallah, N. Moubayed, Survey on the efficiency and the profitability of an hybrid wind/PV system: the Lebanese case, in: 2016 International Conference and Exposition on Electrical and Power Engineering (EPE), Oct. 20–22, 2016, IEEE, Iasi, Romania, 2016, pp. 902–907.
[209] A. Hiendro, R. Kurnianto, M. Rajagukguk, Y.M. Simanjuntak, Techno-economic analysis of photovoltaic/wind hybrid system for onshore/remote area in Indonesia, Energy 59 (2013) 652–657.
[210] J. Prasad, V.K. Vijay, Experimental studies on drying of *Zingiber officinale*, *Curcuma longa* l. and Tinospora cordifolia in solar-biomass hybrid drier, Renew. Energy 30 (14) (2005) 2097–2109.
[211] J.D. Nixon, P.K. Dey, P.A. Davies, The feasibility of hybrid solar-biomass power plants in India, Energy 46 (1) (2012) 541–554.
[212] M. Ayub, A. Mitsos, H. Ghasemi, Thermo-economic analysis of a hybrid solar-binary geothermal power plant, Energy 87 (2015) 326–335.
[213] I. Thain, R. DiPippo, Hybrid geothermal-biomass power plants: applications, designs and performance analysis, in: Proceedings World Geothermal Congress, Melbourne, Australia, April 2015, pp. 19–25.
[214] T. Ma, H. Yang, L. Lu, A feasibility study of a stand-alone hybrid solar–wind–battery system for a remote island, Appl. Energy 121 (2014) 149–158.
[215] Y. Altin, A.Ç. Bedeloğlu, Energy storage textile, Chapter 19, in: S. ul-Islam, B.S. Butola (Eds.), Advances in Functional and Protective Textiles, Woodhead Publishing, Cambridge, UK, 2020, pp. 493–529.
[216] V. Sultan, A. Alzahrani, H. Bitar, B. Hilton, A conceptual framework for battery storage integration into the electric grid, in: 2017 International Conference on Computational Science and Computational Intelligence (CSCI), Dec. 14–16, 2017, IEEE, Las Vegas, NV, USA, 2017, pp. 161–167.
[217] A. Buttler, H. Spliethoff, Current status of water electrolysis for energy storage, grid balancing and sector coupling via power-to-gas and power-to-liquids: a review, Renew. Sustain. Energy Rev. 82 (2018) 2440–2454.
[218] M. Majidi, S. Nojavan, Optimal sizing of energy storage system in a renewable-based microgrid under flexible demand side management considering reliability and uncertainties, J. Oper. Autom. Power Eng. 5 (2) (2017) 205–214.
[219] Y. Yang, S. Bremner, C. Menictas, M. Kay, Battery energy storage system size determination in renewable energy systems: a review, Renew. Sustain. Energy Rev. 91 (2018) 109–125.
[220] P. Holmér, J. Ullmark, L. Göransson, V. Walter, F. Johnsson, Impacts of thermal energy storage on the management of variable demand and production in electricity and district heating systems: a Swedish case study, Int. J. Sustain. Energy 39 (5) (2020) 446–464.
[221] T. Lv, Q. Ai, Interactive energy management of networked microgrids-based active distribution system considering large-scale integration of renewable energy resources, Appl. Energy 163 (2016) 408–422.

[222] A.A. Eladl, M.I. El-Afifi, M.A. Saeed, M.M. El-Saadawi, Optimal operation of energy hubs integrated with renewable energy sources and storage devices considering CO_2 emissions, Int. J. Electr. Power Energy Syst. 117 (2020) 105719.
[223] S. Aznavi, P. Fajri, R. Sabzehgar, A. Asrari, Optimal management of residential energy storage systems in presence of intermittencies, J. Build. Eng. 29 (2020) 101149.
[224] L. Al-Ghussain, R. Samu, O. Taylan, M. Fahrioglu, Sizing renewable energy systems with energy storage systems in microgrids for maximum cost-efficient utilization of renewable energy resources, Sustain. Cities Soc. 55 (2020) 102059.
[225] K. Sharma, A. Arora, S.K. Tripathi, Review of supercapacitors: materials and devices, J. Energy Storage 21 (2019) 801–825.
[226] M.S. Guney, Y. Tepe, Classification and assessment of energy storage systems, Renew. Sustain. Energy Rev. 75 (2017) 1187–1197.
[227] I. Dincer, M. Rosen, Thermal Energy Storage: Systems and Applications, second ed., John Wiley & Sons, Inc., West Sussex, UK, 2011.
[228] X. Luo, J. Wang, M. Dooner, J. Clarke, Overview of current development in electrical energy storage technologies and the application potential in power system operation, Appl. Energy 137 (2015) 511–536.
[229] A.K. Arani, H. Karami, G. Gharehpetian, M. Hejazi, Review of flywheel energy storage systems structures and applications in power systems and microgrids, Renew. Sustain. Energy Rev. 69 (2017) 9–18.
[230] W. He, J. Wang, Optimal selection of air expansion machine in compressed air energy storage: a review, Renew. Sustain. Energy Rev. 87 (2018) 77–95.
[231] T. Sant, D. Buhagiar, R.N. Farrugia, Evaluating a new concept to integrate compressed air energy storage in spar-type floating offshore wind turbine structures, Ocean Eng. 166 (2018) 232–241.
[232] S.P. Koko, K. Kusakana, H.J. Vermaak, Optimal power dispatch of a grid-interactive micro-hydrokinetic-pumped hydro storage system, J. Energy Storage 17 (2018) 63–72.
[233] A.G. Endegnanew, H. Farahmand, D. Huertas-Hernando, Frequency quality in the Nordic power system: wind variability, hydro power pump storage and usage of HVDC links, Energy Procedia 35 (2013) 62–68.
[234] W.-D. Steinmann, Thermal energy storage systems for concentrating solar power (CSP) plants, Chapter 11, in: K. Lovegrove, W. Stein (Eds.), Concentrating Solar Power Technology Principles, Developments and Applications, Woodhead Publishing, Cambridge, UK, 2012, pp. 362–394.
[235] S.T. Revankar, Chemical energy storage, Chapter 6, in: H. Bindra, S. Revankar (Eds.), Storage and Hybridization of Nuclear Energy, Elsevier, 2019, pp. 177–227. Imprint by Academic Press.
[236] N.K.C. Nair, N. Garimella, Battery energy storage systems: assessment for small-scale renewable energy integration, Energ. Buildings 42 (11) (2010) 2124–2130.
[237] Stadler, M. Sterner, Urban energy storage and sector coupling, in: P. Droege (Ed.), Urban Energy Transition: Renewable Strategies for Cities and Regions, second ed., Elsevier, 2018, pp. 225–244.
[238] C. Wolfersdorf, K. Boblenz, R. Pardemann, B. Meyer, Syngas-based annex concepts for chemical energy storage and improving flexibility of pulverized coal combustion power plants, Appl. Energy 156 (2015) 618–627.
[239] D.N. Buckley, C. O'Dwyer, N. Quill, R.P. Lynch, Electrochemical energy storage, in: R.E. Hester, R.M. Harrison (Eds.), Energy Storage Options and Their Environmental Impact, 46, CPI Group Ltd., Croydon, UK, 2018, pp. 115–149.
[240] L. Huibo, X. Jinhua, Y. Jilei, Y. Peng, X. Qingqiang, M. Qian, Study on comprehensive evaluation method of electrochemical energy storage system based on hierarchical analysis, in: 2019 IEEE 2nd International Conference on Automation, Electronics

and Electrical Engineering (AUTEEE), Nov. 22–24, 2019, IEEE, Shenyang, China, 2019, pp. 227–233.
[241] B. Wu, B. Zhang, J. Wang, J. Li, X. Zheng, Y. Liu, B. Mao, Y. Gao, Theoretical research for the application of flow storage battery in demand side management, in: 2010 International Conference on Power System Technology, Oct. 24–28, 2010, IEEE, Hangzhou, China, 2010, pp. 1–6.
[242] S. Xiaoyu, L. Wenyi, L. Xiaolong, Research on ordered charging of battery swapping station based on adaptive genetic algorithm, in: 2019 IEEE Sustainable Power and Energy Conference (iSPEC), Nov. 21–23, 2019, IEEE, Beijing, China, 2019, pp. 792–796.
[243] K. Feng, W. Ahn, G. Lui, H.W. Park, A.G. Kashkooli, G. Jiang, X. Wang, X. Xiao, Z. Chen, Implementing an in-situ carbon network in Si/reduced graphene oxide for high performance lithium-ion battery anodes, Nano Energy 19 (2016) 187–197.
[244] W. Zhuang, S. Lu, H. Lu, Progress in materials for lithium-ion power batteries, in: 2014 International Conference on Intelligent Green Building and Smart Grid (IGBSG) April 23–25, 2014, IEEE, Taipei, Taiwan, 2014, pp. 1–2.
[245] R. Hebner, J. Beno, A. Walls, Flywheel batteries come around again, IEEE Spectr. 39 (4) (2002) 46–51.
[246] L. Fiorini, M. Aiello, D. Poli, P. Pelacchi, Topological considerations on the use of batteries to enhance the reliability of HV-grids, J. Energy Storage 18 (2018) 316–326.
[247] P. Chen, W. Zhou, Z. Xiao, S. Li, Z. Wang, Y. Wang, S. Xie, An integrated configuration with robust interfacial contact for durable and flexible zinc ion batteries, Nano Energy 74 (2020) 104905.
[248] F. Li, R. Li, F. Zhou, Microgrid Technology and Engineering Application, Elsevier, London, UK, 2015. Imprint by Academic Press.
[249] K. Dhanabalan, T. Sadhasivam, S.C. Kim, J.J. Eun, J. Shim, D. Jeon, S.H. Roh, H.Y. Jung, Novel core–shell structure of a lead-activated carbon (Pb@ AC) for advanced lead–acid battery systems, J. Mater. Sci. Mater. Electron. 28 (14) (2017) 10349–10356.
[250] M.A. Zelinsky, J.M. Koch, K.-H. Young, Performance comparison of rechargeable batteries for stationary applications (Ni/MH vs. Ni-Cd and VRLA), Batteries 4 (1) (2018) 1–7.
[251] M. Winter, B. Barnett, K. Xu, Before Li ion batteries, Chem. Rev. 118 (23) (2018) 11433–11456.
[252] B.W. Zhang, T. Sheng, Y.X. Wang, S. Chou, K. Davey, S.X. Dou, S.Z. Qiao, Long-life room-temperature sodium–sulfur batteries by virtue of transition-metal-nanocluster–sulfur interactions, Angew. Chem. 131 (5) (2019) 1498–1502.
[253] C. Choi, S. Kim, R. Kim, Y. Choi, S. Kim, H.Y. Jung, J.H. Yang, H.T. Kim, A review of vanadium electrolytes for vanadium redox flow batteries, Renew. Sustain. Energy Rev. 69 (2017) 263–274.
[254] K. Friedrich, U. Breuer (Eds.), Multifunctionality of Polymer Composites: Challenges and New Solutions, Elsevier, 2015. Imprint by William Andrew.
[255] T. Christen, M.W. Carlen, Theory of Ragone Plots, J. Power Sources 91 (2) (2000) 210–216.
[256] H. Budde-Meiwes, J. Drillkens, B. Lunz, J. Muennix, S. Rothgang, J. Kowal, D.U. Sauer, A review of current automotive battery technology and future prospects, Proc. Inst. Mech. Eng D: J. Autom. Eng. 227 (5) (2013) 761–776.
[257] Z.W. Zhang, H.J. Peng, M. Zhao, J.Q. Huang, Heterogeneous/homogeneous mediators for high–energy–density Lithium–Sulfur batteries: progress and prospects, Adv. Funct. Mater. 28 (38) (2018) 1707536.
[258] Y. Ko, H. Park, B. Lee, Y. Bae, S.K. Park, K. Kang, A comparative kinetic study of redox mediators for high-power lithium–oxygen batteries, J. Mater. Chem. A 7 (11) (2019) 6491–6498.

[259] H. Palneedi, M. Peddigari, G.T. Hwang, D.Y. Jeong, J. Ryu, High-performance dielectric ceramic films for energy storage capacitors: progress and outlook, Adv. Funct. Mater. 28 (42) (2018) 1803665.

[260] I. Sarbu, C. Sebarchievici, A comprehensive review of thermal energy storage, Sustainability 10 (1) (2018) 191.

[261] I. Sarbu, C. Sebarchievici, Solar Heating and Cooling Systems: Fundamentals, Experiments and Applications, Elsevier, 2016. Imprint by Academic Press.

[262] H. Liu, X. Wang, D. Wu, Innovative design of microencapsulated phase change materials for thermal energy storage and versatile applications: a review, Sustainable Energy Fuels 3 (5) (2019) 1091–1149.

[263] A. Andrijanovits, H. Hoimoja, D. Vinnikov, Comparative review of long-term energy storage technologies for renewable energy systems, Elektron. Elektrotech. 118 (2) (2012) 21–26.

[264] D. Bhatnagar, V. Loose, Evaluating Utility Procured Electric Energy Storage Resources: A Perspective for State Electric Utility Regulators, a Study for the DOE Energy Storage Systems Program, Sandia National Laboratories, SAND2012-9422, New Mexico, CA, USA, Nov. 2012.

[265] G.J. May, A. Davidson, B. Monahov, Lead batteries for utility energy storage: a review, J. Energy Storage 15 (2018) 145–157.

[266] H. Chen, T.N. Cong, W. Yang, C. Tan, Y. Li, Y. Ding, Progress in electrical energy storage system: a critical review, Prog. Nat. Sci. 19 (3) (2009) 291–312.

[267] J. Kondoh, I. Ishii, H. Yamaguchi, A. Murata, K. Otani, K. Sakuta, N. Higuchi, S. Sekine, M. Kamimoto, Electrical energy storage systems for energy networks, Energ. Conver. Manage. 41 (17) (2000) 1863–1874.

[268] International Renewable Energy Agency (IRENA), Electricity storage: Technology brief, IEA-ETSAP and IRENA© Technology Policy Brief E18, Jan. 2013. http://www.inship.eu/docs/TES%201%20IRENA-ETSAP%20Tech%20Brief%20E17%20Thermal%20Energy%20Storage.pdf. (Accessed 7 December 2019).

[269] M. Beaudin, H. Zareipour, A. Schellenberglabe, W. Rosehart, Energy storage for mitigating the variability of renewable electricity sources: an updated review, Energy Sustain. Dev. 14 (4) (2010) 302–314.

[270] Hawaiian Electric Company (HEC), Clean Energy—Issues and Challenges—Energy Storage, 2019. http://www.heco.com/heco/Clean-Energy/Issues-and-Challenges/Energy-Storage. (Accessed 1 January 2019).

[271] S.J. Andrepont, Energy storage—thermal energy storage coupled with turbine inlet cooling, in: 14th Annual Electric Power Conference & Exhibition, 2012. http://www.turbineinletcooling.org/resources/papers/Andrepont_2012EP.pdf. (Accessed 6 December 2020).

[272] International Electrotechnical Commission (IEC), Electrical Energy Storage: White Paper, Technical report prepared by electrical energy storage project team, Geneva, Switzerland, 2011.

[273] F.A. Farret, M.G. Simões, Integration of Alternative Sources of Energy, John Wiley & Sons, Inc., Hoboken, NJ, 2006.

[274] M. Finkenrath, S. Pazzi, M. D'Ercole, Status and Technical Challenges of Advanced Compressed Air Energy Storage (CAES) Technology, International Workshop on Environment and Alternative Energy, Munich, Germany, 2009.

[275] R.W.E. Power, ADELE—Adiabatic Compressed-Air Energy Storage (CAES) for Electricity Supply, 2019. http://www.rwe.com/web/cms/mediablob/en/391748/data/235554/1/rwe-power-ag/press/company/Brochure-ADELE.pdf. (Accessed 1 January 2019).

[276] Flowgroup, Compressed Air Batteries, 2019. http://www.flowbattery.co.uk. (Accessed 1 January 2019).

[277] S.M. Shoenung, Characteristics and technologies for long- vs. short-term energy storage: A study by the DOE energy storage systems program, Technical report, SAND2001-0765, Sandia National Laboratories, New Mexico, CA, USA, 2001.
[278] Electricity Storage Association (ESA), Electricity Storage Technology Comparison, 2013. http://www.electricitystorage.org/technology/storage_technologies/technology_comparison. (Accessed 22 November 2013).
[279] W. Smith, The role of fuel cells in energy storage, J. Power Sources 86 (1–2) (2000) 74–83.
[280] Emerson Network Power (ENP), Capacitors Age and Capacitors Have an End of Life, A White Paper From the Experts in Business-Critical Continuity, 2019. http://www.emersonnetworkpower.com/documentation/en-us/brands/liebert/documents/white%20papers/sl-24630.pdf. (Accessed 1 January 2019).
[281] R.J. Miller, Capacitors for power grid storage - multi-hour bulk energy storage using capacitors, in: Trans-Atlantic Workshop on Storage Technologies for Power Grids, JME, Inc. and Case Western Reserve University, Washington, DC, USA, 2010.
[282] M. Winter, R.J. Brodd, What are batteries, fuel cells, and supercapacitors? Chem. Rev. 104 (10) (2004) 4245–4270.
[283] R. Baxter, Energy Storage: A Nontechnical Guide, PennWell, Tulsa, Oklahoma, USA, 2006.
[284] J.P. Barton, D.G. Infield, Energy storage and its use with intermittent renewable energy, IEEE Trans. Energy Convers. 19 (2) (2004) 441–448.
[285] C. Schaber, P. Mazza, R. Hammerschlag, Utility-scale storage of renewable energy, Electr. J. 17 (6) (2004) 21–29.
[286] I. Hadjipaschalis, A. Poullikkas, V. Efthimiou, Overview of current and future energy storage technologies for electric power applications, Renew. Sustain. Energy Rev. 13 (6–7) (2009) 1513–1522.
[287] F. Díaz-González, A. Sumper, O. Gomis-Bellmunt, R. Villafáfila-Robles, A review of energy storage technologies for wind power applications, Renew. Sustain. Energy Rev. 16 (4) (2012) 2154–2171.
[288] J.K. Kaldellis, D. Zafirakis, Optimum energy storage techniques for the improvement of renewable energy sources-based electricity generation economic efficiency, Energy 32 (12) (2007) 2295–2305.
[289] C.J. Rydh, B.A. Sanden, Energy analysis of batteries in photovoltaic Systems—part I: performance and energy requirements, Energ. Conver. Manage. 46 (11–12) (2005) 1957–1979.
[290] Department of Trade and Industry (DTI), Review of Electrical Energy Storage Technologies and Systems and of Their Potential for the UK, DTI Report, DG/DTI/00055/00/00, UK Department of Trade and Industry, Durham, UK, 2004.
[291] D. Rastler, Electricity Energy Storage Technology Options: A White Paper Primer on Applications, Costs, and Options, Technical Report, 1020676, Electric Power Research Institute (EPRI), Inc., Washington, DC, USA, Dec. 2010.
[292] M.P. Moghaddam, M R. Haghifam, G.R. Yousefi, Electric energy storage systems in a market-based economy: comparison of emerging and traditional technologies, Renew. Energy 34 (12) (2009) 2630–2639.
[293] S. Schoenung, Energy Storage Systems Cost Update: A Study for the DOE Energy Storage Systems Program, Technical Report, Sandia National Laboratories, New Mexico, CA, USA, 2011.
[294] J. McDowall, Integrating energy storage with wind power in weak electricity grids, J. Power Sources 162 (2) (2006) 959–964.
[295] C. Ponce de León, A. Frías-Ferrer, J. González-García, D.A. Szánto, F.C. Walsh, Redox flow cells for energy conversion, J. Power Sources 160 (1) (2006) 716–732.

[296] S.J. Amodeo, H.G. Chiacchiarini, J.A. Solsona, C.A. Busada, High-performance sensorless nonlinear power control of a flywheel energy storage system, Energ. Conver. Manage. 50 (7) (2009) 1722–1729.
[297] A.Z. Weber, M.M. Mench, J.P. Meyers, P.N. Ross, J.T. Gostick, Q. Liu, Redox flow batteries: a review, J. Appl. Electrochem. 41 (10) (2011) 1137–1164.
[298] B. Paul, The Future of Electrical Energy Storage: The Economics and Potential of New Technologies, Business Wire, Inc., Report, Business Insights (Energy), 2009.
[299] P. Taylor, R. Bolton, D. Stone, X.-P. Zhang, C. Martin, P. Upham, Pathways for energy storage in the UK, The Centre for Low Carbon Futures, White Rose University Consortium, New York, UK, March 2012, p. 56.
[300] R. Dan, Overview of Energy Storage Options for the Electric Enterprise, Electric Power Research Institute (EPRI), Washington, DC, USA, 2009.
[301] S. Mekhilef, R. Saidur, A. Safari, Comparative study of different fuel cell technologies, Renew. Sustain. Energy Rev. 16 (1) (2012) 981–989.
[302] A. Barin, L.N. Canha, K. Magnago, A. da Rosa Abaide, Fuzzy multi-sets and multi-rules: analysis of hybrid systems concerning renewable sources with conventional and flow batteries, in: 2009 15th International Conference on Intelligent System Applications to Power Systems, IEEE, Curitiba, Brazil, 2009, pp. 1–6.
[303] K.C. Divya, J. Østergaard, Battery energy storage technology for power systems—an overview, Electr. Pow. Syst. Res. 79 (4) (2009) 511–520.
[304] J. Baker, New technology and possible advances in energy storage, Energy Policy 36 (12) (2008) 4368–4373.
[305] S. Cole, D. Van Hertem, L. Meeus, R. Belmans, SWOT analysis of utility side energy storage technologies, in: Proceedings of the 5th WSEAS/IASME International Conference on Electric Power Systems, High Voltages, Electric Machines, Tenerife, Spain, 2005, pp. 16–8.
[306] European Commission (EC), D4.1 SWOT Analysis, Executive Agency for Competitiveness and Innovation (EACI), EIE/07/159/SI2, 466845, 2010. https://ec.europa.eu/energy/intelligent/projects/sites/iee-projects/files/projects/documents/stories_swot_analysis.pdf. (Accessed 7 December 2019).
[307] C.-C. Liu, S. McArthur, S.-J. Lee (Eds.), Smart Grid Handbook, vol. 1, John Wiley & Sons, Inc., West Sussex, UK, 2016.
[308] S. Sabihuddin, A.E. Kiprakis, M. Mueller, A numerical and graphical review of energy storage technologies, Energies 8 (1) (2015) 172–216.
[309] T.B. Reddy, D. Linden, Linden's Handbook of Batteries, fourth ed., McGraw Hill Professional, New York, NY, 2010.
[310] E. Ela, C. Wang, S. Moorty, K. Ragsdale, J. O'Sullivan, M. Rothleder, B. Hobbs, Electricity markets and renewables: a survey of potential design changes and their consequences, IEEE Power Energ. Mag. 15 (6) (2017) 70–82.
[311] M.E. Amiryar, K.R. Pullen, A review of flywheel energy storage system technologies and their applications, Appl. Sci. 7 (3) (2017) 286.
[312] H.A. Kiehne (Ed.), Battery Technology Handbook, second ed., Expert Verlag, Renningen, Germany, 2003.
[313] M.C. McManus, Environmental consequences of the use of batteries in low carbon systems: the impact of battery production, Appl. Energy 93 (2012) 288–295.
[314] M.A. Rahman, X. Wang, C. Wen, High energy density metal-air batteries: a review, J. Electrochem. Soc. 160 (10) (2013) A1759–A1771.
[315] C. Zhang, Y.-L. Wei, P.-F. Cao, M.-C. Lin, Energy storage system: current studies on batteries and power condition system, Renew. Sustain. Energy Rev. 82 (2018) 3091–3106.
[316] M. Medrano, M.O. Yilmaz, M. Nogués, I. Martorell, J. Roca, L.F. Cabeza, Experimental evaluation of commercial heat exchangers for use as PCM thermal storage systems, Appl. Energy 86 (10) (2009) 2047–2055.

[317] M. Noro, R.M. Lazzarin, F. Busato, Solar cooling and heating plants: an energy and economic analysis of liquid sensible vs phase change material (PCM) heat storage, Int. J. Refrig. 39 (2014) 104–116.
[318] M.M.A. Khan, R. Saidur, F.A. Al-Sulaiman, A review for phase change materials (PCMs) in solar absorption refrigeration systems, Renew. Sustain. Energy Rev. 76 (2017) 105–137.
[319] L.A. Chidambaram, A.S. Ramana, G. Kamaraj, R. Velraj, Review of solar cooling methods and thermal storage options, Renew. Sustain. Energy Rev. 15 (6) (2011) 3220–3228.
[320] A. Sharma, V.V. Tyagi, C.R. Chen, D. Buddhi, Review on thermal energy storage with phase change materials and applications, Renew. Sustain. Energy Rev. 13 (2) (2009) 318–345.
[321] D. Zhou, C.Y. Zhao, Y. Tian, Review on thermal energy storage with phase change materials (PCMs) in building applications, Appl. Energy 92 (2012) 593–605.
[322] P. Moreno, C. Solé, A. Castell, L.F. Cabeza, The use of phase change materials in domestic heat pump and air-conditioning systems for short term storage: a review, Renew. Sustain. Energy Rev. 39 (2014) 1–13.
[323] M. Liu, W. Saman, F. Bruno, Review on storage materials and thermal performance enhancement techniques for high temperature phase change thermal storage systems, Renew. Sustain. Energy Rev. 16 (4) (2012) 2118–2132.
[324] X.Q. Zhai, X.L. Wang, T. Wang, R.Z. Wang, A review on phase change cold storage in air-conditioning system: materials and applications, Renew. Sustain. Energy Rev. 22 (2013) 108–120.

CHAPTER 2

Natural production of ice

2.1 Introduction

Before the appearance of artificial cold production via refrigeration systems, a number of traditional techniques and environmentally friendly technologies allowed people to drink and enjoy cold water and/or a wide variety of cold beverages throughout the hot/warm months of the year. Physiologically, drinking cold fluids on hot summer days is more refreshing and pleasant than drinking lukewarm ones, and the fluid can be absorbed more swiftly by the body, thereby reducing the risk of dehydration [1]. The following paragraphs introduce and describe some traditional approaches and eco-friendly technologies that have used to supply cold water and produce ice naturally in past times:

1. Water was stored in cisterns, or underground water reservoirs, during the winter and sometimes early spring for consumption all through the hot/warm months of the year [2–5]. For this purpose, various types of cisterns (public, private, and public-private) were constructed in cities, villages, caravanserais[a] (Fig. 2.1), en routes, and so on. They were then filled with water from different sources, including Qanats[b] (Fig. 2.2), springs, wells, rainwater, seasonal rivers, and spring floods, to provide cold water for one or two years of summertime use [4]. Fig. 2.3 shows in cross-section the major components of a typical cistern located in a

[a] A caravanserai (or caravansary) was a roadside inn, set within a secure rectangular or square (mostly) building (Fig. 2.1), where travelers (caravaners) could rest overnight and recover from the day's journey [7, 8]. Caravanserais typically consisted of a central courtyard surrounded by lodgings, storerooms, and stables, which were employed sometimes as bazaars (or markets) [8, 9]. They supported the flow of commerce, information, and people across the network of trade routes covering North Africa, South-East Europe, and the Aegean to India and Western China (most notably the "Silk Road"), even though most of them were located between Persia (Iran) and the Ottoman Empire (Turkey) [10]. The word *caravanserai* is a compound noun that originated in Persia but was globally adopted. *Caravan* means *group of travelers*, and *serai* means *house and place to stay*.

[b] A Qanat (or Karez) is a gently sloping underground canal used to transport water over long distances from an aquifer, or water well to the surface, for drinking and irrigation (Fig. 2.2) [4, 13].

Ice-Houses
https://doi.org/10.1016/B978-0-12-822275-1.00007-2
Copyright © 2021 Elsevier Inc.
All rights reserved.

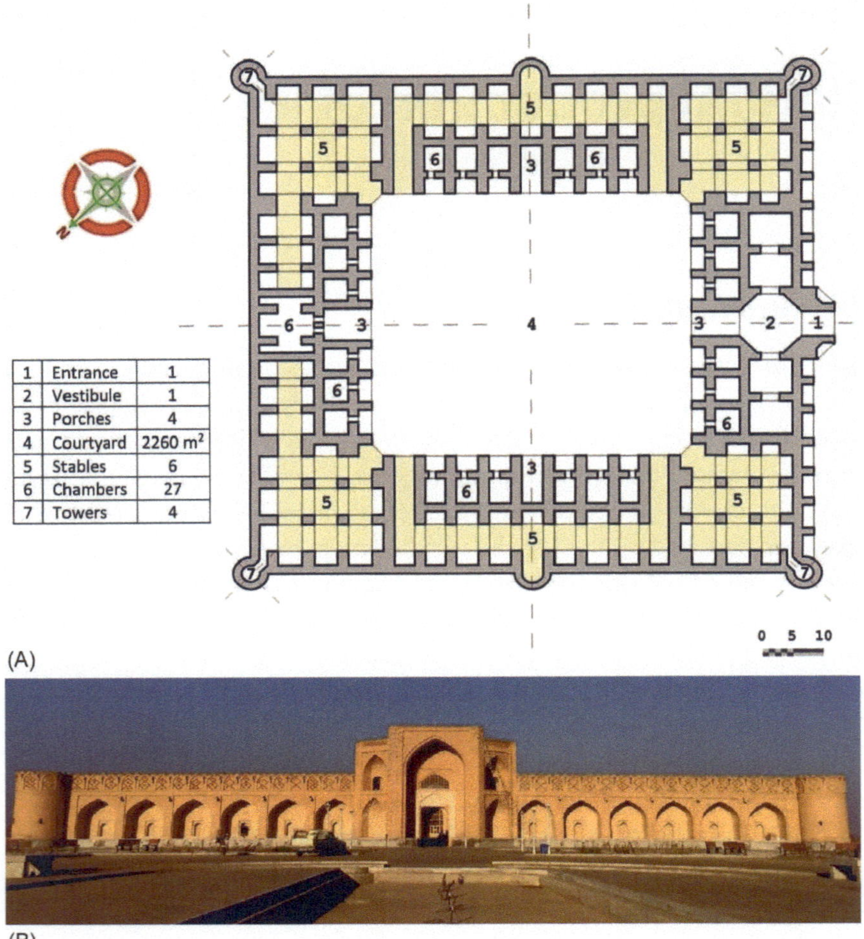

Fig. 2.1 (A) A schematic plan and (B) front view of "*Madar-e-Shah*" Caravansary in Isfahan, Iran [6].

hot and dry area of Iran. People accessed the cold water through a tap positioned near the bottom of the reservoir. It should be noted that the construction of these structures varied according to local climate and geological conditions, economic levels, water sources, intended use, and so on. Fig. 2.4 depicts a cistern with components such as an entrance, wind towers,[c] a domed roof, and staircase, situated in a hot and arid

[c]Wind towers (or wind-catchers), called "*Baudgeers*" in Persian, are small towers installed on top of buildings and are of different shapes and structures. For centuries, wind towers have been used in the Middle East (especially in Iran, the origin of this technology) and Egypt for natural ventilation and passive cooling of buildings in hot/warm climates. Wind towers were also used to maintain natural air circulation within cisterns and keep the stored water cold [5].

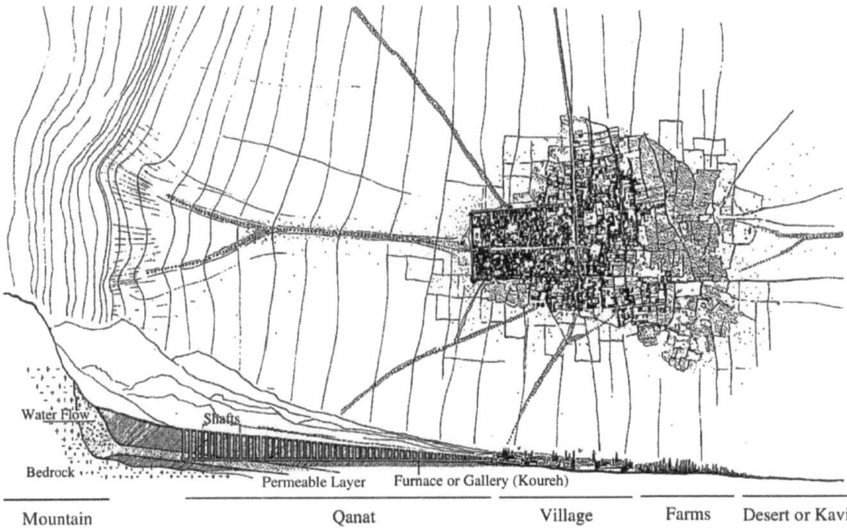

Fig. 2.2 Schematic illustration of a Qanat directing water from a foothill to a village [11, 12].

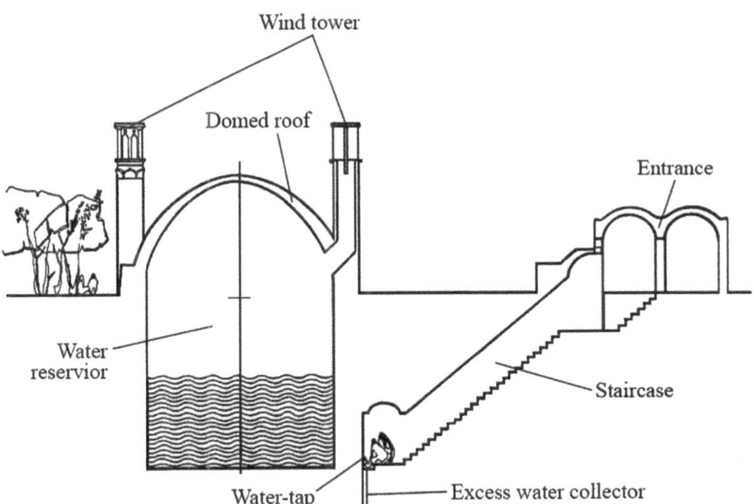

Fig. 2.3 Schematic illustration of *"Panjeh-Shah"* Cistern in the city of Kashan, Isfahan, Iran [4, 11].

climate, which was filled with water from a Qanat in winter. In Fig. 2.5, several cisterns of different shapes (rectangular and circular) located in a hot and humid climate are indicated; they were filled from rainwater and spring flood water. In the old photo reproduced in Fig. 2.6, Iranian children are carrying water jugs they have filled with cold water from the water reservoir. According to Dehghani-Sanij et al. [4, 16–19],

Fig. 2.4 A view of the entrance, wind towers, domed roof, and staircase of the cistern called locally "*Six-Baudgeer Aub-anbar*" in the city of Yazd, Yazd, Iran [4, 14].

Fig. 2.5 Rectangular and a circular cisterns in the village of Dezhgan, the central district of Bandar-e Lengeh county, Hormozgan, Iran [4].

for a cistern situated in the city of Yazd (a hot and dry region), Iran, the temperature in the lower layer of water inside the reservoir (the position of the tap) would change over the discharge time from 11.5°C in mid-April to 13.1°C in mid-October.[d] In contrast, the average am-

[d]Drinking water temperatures ranged from 5°C to 15°C, the most favored being between 8°C and 13°C [4, 20].

Fig. 2.6 Filling water jugs from a cistern to supply potable water to the home, taken by Antoin Sevruguin (1851–1933) [15].

bient temperature would have ranged from 23°C to 38°C. Accordingly, the studied cistern would have been able to provide cold water for the people living around it for the duration of its six-month discharge cycle, even though the demand for cold water was high and the ambient temperature reached 42°C or more on some afternoons [4].

2. Ice created in freezing lakes and rivers was stored in underground reservoirs during the winter, either for local use in the summer or for transfer to hot/warm regions. Before the invention and use of mechanical and artificial refrigeration systems, the "ice trade," also known as "the frozen water trade," and the ice transfer were thriving businesses based in Great Britain, North America, and even Norway, Sweden, and France [21, 22]. According to Jahren and Sui [23], the years between 1890 and 1900 were the peak decade of the trade and export of natural ice. For example, in the United States, natural ice export was the second major export after cotton in terms of revenue in the second half of the 19th century [22]. Traders shipped ice to countries with hot/warm climates around the world, such as the Caribbean (e.g., Cuba, Puerto Rico), Brazil, Argentina, Peru, India (Fig. 2.7), the Persian Gulf, Australia, New Zealand, China, Hong Kong, and the Philippines (Fig. 2.8) [25–30]. Fig. 2.9 shows ice trade activities around New York City, USA. Starting from the top, this figure depicts icehouses on the Hudson River; ice barges being towed to New York; barges being unloaded; an ocean steamship being supplied; ice being weighed;

Fig. 2.7 Taking on ice on board the British troop ship HMS Serapis prior to her journey to India, 1875 [24].

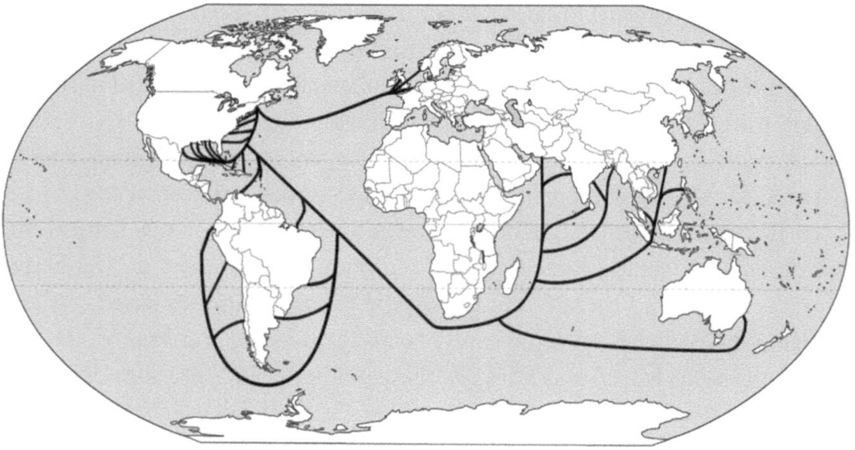

Fig. 2.8 Map showing the ice trade at its widest extent [25].

small customers being sold ice; the "uptown trade" to wealthier customers; and an ice cellar being filled. In the past, in many regions (e.g., Russia, Iran, Iraq, Turkey, Syria), people stored the winter's snow and ice for later consumption in the summer months [2, 27, 32]. For instance, for many years, Russians gathered ice along the Neva River throughout the cold months

Fig. 2.9 The ice trade in and around New York City, USA; by F. Ray, *Harper's Weekly*, 30 August 1884 [31].

for use in Saint Petersburg [27]. In past decades, Iranians used "*Sangab*"—large stone pots or troughs of different shapes (for example, hemispherical, rectangular, square)—to provide public sources of cold water for thirsty people next to mosques, shrines, bazaars, passageways, the intersections of alleys, and similar in the hot/warm summer days [4]. The water inside the Sangab was cooled to a pleasant temperature with lumps of ice. Fig. 2.10 shows a beautifully carved hemispherical Sangab at the entrance of *Imam* Mosque in the city of Isfahan, Iran. People drank water from the Sangab using a small beautifully decorated metal bowl. A thin screen was placed

Fig. 2.10 A view of the Sangab at the entrance of *Imam* Mosque in Isfahan City, Isfahan, Iran [4].

on top of the Sangab to prevent the contamination of its water by dust and insects, for example.

3. In the winters of past centuries, natural techniques were used to manufacture ice in some hot/warm regions of the world such as Iran, India, and Egypt [33, 34]. For instance, in the 19th century, in India, ice was produced naturally in small amounts during the winter [25]. Porous clay pots containing boiled, chilled water were laid on top of straw in shallow pits (Fig. 2.11). On cold winter nights, thin ice layers that formed on the water's surface were collected and compressed for later sale [28]. There were manufacturing locations at Hugli-Chuchura and Allahabad (Fig. 2.11), India, so this kind of ice, called "hooghly ice," was only available in limited quantities and considered of poor quality because it often resembled soft slush instead of hard crystals [27–29]. The ice produced in old ice-houses (known as Yakhchals) in Iran tended to be of much higher quality and quantities, as will be explained in the following section.

This chapter discusses and analyzes the method of making ice naturally via traditional Iranian ice-houses that were employed in past centuries. Based on this chapter, a new approach for producing ice by means of low-energy technology in areas with cold winters and storing it in

Fig. 2.11 Natural ice making through skimming ice from water-filled pots, near Allahabad, India, in 1828 [35].

reservoirs for different applications (for example, space cooling) in summer will be introduced and described in Chapter 6.

2.2 Natural production of ice during winter nights in shallow ponds

Since olden times, ice has been made naturally on cold and clear winter nights and stored for summer use in underground reservoirs or pits in different villages and cities of Iran, especially in arid and semi-arid regions [3, 36–38]. A traditional ice-house included one or more shallow ponds, on average, around 30 cm deep, 10 m wide, oriented north–south, and a few decameters length running east–west [2, 39]. In some regions of Iran, in particular central areas such as Yazd and Kerman, a high and long wall was built along the southern edge of the shallow ponds to block the sun's rays from hitting the surface of ponds during winter days. In the eastern and western parts, shorter walls were built to prevent solar radiation hitting the ponds' surfaces in the early morning and the late afternoon. These protective adobe walls also helped to decrease wind impacts on the shallow ponds. In Chapter 5, the architectural features and performance details of various types of ice-houses in different areas of Iran are described in detail.

To produce ice on cold winter nights when there was a low enough ambient air temperature and relative humidity and clear skies, the shallow ponds were first cleaned, then filled with freshwater to a level of ~1 mm. After that layer froze, another layer of water was poured on the ice to be frozen. These actions were repeated throughout consecutive nights until, gradually, an ice cover of about 30 cm formed. It should be noted that the thickness of the ice produced in different regions of Iran varied due to weather and geographical conditions. For instance, it has been reported that the ice thickness in the city of Shiraz reached 20–25 cm when winters were very cold [40]. This thick ice was then cut into pieces and saved in storage pits for use in summer months [2, 3, 39, 41].

At the present time, due to the widespread use of mechanical and artificial refrigeration systems (for instance, refrigerators and freezers), natural production of ice through these structures has been completely abandoned. Their abandonment was also hastened by the lack of adequate sanitation during the production process and transportation from ice-houses to desired locations, as well as the cost of constructing new ice-houses and maintaining old ice-houses. Consequently, natural ice-making facilities, or traditional ice-houses, have been destroyed or left to deteriorate. Today, the shadowing adobe walls and the domes of a number of ice-houses remain in various villages and cities in Iran (Figs. 2.12 and 2.13, as an example), and some are gradually being restored as tourist attractions.

Fig. 2.12 An old picture of the Arg-e Bam ice-house's shadowing adobe wall and dome located in the city of Bam, Kerman, Iran.

Fig. 2.13 Picture of shadowing adobe wall taken approximately forty-five years ago in Yazd City, Yazd, Iran. The structure has since been completely destroyed.

2.3 Governing equations for natural production of ice

This section describes and analyzes the method of producing natural ice that would have been used by a typical traditional ice-house during winter. Fig. 2.14 shows an ice-house configuration in cross-section, with the shallow pond, an ice block, and the exchange of energy through a layer of water located on the ice. An ice layer of a certain height is considered to

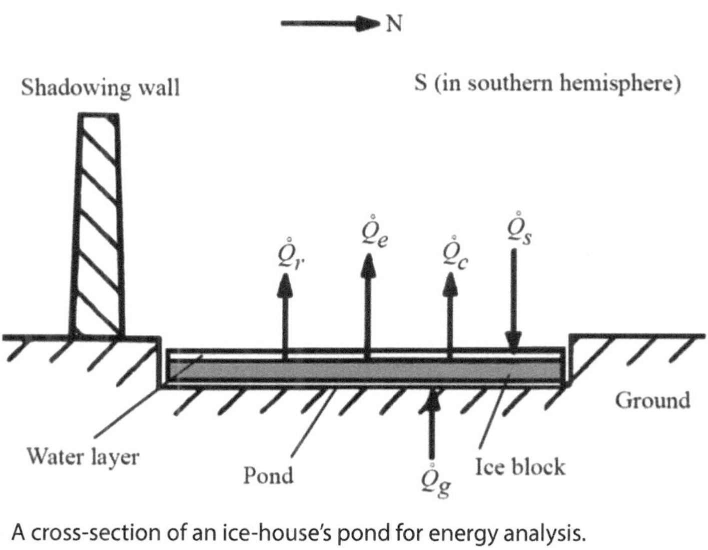

Fig. 2.14 A cross-section of an ice-house's pond for energy analysis.

preexist on the pond bottom, and an S thickness layer of water is poured on it. To determine the icing rate and obtain the amount of ice produced, an energy balance is required for the thermodynamic process. For the layer of water considered in Fig. 2.14, the following equations can be written as in [42, 43]:

(a) To cool the water until it reaches the freezing point:

$$\left(\dot{Q}_{r1} + \dot{Q}_{e1} + \dot{Q}_{c1} - \dot{Q}_{s1}\right)\Delta t_1 = \rho_w A S C_w \left(T_{wi} - 0\right) \tag{2.1}$$

(b) To freeze the water layer at 0°C:

$$\left(\dot{Q}_{r2} + \dot{Q}_{e2} + \dot{Q}_{c2} - \dot{Q}_{s2}\right)\Delta t_2 = \rho_w A S h_{if} \tag{2.2}$$

In the above equations, $\dot{Q}$ is the transferred heat (absorbed or lost), in units ofWatt; the subscripts c, r, e, and s, respectively, refer to convection, radiation, evaporation, and solar radiation; A is the surface area of the pond; ρ_w is the water density; S is the thickness of water poured on the ice; C_w is the specific heat capacity of water; h_{if} is the enthalpy of the solidification of water; Δt_1 and Δt_2 are the times needed to bring water from the initial temperature, T_{wi}, to zero and then freeze it at 0°C, respectively.

The heat loss through convection, $\dot{Q}_c$, between the pond and the ambient air is obtained by:

$$\dot{Q}_c = h_c A_w \left(T_w - T_a\right) \tag{2.3}$$

Here, T_w is the temperature of the water (or ice), T_a is the ambient air temperature, A_w is the surface area of water (or ice) in the pond, and h_c is the convective heat transfer coefficient, which is dependent on the surface geometry, the type of fluid motion (laminar or turbulent), and a set of fluid thermodynamic and transport properties [44]. The convective heat transfer coefficient for the planar surfaces can be determined by [45–47]:

$$h_c = \frac{Nu k_a}{L} = C Re^{0.8} Pr^{0.33} \tag{2.4}$$

where Nu is the Nusselt number, k_a is the air thermal conductivity, L is the length of the pond surface, Re is the Reynolds number, and Pr is the Prandtl number. According to Rohsenow and Choi [45], the parameter of C, on average, is equal to 0.037 for a flat surface (like a plate), over a length L, and in a turbulent flow parallel to the surface.

The thermal conductivity of the air, k_a, is expressed by [48–50]:

$$k_a = 0.024577 + 9.027 \times 10^{-5} T_a \tag{2.5}$$

The Reynolds number, Re, and the Prandtl number, Pr, respectively, can be defined by the following formulae:

$$Re = \frac{UL}{\upsilon_a} \tag{2.6}$$

$$Pr = \frac{\mu_a}{\rho_a \alpha_a} \tag{2.7}$$

Here, U is the wind velocity, and υ_a, α_a, and μ_a represent the kinematic viscosity, the thermal diffusivity, and the dynamic viscosity of the air, respectively.

The kinematic viscosity of the air, υ_a, is given by [50, 51]:

$$\upsilon_a = \frac{1}{(80711.7 - 766.15T_a)} \tag{2.8}$$

The thermal diffusivity of the air, α_a, can be computed by [48–50]:

$$\alpha_a = \frac{1}{(57736 - 585.78T_a)} \tag{2.9}$$

The dynamic viscosity of the air, μ_a, can be expressed by [50, 52]:

$$\mu_a = \mu_o \left[\frac{T_o + 120}{T + 120} \left(\frac{T}{T_o} \right)^{1.5} \right] \tag{2.10}$$

where $\mu_o = 1.8325 \times 10^{-5}$Pa s, $T_o = 296.16$K, and $T = (T_a + 273.15)$K. Additionally, the density of the air, ρ_a, is given by [50, 52]:

$$\rho_a = \frac{P}{287.04(T_a + 273.15)} \tag{2.11}$$

where P is the atmospheric air pressure and T_a is in units of °C.

In Eqs. (2.1) and (2.2), $\dot{Q}_r$ represents the heat losses through radiation, which can be divided into two parts: (1) the radiation heat exchange between the surface of the water (or ice) in the pond and the sky ($\dot{Q}_{r,sky}$), and (2) the heat transfer by radiation between the pond and miscellaneous objects surrounding the ice-house's pond, such as the adobe protection walls built in the south, east, and west directions ($\dot{Q}_{r,mis}$).

The studies conducted indicate that the operation of traditional ice-houses primarily relied on the heat radiation losses of the pond surface to the sky [2, 53]. The heat exchange through radiation between the surface of the water (or ice) in the pond and the sky ($\dot{Q}_{r,sky}$) can be calculated if the equivalent temperature of the sky is known, which can be determined by various empirical relationships, plus knowing the shape or view factors between them. This exchange of heat can be defined by the following formula:

$$\dot{Q}_{r,sky} = \varepsilon_w A_w F_{w-s} \sigma \left(T_w^4 - T_{sky}^4 \right) \tag{2.12}$$

Here, ε_w is the emissivity of water, F_{w-s} is the shape factor between the water (or ice) surface and the sky, σ is the Stefan-Boltzmann constant ($5.67 \times 10^{-8} \mathrm{W/m^2 K^4}$), and T_{sky} is the equivalent temperature of the sky.

To estimate the equivalent black-body sky temperature, T_{sky}, Bliss [54] presented the following relationship:

$$T_{sky} = T_a \left(0.8 + 0.004 T_{dp} \right)^{0.25} \tag{2.13}$$

Berdahl and Martin [55] also proposed another relationship:

$$T_{sky} = T_a \left[0.711 + 0.0056 T_{dp} + 0.000073 T_{dp}^2 + 0.013 \cos\left(2\pi \frac{t_m}{24} \right) \right]^{0.25} \tag{2.14}$$

In the above equations, T_{dp} is the dew point temperature of the air (°C), t_m the time after midnight (h), and T_{sky} and T_a are in units of Kelvin. The dew point temperature (T_{dp}) is equal to the saturation temperature of water-vapor in the air. Once the air temperature and the relative humidity are known, the partial pressure of the existing water-vapor in the air can be determined. The water-vapor saturation temperature at the partial pressure can be defined by referring to water-vapor tables. Another method is to enter the air temperature and the relative humidity into a psychrometric chart, and to determine T_{dp}.

The radiation heat loss between the pond and miscellaneous objects surrounding the ice-house's pond such as the adobe protection walls built in the south, east, and west directions ($\dot{Q}_{r,mis}$) can be computed by:

$$\dot{Q}_{r,mis} = \sum_{i=1}^{n} \frac{\sigma \left(T_w^4 - T_{o_i}^4 \right)}{\dfrac{1-\varepsilon_w}{\varepsilon_w A_w} + \dfrac{1}{A_w F_{w-o_i}} + \dfrac{1-\varepsilon_{o_i}}{\varepsilon_{o_i} A_{o_i}}} \tag{2.15}$$

where T_{o_i}, ε_{o_i}, and A_{o_i} represent the temperature, emissivity, and area of object i surrounding the pond, respectively; and F_{w-o_i} is the shape factor of the water (or ice) surface with respect to the object i. Note that the pond's shape factor relative to the walls can be determined using the width of the pond and the height of the walls. In addition, the radiation transfer between these surfaces can be obtained once the temperature of the internal surfaces of the walls is known. The temperature of the internal surfaces of the walls can be computed if the height, thickness, and materials used in the wall structure, plus climate conditions such as solar radiation, air temperature, relative humidity of the air, wind speed and its direction, and the sky temperature are known. It can be assumed that the temperature of the internal surfaces of the walls is approximately equal to the air temperature at any moment [2, 42].

The heat loss because of evaporation from the surface of the pond to the ambient air, $\dot{Q}_e$, can be determined by the following formula [47, 50, 56]:

$$Q_e = h_c \left(\frac{Pr}{Sc} \right)^{0.63} \frac{\in L_v}{PC_{pa}} \left(e_s(T) - \phi\, e_s(T_a) \right) \tag{2.16}$$

Here, Sc is the Schmidt number, $\in$ is the ratio of the molecular weight of water vapor and dry air (considered equal to 0.622 [57, 58]), L_v is the latent heat of vaporization of water at the surface temperature (T_w), C_{pa} is the specific heat capacity of dry air at a constant pressure (~1005 J/kg K [50, 52]), ϕ is the relative humidity of the air, and $e_s(T)$ is the saturated vapor pressure, which is linearized by [47, 59]:

$$e_s(T) \approx E_0 + e_0 T \tag{2.17}$$

where $e_0 = 27.03$ Pa/K and $E_0 = -6803$ Pa. The Schmidt number is given by:

$$Sc = \frac{\mu_a}{\rho_a D_{ab}} \tag{2.18}$$

Here, D_{ab} is the air-water vapor diffusivity, which can be calculated by [48–50]:

$$D_{ab} = 2.227 \times 10^{-5} \left(\frac{T_a + 273}{273} \right)^{1.81} \tag{2.19}$$

In Eqs. (2.1) and (2.2), $\dot{Q}_s$ represents the diffuse solar radiation absorbed by the surface of water (or ice) in the pond. To calculate the radiation of the sun (both direct and diffuse), the first step is to estimate the length of the day and determine the rate of solar radiation at different hours of the day. To compute the length of day, it is necessary to specify the solar noon, the hours of the sunrise and the sunset relative to a common time reference. In every country, time is determined by the origin longitude in Greenwich, England. To simplify, Iran is considered to lie in one time zone because the time difference between the farthest points east and west is only one hour, and the country's time is three-and-a-half hours before Greenwich Mean Time (GMT). Iran's longitude associated with this time is equal to $3.5 \times \frac{15}{360} = 52.5°$.

The apparent solar time (AST), or the local solar time, can be calculated using the following relationship [5, 60]:

$$AST = LST \pm \left(4\frac{min}{deg}\right)\left(L_{std} - L_{loc}\right) + EoT \tag{2.20}$$

For all locations east of Greenwich (such as Iran), the above relationship is represented with a minus sign. In Eq. (2.20), LST is the local standard time (or local clock time), L_{std} is the local longitude of the standard time meridian (in degrees), L_{loc} is the local longitude at the position of interest (in degrees), and EoT is the equation of time,[e] which can be estimated by [5, 60]:

$$EoT = 9.87 sin2\psi - 7.53 cos\psi - 1.5 sin\psi \tag{2.21}$$

$$\psi = \frac{360^{\circ} n}{365} \tag{2.22}$$

and n is the number of the day in the year. For instance, $n = 1$ for the first day of the first Iranian month ("Farvardin") that is generally equal to March 21, and $n = 187$ for the first day of the seventh Iranian month ("Mehr"), which is usually equal to September 23.

As a result of passing through the Earth's atmosphere, the sunlight is broken down into two components: direct solar radiation (sometimes called direct beam radiation) and diffuse solar radiation. Direct solar radiation travels directly from the sun to the surface of the Earth without scattering by the atmosphere. This radiation can produce a recognizable shadow when interrupted by an opaque object. In contrast, diffuse solar radiation cannot

[e]The difference between the apparent solar time (indicated by sundials) and the mean solar time (as shown by clocks) varies continuously day-to-day in an annual cycle; this amount is known as the equation of time [61, 62].

reach the Earth's surface directly due to molecules or particulates in the atmosphere such as water vapor, dust particles, or pollution, and also cannot lead to a shadow. The sum of both is called global solar radiation.

According to Fig. 2.15, the solar incidence angle (θ) for the planar surface, defined as the angle between the rays of the sun and the normal on a surface [63], depends on the latitude of location (φ); day of the year, characterized by the declination angle (δ); the slope of the planar surface to the horizon (β); surface azimuthal angle (γ); and solar hour angle (ω) [64], can be calculated by [60, 65]:

$$\begin{aligned} cos\theta &= sin\varphi\ sin\delta\ cos\beta - sin\delta\ cos\varphi\ cos\gamma\ sin\beta + cos\varphi\ cos\delta\ cos\omega\ cos\beta \\ &+ cos\delta\ sin\varphi\ sin\beta\ cos\gamma\ cos\omega + cos\delta\ sin\beta\ sin\gamma\ sin\omega \end{aligned} \tag{2.23}$$

The various angles mentioned in the above equation are described as follows:

1. *Location's latitude* (φ) is the angle created through the projection of the line on the plane of the equator with the radial line joining the location to the center of the Earth. It varies between $-90°$ and $+90°$ (for example, $\varphi = +35.5°$ for Tehran) and is positive (negative) for northern latitudes (southern latitudes).

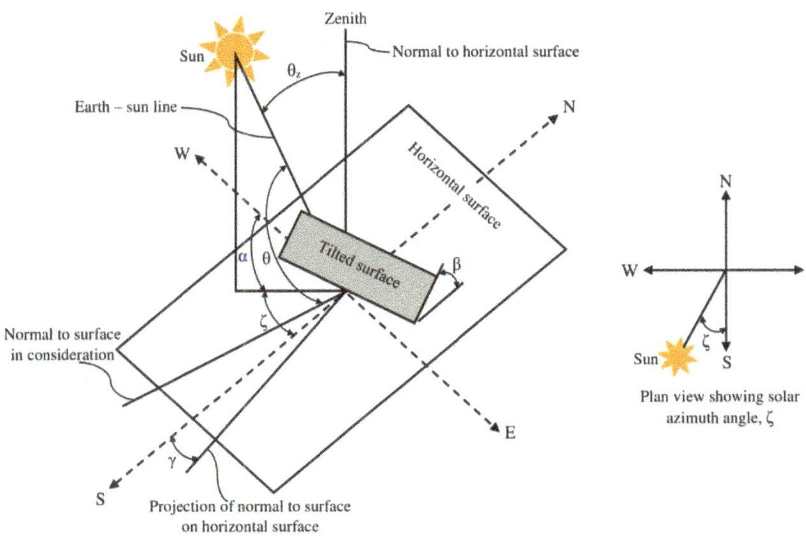

Fig. 2.15 Sketch showing the solar angles [63].

2. *Solar declination angle* (δ) is the angle between the sun's rays and the Earth's equatorial plane. It ranges between −23.45° (winter solstice in the Northern Hemisphere) and +23.45° (summer solstice in the Northern Hemisphere) during its annual cycle, and can be obtained by:

$$\delta = 23.45^\circ \, sin\left(\frac{360^\circ n}{365}\right) \tag{2.24}$$

3. *Tilt angle* (β) is the angle between the planar surface and the horizon. The angle β is equal to 0° and 90° for the horizontal and vertical surfaces, respectively.
4. *Surface azimuth angle* (γ) is the angle between south and the horizontal projection of the surface normal. This angle is measured from the south, so that γ is equal to zero for a surface completely south-facing and negative (positive) for a surface that faces east of south (west of south).
5. *Solar hour angle* (ω) is defined based on the rotation of the Earth around its polar axis, leading to the day and night cycle, which is an angular measure of time. It is determined relative to solar noon, so that ω is equal to zero at solar noon and varies by 15° per hour from noon, which means +15° per hour during the morning and −15° in the afternoon. The hour angle is given by:

$$\omega = \pm 15^\circ (AST - 12) \tag{2.25}$$

For example, at 8 am solar time, we have: $\omega = 15^\circ (8-12) = -60^\circ$.

For the horizontal surfaces, $\beta = 0^\circ$ and the solar incidence angle (θ) and the solar altitude angle (α) are identical. Therefore, Eq. (2.23) becomes:

$$cos\theta = sin\alpha = sin\delta \; sin\varphi + cos\delta \; cos\varphi \; cos\omega \tag{2.26}$$

The solar altitude angle (α) is defined as the vertical angle positioned between the horizon and an imaginary line connecting to the sun. It is zero at sunrise and sunset. In addition, $sin\alpha = cos\theta_z$ ($\alpha + \theta_z = 90^\circ$), in which θ_z is the solar zenith angle (the angle between the local vertical and the instantaneous position of the sun).

Using Eq. (2.26), the hour angle, ω_s, at either sunrise (when a negative quantity is taken) or sunset (when a positive quantity is taken), as well as the length of the day, N (h), can be determined by the following relationships:

$$cos\omega_s = -tan\delta \; tan\varphi \tag{2.27}$$

$$N = \frac{2}{15} Arccos(-tan\delta \; tan\varphi) \tag{2.28}$$

By means of Eqs. (2.23) and (2.26), the intensity of the solar radiation on an inclined surface (G_{bt}) and also on a horizontal surface (G_{bh}) at any moment, respectively, can be calculated by:

$$G_{bt} = G_{bn} \, cos\theta \tag{2.29}$$

$$G_{bh} = G_{bn} \, cos\theta_z \tag{2.30}$$

$$R_b = \frac{G_{bt}}{G_{bh}} = \frac{cos\,\theta}{cos\,\theta_z} \tag{2.31}$$

In the above equations, G_{bn} represents the intensity of the solar radiation on a surface normal to the direction of the sun. Moreover, G_{bt}, G_{bh}, and G_{bn} are in units of W/m^2.

A number of methods can be used to determine the amount of solar radiation. One of these is to use the clearness index (or atmospheric transmittance), K_T, which is empirically obtained through continuous measurements of solar energy. The clearness index is defined based on the ratio of the instantaneous global solar radiation at ground level on a horizontal surface to the corresponding quantity outside the Earth's atmosphere [66]. It is a stochastic parameter and function of the time of year, season, climatic conditions, and geographical situation of a place [67–69]. The daily clearness index is expressed by [54]:

$$K_T = \frac{H}{H_{oh}} \tag{2.32}$$

The parameters H and H_{oh} are in units of MJ/m^2d. Similarly, the hourly clearness index is given by [60]:

$$k_T = \frac{I_h}{I_{oh}} \tag{2.33}$$

Here, I_h is the hourly global solar radiation on a horizontal surface, and I_{oh} is the hourly extraterrestrial radiation. The parameters K_T and k_T are related to each other by the following relationship [70]:

$$k_T = \left[a + bcos\dot{\omega}(t - 12)\right]K_T \tag{2.34}$$

where t is the time (12 for solar noon), and $\dot{\omega}$, a, and b can be determined by:

$$\dot{\omega} = \frac{2\pi}{24} \quad (2.35)$$

$$a = 0.409 + 0.5016 sin(\omega_s - 60) \quad (2.36)$$

$$b = 0.6609 - 0.4767 sin(\omega_s - 60) \quad (2.37)$$

By means of meteorological data, the parameter K_T can be found. Table 2.1 shows the clearness index for a number of cities in Iran for various months.

The parameter I_{oh} can be calculated by [60]:

$$I_{oh} = \frac{12 \times 3600}{\pi} G_{sc} \left(1 + 0.033 cos \frac{360(n+81)}{365} \right) \times \left[cos\varphi \, cos\delta \left(sin\omega_2 - sin\omega_1 \right) + \frac{\pi \left(\omega_2 - \omega_1 \right)}{180} sin\varphi \, sin\delta \right] \quad (2.38)$$

Here, G_{sc} is a solar constant (1367W/m^2 [60]), and ω_1 and ω_2 are, respectively, the hour angle at the beginning and end of the time interval. Note that all angles in Eq. (2.38) are given in degrees.

According to Orgill and Hollands' model [73], the ratio of the hourly diffuse solar radiation on the horizontal surface, I_{dh}, to the total (global) solar radiation, based on k_T, for three categories, is defined by:

$$\frac{I_{dh}}{I_h} = \begin{cases} (1 - 0.249k_T) \text{ for } 0 < k_T < 0.35 \\ (1.577 - 1.84k_T) \text{ for } 0.35 \le k_T \le 0.75 \\ (0.177) \text{ for } 0.75 < k_T \le 1 \end{cases} \quad (2.39)$$

where

$$I_{dh} = I_h - I_{bh} \quad (2.40)$$

Here, I_{bh} is the hourly direct solar radiation on the horizontal surface. It can be computed by the following formula [69]:

$$I_{bh} = I_N cos\theta_z \quad (2.41)$$

where I_N is the direct normal solar radiation on horizontal surfaces, which can be measured by an instrument called a "pyrheliometer." It can also be determined through a number of theoretical models.

Table 2.1 The sky clearness index, K_T, for some cities of Iran in different months [71, 72].

City	Month											
	April	May	June	July	Aug.	Sep.	Oct.	Nov.	Dec.	Jan.	Feb.	March
Hamadan	0.5	0.52	0.58	0.59	0.60	0.60	0.56	0.50	0.46	0.52	0.51	0.49
Isfahan	0.61	0.63	0.66	0.66	0.67	0.67	0.66	0.63	0.60	0.59	0.62	0.59
Kerman	0.61	0.64	0.67	0.70	0.71	0.70	0.68	0.65	0.61	0.56	0.59	0.57
Mashhad	0.58	0.61	0.69	0.69	0.70	0.70	0.66	0.58	0.56	0.51	0.57	0.55
Shiraz	0.60	0.64	0.68	0.65	0.67	0.68	0.67	0.63	0.62	0.59	0.60	0.59
Tehran	0.55	0.58	0.64	0.64	0.65	0.65	0.61	0.55	0.50	0.51	0.53	0.53
Tabriz	0.50	0.56	0.64	0.48	0.70	0.68	0.64	0.57	0.47	0.41	0.48	0.47
Yazd	0.61	0.68	0.74	0.73	0.77	0.76	0.74	0.65	0.64	0.60	0.62	0.59
Zanjan	0.51	0.54	0.60	0.60	0.61	0.62	0.56	0.52	0.49	0.51	0.49	0.49

The total radiation on a tilted surface (I_T) can be calculated by [74]:

$$I_T = I_{bh} R_b + I_{dh} R_d + \rho_g I_h R_r \tag{2.42}$$

Here, R_b, R_d, and R_r represent conversion factors for direct, diffuse, and reflected components, respectively, and ρ_g is the reflection coefficient of the ground. The factor R_b is obtained by Eq. (2.31). The factors R_d and R_r, respectively, are expressed as follows:

$$R_d = \frac{1 + \cos\beta}{2} \tag{2.43}$$

$$R_r = \frac{1 - \cos\beta}{2} \tag{2.44}$$

The heat transfer (or heat gain) from the ground to the ice, $\dot{Q}_g$, has been neglected in Eqs. (2.1) and (2.2). The effect of this heat exchange is to melt a part of the ice, which establishes a thin layer of water under the ice. This layer of water floats the ice block, so that it no longer sticks to the bottom of the pond, making it easier to break the ice block into smaller pieces [2, 42].

The amount of ice production per unit time (or the rate of icing) can be calculated by:

$$\dot{S}_i = \frac{\rho_w}{\rho_i} \frac{S}{\Delta t} \tag{2.45}$$

where

$$\Delta t = \Delta t_1 + \Delta t_2 \tag{2.46}$$

In the above equations, the heat transfer from the water layer to the ice beneath it has been neglected. When the water is poured on the ice, heat transfers from the water to the ice and melts some of it. This procedure reduces the time of Δt_1. Instead, the time of Δt_2 increases to freeze the water poured on the ice, because it is necessary to freeze the initial layer of water with the thickness of S (mm) and the melted ice as well. However, $\Delta t = \Delta t_1 + \Delta t_2$ does not change much.

2.4 Solving governing equations and determining ice production rate for different climatic conditions

By calculating and substituting the $\dot{Q}_r$, $\dot{Q}_c$, $\dot{Q}_e$, and $\dot{Q}_s$ in Eqs. (2.1) and (2.2), and then, solving these equations along with Eqs. (2.45) and (2.46) for different climate conditions, the rate of natural ice production per hour, day, month, or season can be determined. Table 2.2 shows the dew point temperature and the equivalent temperature of the sky (using Eq. (2.13)) for different temperatures and relative humidities of the air. Table 2.3 indicates

Table 2.2 Saturation pressure (P_{vs}), dew point temperature (T_{dp}), and the equivalent temperature of the sky (T_{sky}) at different air temperatures (T_a) and relative humidities (ϕ).

		ϕ=100%		ϕ=80%		ϕ=50%		ϕ=20%		ϕ=10%		ϕ=5%	
T_a (°C)	P_{vs} (kPa)	T_{dp} (°C)	T_{sky} (°C)	T_{dp} (°C)	T_{sky} (°C)	T_{dp} (°C)	T_{sky} (°C)	T_{dp} (°C)	T_{sky} (°C)	T_{dp} (°C)	T_{sky} (°C)	T_{dp} (°C)	T_{sky} (°C)
20	2.34	20	10.8	16.6	9.7	9.1	7.2	−3.1	3	−11	0.21	−18.8	−2.7
16	1.82	16	5.6	12.5	4.5	5.7	2.25	−6.1	−1.8	−14	−4.6	−21.2	−7.2
12	1.4	12	0.5	8.7	−0.6	2.0	−2.8	−9.1	−6.6	−16.5	−9.2	−24	−11.9
8	1.07	8	−4.6	4.6	−5.7	−1.5	−7.75	−12	−11.3	−19.5	−14	−26.5	−16.5
4	0.81	4	−9.7	0.96	−11.3	−5	−12.7	−15	−16.1	−22.5	−18.7	−29.2	−21.2
0	0.61	0	−14.8	−3	−15.8	−8.5	−17.6	−18.2	−20.9	−25.5	−23.5	−32	−25.8
−4	0.44	−4	−19.9	−7	−20.9	−12	−22.5	−21.5	−25.7	−28.5	−28.2	−35	−30.5
−8	0.31	−8	−24.9	−10.5	−25.7	−16	−27.5	−25.2	−30.7	−32	−33.1	−38.2	−35.3
−12	0.22	−12	−30	−14.5	−30.8	−19.5	−32.4	−28.5	−35.5	−35	−37.8	−41.8	−40.1
−16	0.151	−16	−35	−18.5	−35.8	−23.1	−37.3	−32.5	−40.5	−38.5	−42.6	−44.5	−44.8
−20	0.103	−20	−40	−22.5	−40.1	−27	−42.2	−35.5	−45.1	−42	−47.4	−47.5	−49.4

Note that T_{sky} is determined by Eq. (2.13).

Table 2.3 The ratio of the equivalent temperature of the sky (T_{sky}) to the air temperature (T_a) at different dew point temperatures (T_{dp}) and day and night hours [2].

Hour	T_{dp} (°C)	T_{sky}/T_a Eq. (2.13)	T_{sky}/T_a Eq. (2.14)
24	10	0.957	0.942
	0	0.946	0.922
	−10	0.934	0.907
	−20	0.921	0.895
18 or 6	10	0.957	0.938
	0	0.946	0.918
	−10	0.934	0.902
	−20	0.921	0.980
12	10	0.957	0.934
	0	0.946	0.914
	−10	0.934	0.898
	−20	0.921	0.886

Note that T_{sky} and T_a are in units of Kelvin.

the ratio of the equivalent temperature of the sky (T_{sky}) to the air temperature (T_a) at different dew point temperatures and the day and night hours using Eqs. (2.13) and (2.14).

As illustrated in Table 2.3, Eq. (2.14) provides a relatively lower result for the equivalent temperature of the sky at various dew point temperatures. For example, for the dew point temperature and air temperature of 10°C, Eq. (2.13) provides a T_{sky} equal to −2.17°C throughout the day, and Eq. (2.14) provides the T_{sky} of −6.14, −7.55 and −8.68°C for 24 pm (midnight), 6 and 12 pm (noon), respectively. In addition, Eq. (2.13) represents the T_{sky} equal to −14.74°C for the whole day at the dew point temperature and air temperature of 0°C, but in Eq. (2.14), the T_{sky} at 24 pm, 6 and 12 pm is equal to −21.29°C, −22.39°C and −23.48°C, respectively. For the dew point temperature and air temperature of −10°C, Eq. (2.13) shows the T_{sky} equal to −27.36°C at all hours of the day, and in Eq. (2.14), the T_{sky} at 24 pm, 6 and 12 pm is equal to −34.46°C, −35.77°C and −36.83°C, respectively.

Table 2.4 and Fig. 2.16 indicate the rate of natural production of ice for different situations using Eq. (2.13). Table 2.5 shows the results of using Eq. (2.14) to determine the rate of natural production of ice for 16 cities in Iran based on their climatic conditions.

As illustrated in Fig. 2.16, natural ice can be produced at temperatures slightly above 0°C. An experimental study in Shiraz, Iran [53], confirmed

Table 2.4 Natural production of ice in a traditional ice-house's pond for sea-level regions at latitude 32°, under perfectly clear skies [2, 42].[a]

$\dot{Q}_s$ (W/m^2)	h_c (W/m^2°C)	T_a (°C)	$T_{wi}=5°C$ $\phi=5\%$	10%	20%	50%	80%	100%	$T_{wi}=15°C$ 5%	10%	20%	50%	80%	100%
0	5	−12	2.51	2.45	2.37	2.24	2.13	2.08	2.35	2.29	2.22	2.1	2.01	1.95
		−8	2.14	2.07	1.98	1.81	1.68	1.6	2.01	1.94	1.87	1.71	1.59	1.51
		−4	1.79	1.68	1.57	1.35	1.18	1.07	1.66	1.59	1.49	1.13	1.13	1.03
		0	1.36	1.27	1.14	0.87	0.64	0.49	1.3	1.21	1.09	0.84	0.62	0.48
		4	0.96	0.85	0.7	0.35	0.05	−0.12	0.93	0.82	0.68	0.35	−0.05	0.12
		8	0.55	0.42	0.23	−0.18	−0.52	−0.73	0.54	0.41	0.23	−0.18	−0.52	−0.73
		12	0.11	−0.40	−0.24	−0.61	−1.1		0.11	−0.04	−0.24	−0.71		
		16	−0.3	−0.45	−0.45	−0.7			−0.3	−0.45	−0.69			
	15	−12	5.02	4.94	4.82	4.56	4.33	4.19	4.72	4.65	4.45	4.3	4.1	3.96
		−8	4.2	4.1	3.95	3.6	3.29	3.09	3.97	3.88	3.74	3.42	3.13	2.95
		−4	3.36	3.23	3.04	2.57	2.15	1.87	3.2	3.1	2.91	2.47	2.08	1.81
		0	2.5	2.34	2.1	1.47	0.88	0.5	2.4	2.26	2.03	1.44	0.87	0.5
		4	1.63	1.44	1.13	0.31	−0.42	−0.87	1.59	1.41	1.11	0.31	−0.42	−0.87
		8	0.74	0.50	0.1	−0.85	−1.7	−2.3	0.73	0.50	0.10	−0.85	−1.7	−2.3
		12	−0.17	−0.42	−0.86	−2.1			−0.17	−0.42	−0.86	−2.1		
	25	−12	7.53	7.43	7.27	6.9	6.53	6.30	7.1	7.0	6.86	6.50	6.18	5.97
		−8	6.25	6.13	5.92	5.39	4.9	4.58	5.93	5.81	5.62	5.13	4.68	4.38
		−4	4.95	4.79	4.52	3.8	3.11	2.67	4.73	4.58	4.33	3.66	3.02	2.6
		0	3.63	3.42	3.06	2.07	1.13	0.50	3.5	3.31	2.97	2.03	1.12	0.5
		4	2.3	2.03	1.56	0.25	−0.89	−1.6	2.25	2.0	1.54	0.25	−0.89	−1.6
		8	0.93	0.59	−0.03	−1.5	−3.0	−3.9	0.93	0.59	−0.03	−1.5	−3.0	−3.9
		12	−0.44	−0.81	−1.5	−3.4			−0.44	−0.81	−1.5	−3.4		

Continued

Table 2.4 Natural production of ice in a traditional ice-house's pond for sea-level regions at latitude 32°, under perfectly clear skies—cont'd

$\dot{Q}_s$ (W/m^2)	h_c (W/m^2°C)	T_a (°C)	$T_{wi}=5$°C $\phi=5\%$	10%	20%	50%	80%	100%	$T_{wi}=15$°C 5%	10%	20%	50%	80%	100%
60	5	−12	1.84	1.78	1.71	1.57	1.47	1.41	1.74	1.68	1.62	1.49	1.4	1.34
		−8	1.47	1.40	1.32	1.14	1.01	0.93	1.4	1.34	1.26	1.1	0.97	0.90
		−4	1.09	1.01	0.90	0.69	0.51	0.4	1.05	0.97	0.87	0.67	0.50	0.4
		0	0.69	0.6	0.47	0.19	−0.04	−0.16	0.68	0.59	0.46	0.19	−0.04	−0.17
		4	0.29	0.18	0.02	−0.29	−0.55	−0.72	0.29	0.18	0.02	−0.29	−0.55	−0.72
		8	−0.12	−0.23	−0.04	−0.77			−0.12	−0.23	−0.4	−0.77		
		12	−0.5	−0.63	−0.83				−0.5	−0.63	−0.83			
	15	−12	4.36	4.27	4.16	3.9	3.67	3.52	4.11	4.04	3.93	3.69	3.48	3.35
		−8	3.53	3.43	3.28	2.93	2.62	2.42	3.36	3.27	3.13	2.81	2.52	2.33
		−4	2.69	2.57	2.38	1.91	1.48	1.2	2.58	2.47	2.29	1.85	1.44	1.18
		0	1.83	1.68	1.43	0.8	0.20	−0.17	1.78	1.63	1.4	0.79	0.20	−0.17
		4	0.96	0.77	0.45	−0.33	1.0	−1.56	0.95	0.76	0.45	−0.33	−1.0	−1.5
		8	0.05	−0.16	−0.51	−1.4	−2.3		0.05	−0.16	−0.51	−1.4	−2.3	
		12	−0.76	−0.1	−1.5				−0.76	−1.0	−1.5			
	25	−12	6.87	6.76	6.61	6.22	5.87	5.64	6.49	6.4	6.25	5.89	5.57	5.36
		−8	5.59	5.46	5.25	4.72	4.23	3.91	5.32	5.2	5.01	4.52	4.06	3.76
		−4	4.29	4.12	3.85	3.13	2.45	2.0	4.11	3.96	3.71	3.03	2.39	1.96
		0	2.97	2.75	2.39	1.4	0.45	−0.17	2.88	2.68	2.33	1.38	0.48	−0.17
		4	1.63	1.36	0.89	−0.38	−1.5	−2.2	1.60	1.34	0.88	−0.38	−1.5	−2.2
		8	0.25	−0.09	−0.62	−2.1	−3.5		0.25	−0.09	−0.62	−2.1	−3.6	
		12	−0.1	−1.4	−2.1				−1.0	−1.4	−2.1			

Note that the rate of ice production is in mm/h.

[a] These amounts do not contain heat gained from the ground, and the equivalent temperature of the sky is determined from Eq. (2.13) [2, 42]. The amount of ice for different climate conditions between the given amounts can be obtained by averaging.

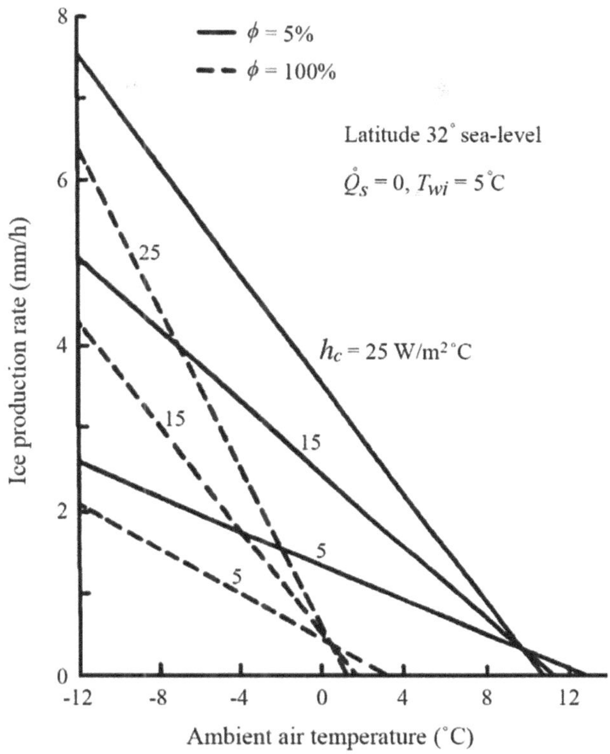

Fig. 2.16 The rate of ice production in an ice-house's shallow pond as a function of ambient air temperature (T_a) and relative humidity (ϕ), as well as the convective heat transfer coefficient (h_c).

Table 2.5 The rate of ice production for different cities in Iran [2, 75, 76].

	Month					
City	**November mm/month**	**December mm/month**	**January mm/month**	**February mm/month**	**March mm/month**	**Total mm/year**
Arak	96.4	363.2	613.2	426	86.4	1585
Isfahan	103.2	316.8	302.4	186.4	100.4	1009
Urmia	192	255.6	372	284.8	173.6	1278
Tabriz	90	396	492.8	324	145.6	1448
Tehran	0	192	277.2	171.2	87.6	728
Rasht	0	0	58	46.4	0	104
Zahedan	28.1	108.4	150	95.6	0	382
Zanjan	93.6	358.4	582.4	321.6	247.2	1603
Sanandaj	191.2	255.6	292.8	276.8	164	1180
Shiraz	0	101.6	295.2	93.6	24.5	514
Kashan	0	98	273.6	86.4	0	458
Kerman	106.8	398.4	372.8	172	92.4	1142
Kermanshah	23.1	172.8	243.6	225.6	86.4	751
Gorgan	0	0	83.6	0	0	83
Mashhad	205.6	285.6	376	237.6	72.8	1171
Hamedan	192	537.6	576.8	585.2	179.2	2070

the possibility of producing ice naturally on clear winter nights with an air temperature as high as 4°C.

Example 2.1 In a natural ice production factory near the city of Tehran, Iran, the ambient air temperature on a clear winter night is −8°C, the relative humidity of air is 35%, and due to the wind speed, the convective heat transfer coefficient is equal to $15 W/m^2 °C$ (case 1). The initial temperature of water in the pond is 5°C. Determine:

(a) The rate of ice production in mm/h.

(b) If the ambient air temperature is 4°C and other conditions are the same as before (case 2), what the rate of ice production is.

Solution: Referring to Table 2.4 and averaging (linearly) between the relative humidities of 20% and 50%, for case 1, the rate of ice production is equal to 3.775 mm/h, and for case 2, it is equal to 0.72 mm/h.

Example 2.2 Considering the climate conditions of Tehran and Hamedan in Iran, determine the rate of ice that can be produced naturally in these two cities during a year.

Solution: Referring to Table 2.5, the rates of ice production in Tehran and Hamedan are 728 and 2070 mm/year, respectively.

Examples 2.1 and 2.2 indicate the rate of natural production of ice under the different climatic conditions of Iran's cities. In the past, the air temperatures and relative humidities in these cities were lower, and the equivalent temperature of the sky was also lower because of lower amounts of (a) water vapor and (b) solid particles in the air. The rates of ice production in these cities were therefore higher than those calculated and shown in Tables 2.4 and 2.5.

2.5 Conclusion

Although traditional ice-houses are no longer used for a variety of reasons (for instance, health problems and the emergence and common usage of artificial refrigeration systems), the sustainable and eco-friendly technology employed in these ice-houses for producing ice naturally can be utilized today for various applications such as space cooling, ventilation, and air-conditioning (AC) purposes. This green and sustainable technology can be considered an effective, economical, and practical solution not only for meeting the thermal comfort requirements of inhabitants, but also for reducing the environmental threats and concerns (for example, CO_2 emissions and resulting global warming) related to electrical energy generation through fossil-fuel power plants and energy consumed for cooling and AC in buildings. As a result, the use of this environmentally friendly and cost-effective technology can help achieve sustainable development and energy management by diminishing the energy consumption and costs of refrigeration and AC systems in buildings.

References

[1] J.K.W. Lee, S.M. Shirreffs, R.J. Maughan, Cold drink ingestion improves exercise endurance capacity in the heat, Med. Sci. Sports Exerc. 40 (2008) 1637–1644.

[2] M.N. Bahadori, M. Yaghoubi, Ventilation and Natural Cooling System in Traditional Buildings of Iran, Markaz-e Nashr-e Daneshgahi, Tehran, Iran, 2006 (in Farsi).

[3] M.N. Bahadori, Passive cooling systems in Iranian architecture, Sci. Am. 238 (2) (Feb. 1978) 144–154.

[4] A.R. Dehghani-Sanij, in: A. Sayigh (Ed.), Cisterns: Sustainable Development, Architecture and Energy, River Publishers, Denmark, 2016.

[5] M.N. Bahadori, A.R. Dehghani-Sanij, in: A. Sayigh (Ed.), Wind Towers: Architecture, Climate and Sustainability, Springer International Publishing, New York, NY, 2014.

[6] Karavansera, Madar-e-Shah Caravansary, 2020. http://www.karavansera.com/madar-e-shah.html. (Accessed 20 September 2020).

[7] H. Chisholm (Ed.), Caravanserai, Encyclopædia Britannica, eleventh ed., Cambridge University Press, Cambridge, UK, 1911.

[8] M. Siroux, Caravansérails d'Iran et petites constructions routières, de l'Institut Français d'Archéologie Orientale, Série t. 81, Le Caire, Egypt, 1949 (in French).

[9] J.M. Bloom, S.S. Blair (Eds.), Grove Encyclopedia of Islamic Art & Architecture: Three-Volume Set, Oxford University Press, Oxford, UK, 2009.

[10] E. Edgü, A. Ünlü, M.E. Şalgamcıoğlu, A. Mansouri, Traditional shopping: a syntactic comparison of commercial spaces in Iran and Turkey, in: M. Greene, J. Reyes, A. Castro (Eds.), Proceedings of 8th International Space Syntax Symposium, Paper Ref.# 8099, PUC, Santiago de Chile, 2012.

[11] M.R. Moghtader, Ab-anbar: Conservation de l'eau Sur le Plateau Iranien, UNESCO, Paris, France, Dec. 1982 (in French).

[12] A.R. Dehghani, Water in the Plateau of Iran: Qanat, Cistern and Ice-House, Yazda Publications, Tehran, Iran, Oct. 2009 (in Farsi).

[13] A.A. Semsar Yazdi, M. Labbaf Khaneiki, Qanat Knowledge: Construction and Maintenance, Springer, The Netherlands, 2017.

[14] J. Ghazbanpour, New Life—Old Structure: A Selection of Valuable Historical Structures, second ed., Ministry of Housing and City Development with the Cooperation of Iranian Cultural Heritage Organization, Tehran, Iran, Aug. 1995 (in Farsi).

[15] Youngs Photographers Club, A Look at the Works and Photographic Approach of Antoin Sevruguin, 2020. https://news.yphc.ir/85487/. (Accessed 1 November 2020).

[16] M. Razavi, A.R. Dehghani-Sanij, M.R. Khani, M.R. Dehghani, Comparing meshless local Petrov-Galerkin and artificial neural networks methods for modeling heat transfer in cisterns, Renew. Sustain. Energy Rev. 43 (March 2015) 521–529.

[17] A.A. Dehghan, A.R. Dehghani, Experimental and theoretical investigation of thermal performance of underground cold-water reservoirs, Int. J. Therm. Sci. 50 (May 2011) 816–824.

[18] H.R. Ameri-Siahoui, A.R. Dehghani, M. Razavi, M.R. Khani, Investigation of thermal stratification in cisterns using analytical and artificial neural networks methods, J. Energy Conver. Manage. 52 (1) (Jan. 2011) 505–511.

[19] A. Arefmanesh, A.A. Dehghan, A.R. Dehghani, Thermal characteristics of an underground cold water reservoir: analytical and experimental studies, Appl. Therm. Eng. 29 (14–15) (Oct. 2009) 3261–3265.

[20] C. Dahlhaus, H. Damrath, Wasserversorgung, Abschnitt 3, Stuttgart, Germany, 1982 (in German).

[21] V. Udchan, LPG Refrigeration System with Zero Operating Cost, PhD Thesis, Department of Mechanical Engineering, Visvesvaraya Technological University, Belgaum, India, 2018.
[22] S. Reif-Acherman, Domestic ice-making machines 1830–1930, in: K. Gavroglu (Ed.), History of Artificial Cold, Scientific, Technological and Cultural Issues, Boston Studies in the Philosophy and History of Science, vol. 299, Springer, Dordrecht, The Netherlands, 2014, pp. 135–170, Chapter 8.
[23] P. Jahren, T. Sui, Snow and ice, in: P. Jahren, T. Sui (Eds.), How Water Influences Our Lives, Springer, Singapore, 2016, pp. 39–57, Chapter 3.
[24] Taking on ice on HMS Serapis, London Illustrated News, Oct. 2, 1875. https://commons.wikimedia.org/wiki/File:Taking_on_ice_on_HMS_Serapis.jpg. (Accessed 25 September 2020).
[25] D.G. Dickason, The nineteenth-century indo-American ice trade: an hyperborean epic, Mod. Asian Stud. 25 (1) (1991) 53–89.
[26] B.B. Blain, Melting markets: the rise and decline of the Anglo-Norwegian ice trade, 1850–1920, in: Working Papers of the Global Economic History Network (GEHN) (20/06), London School of Economics and Political Science, London, UK, Feb. 2006.
[27] R.O. Cummings, The American Ice Harvests: A Historical Study in Technology, 1800–1918, California University Press, Berkeley, CA, 1949.
[28] M.W. Herold, Ice in the tropics: the export of 'Crystal blocks of Yankee coldness' to India and Brazil, Rev. Esp. Acad. 11 (126) (2011) 162–177.
[29] P.C.F. Smith, Concentrated Wenham: New England Ice in Albion, in: D.V. Proctor (Ed.), Ice Carrying Trade at Sea, National Maritime Museum, London, UK, 1982.
[30] G. Weightman, The Frozen Water Trade: How Ice from New England Lakes Kept the World Cool, Harper Collins, London, UK, 2003.
[31] F. Ray, Harper's Weekly, New York Ice Trade, 30 August 1884. https://upload.wikimedia.org/wikipedia/commons/c/cd/New_York_Ice_trade_colour.jpg. (Accessed 21 July 2020).
[32] E. Morofsky, History of thermal energy storage, in: H.O. Paksoy (Ed.), Thermal Energy Storage for Sustainable Energy Consumption: Fundamentals, Case Studies and Design, NATO Science Series, Springer and NATO Public Diplomacy Division, pp. 3–22, 2005.
[33] C. Gantz, Refrigeration: A History, McFarland & Company, Inc., Jefferson, NC, 2015.
[34] J.M. Scarcelli, Disaster Preparedness in Food Service: Government Agency Recommendations, Industry Behavior, and the Impact of a Power Loss on Refrigeration, PhD Thesis, Purdue University, West Lafayette, IN, Aug. 2010.
[35] F.P. Parlby, Wanderings of a Pilgrim in Search of the Picturesque: During Four-and-Twenty Years in the East, vol. 1, Pelham Richardson, London, UK, 1850.
[36] M.N. Bahadori, Production and storage of ice for cooling, in: Proceedings of the Second International Passive and Low Energy Architecture Conference, Pergamon Press, Crete, Greece, June 1983.
[37] M.N. Bahadori, D.K. Pugsley, Long-term storage of coolness through a gradual production of ice in winter, in: Proceedings of the First U.S.—China Conference on Energy, Resources and Environment, Pergamon Press, Beijing, China, Nov. 1982.
[38] M.N. Bahadori, E. Hajidavalloo, A feasibility study of passive solar heating and cooling under Iranian conditions, in: Proceedings of the 20th National Passive Solar Conference, American Solar Energy Society, Minneapolis, MN, 1995, pp. 33–38, July 15–20.
[39] M.N. Bahadori, Natural cooling in hot arid regions, in: A.A.M. Sayigh (Ed.), Solar Energy Application in Buildings, Academic Press, Inc., New York, NY, 1979, pp. 195–225.
[40] P. Javaheri, M. Javaheri, Challenge for Water in Ancient Fars, vol. 2, Iranian National Committee on Irrigation & Drainage (IRNCID) and Water National Museum of Iran (WNMI), Tehran, Iran, 2000 (in Farsi).

[41] V. Ghobadian, A Climatic Analysis of the Traditional Buildings in Iran, University of Tehran Press, Tehran, Iran, 1994 (in Farsi).
[42] M.N. Bahadori, T.P. Jantzi, Natural production of ice in shallow ponds, in: Proceedings of Solar World Congress, vol. 1, International Solar Energy Society, Perth, Australia, Aug. 1983, pp. 51–56.
[43] M.N. Bahadori, Natural production, storage, and utilization of ice in deep ponds for summer air conditioning, Sol. Energy J. 34 (2) (1985) 143–149.
[44] T.L. Bergman, A.S. Lavine, F.P. Incropera, D.P. DeWitt, Fundamentals of Heat and Mass Transfer, seventh ed., John Wiley & Sons, Inc., New York, NY, 2011.
[45] W.M. Rohsenow, H.Y. Choi, Heat, Mass, and Momentum Transfer, Englewood Cliffs, Prentice-Hall, NJ, 1961.
[46] E.P. Lozowski, K. Szilder, L. Makkonen, Computer simulation of marine ice accretion, Phil. Trans. R. Soc. Lond. A 358 (2000) 2811–2845.
[47] A.R. Dehghani-Sanij, S.R. Dehghani, G.F. Naterer, Y.S. Muzychka, Sea spray icing phenomena on marine vessels and offshore structures: review and formulation, Ocean Eng. 132 (2017) 25–39.
[48] ASHRAE, ASHRAE Handbook & Product Directory: Fundamentals, American Society of Heating, Refrigerating and Air Conditioning Engineers (ASHRAE), Inc., New York, NY, 1977.
[49] B.V. Karlekar, R.M. Desmond, Engineering Heat Transfer, West Publishing Co., Minneapolis, MN, 1977.
[50] A.R. Dehghani-Sanij, S. MacLachlan, G.F. Naterer, Y.S. Muzychka, R.D. Haynes, V. Enjilela, Multistage cooling and freezing of a saline spherical water droplet, Int. J. Therm. Sci. 147 (2020) 106095.
[51] J.P. Zarling, Heat and Mass Transfer from Freely Falling Drops at Low Temperatures, US Army Cold Regions Research and Engineering Laboratory, Hanover, NH, Aug. 1980. CRREL Report BO-1B.
[52] C.R. Tracy, W.R. Welch, W.P. Porter, Properties of Air, A Manual for Use in Biophysical Ecology, third ed., University of Wisconsin, Madison, WI, 1980, Technical report No. 1.
[53] M.N. Bahadori, A. Kosari, Performance of natural ice makers, in: Proceedings of the International Solar Energy Society Congress, vol. 3, Jan. 1978, pp. 1976–1978, New Delhi, India.
[54] R.W. Bliss, Atmospheric radiation near the surface of the earth: a summary for engineers, Sol. Energy 5 (3) (1961) 103–120.
[55] P. Berdahl, M. Martin, Emissivity of clear skies, Sol. Energy 32 (5) (1984) 663–670.
[56] T.L. Bergman, A.S. Lavine, F.P. Incropera, D.P. DeWitt, Introduction to Heat Transfer, sixth ed., John Wiley & Sons, Inc., New York, NY, 2011.
[57] L. Makkonen, Estimation of wet snow accretion on structures, Cold Reg. Sci. Technol. 17 (1989) 83–88.
[58] R. Kato, Modeling of Ship Superstructure Icing: Application to Ice Bridge Simulators, M.Sc. Thesis, Department of Marine Technology, Norwegian University of Science and Technology, Trondheim, Norway, 2012.
[59] T.G. Myers, Extension to the Messinger model for aircraft icing, AIAA J. 39 (2) (2001) 211–218.
[60] J.A. Duffie, W.A. Beckman, Solar Engineering of Thermal Processes, fourth ed., John Wiley & Sons, Inc., New York, NY, 2013.
[61] M. Benghanem, Optimization of tilt angle for solar panel: case study for Madinah, Saudi Arabia, Appl. Energy 88 (4) (April 2011) 1427–1433.
[62] K. Hamilton, T. Sakazaki, A note on apparent solar time and the seasonal cycle of atmospheric solar tides, Q. J. Roy. Meteorol. Soc. 143 (July 2017) 2310–2314.
[63] S.A. Kalogirou, Environmental characteristics, in: S.A. Kalogirou (Ed.), Solar Energy Engineering, Processes and Systems, Elsevier, Imprint by Academic Press, 2009, pp. 49–119, Chapter 2.

[64] N.D. Kaushika, A. Mishra, A.K. Rai, Solar radiation characteristics, in: N.D. Kaushika, A. Mishra, A.K. Rai (Eds.), Solar Photovoltaics: Technology, System Design, Reliability and Viability, Springer International Publishing, 2018, pp. 15–26, Chapter 2.
[65] S.P. Sukhatme, Solar Energy Principles of Thermal Collection and Storage, Tata McGraw Hill, New Delhi, London, 1984.
[66] A.M. El-Nashar, Solar radiation characteristics in Abu Dhabi, Sol. Energy 47 (1) (1991) 49–55.
[67] R. Kumar, L. Umanand, Estimation of global radiation using clearness index model for sizing photovoltaic system, Renew. Energy 30 (15) (Dec. 2005) 2221–2233.
[68] L.M. Halabi, S. Mekhilef, M. Hossain, Performance evaluation of hybrid adaptive neuro-fuzzy inference system models for predicting monthly global solar radiation, Appl. Energy 213 (March 2018) 247–261.
[69] S.A. Mousavi Maleki, H. Hizam, C. Gomes, Estimation of hourly, daily and monthly global solar radiation on inclined surfaces: models re-visited, Energies 10 (134) (2017), 1–28.
[70] M.N. Bahadori, M.S. Chamberlain, A simplification of weather data to evaluate daily and monthly energy needs of residential buildings, Sol. Energy 36 (6) (1986) 499–507.
[71] S.A. Mirhosseini, Determination of the Sky Clearness Index for Different Cities of Iran, M.Sc. Thesis, School of Mechanical Engineering, Sharif University of Technology, Tehran, Iran, 2003 (in Farsi).
[72] M.N. Bahadori, S.A. Mirhosseini, Sky Clearness Index for Iranian Cities, Paper No. 1120, ISES 2005, Solar World Congress, Orlando, FL, Aug. 2005.
[73] J.F. Orgill, K.G.T. Hollands, Correlation equation for hourly diffuse radiation on a horizontal surface, Sol. Energy 19 (1977) 357–359.
[74] B.Y.H. Liu, R.C. Jordan, Daily insolation on surfaces tilted towards the equator, ASHRAE Trans. 67 (1962) 526–541.
[75] B. Sadri, Study of Cold Storage in Winter to Provide the Cold Needs of Buildings, B. Sc. Thesis, School of Mechanical Engineering, Sharif University of Technology, Tehran, Iran, June 1989 (in Farsi).
[76] M.N. Bahadori, B. Sadri, Potential of natural ice making for cooling of buildings in Iran, in: Proceedings of the 14th National Passive Solar Conference, American Solar Energy Society, Denver, CO, 1989, pp. 303–308, June 19–22.

CHAPTER 3

Artificial production of ice

3.1 Introduction

The invention and development of different types of refrigeration and air-conditioning (AC) systems have made an enormous improvement in people's living standards and currently play a significant role in the modern world. Refrigeration is generally defined as any artificial process whereby heat is withdrawn from a confined (or an enclosed) place, system, or substance so as to bring its temperature to below the surrounding temperature. It then preserves that temperature at a level suitable for the intended purposes and transfers the removal heat elsewhere (e.g., to another place, system, or substance) at a higher temperature [1–5]. In simple terms, artificial or human-made cooling through removing heat is called refrigeration [6]. In another definition, from the American Society of Refrigerating Engineers (ASRE) [7], refrigeration means "*the science of providing and maintaining temperatures below the ambient temperatures.*"

Refrigeration is solely involved with cooling for a variety of applications, whereas AC, which means the process of air treatment to simultaneously control temperature, humidity,[a] purity,[b] and circulation to fulfill the thermal needs of an enclosed space, deals with both cooling and heating. As shown in Fig. 3.1, refrigeration applications can generally be classified into four broad categories. In the past, the principal purpose of refrigeration was ice making. Nowadays, refrigeration is broadly utilized in maintaining, transferring, and storing (short- or long-term) perishables. It can also be applied to various industries, such as petroleum and gas refining, pulp and paper, pharmaceutical, chemical, and petrochemical industries that involve low-temperature processes [8–12], for example, for the liquefaction and purification of natural gas [12–14].

[a]Control of humidity is to increase (in winter) or decrease (in summer) the moisture content in the air.
[b]Purity (or cleanliness) control means filtering the air to remove dust and other undesirable particles.

Ice-Houses
https://doi.org/10.1016/B978-0-12-822275-1.00010-2
Copyright © 2021 Elsevier Inc.
All rights reserved.

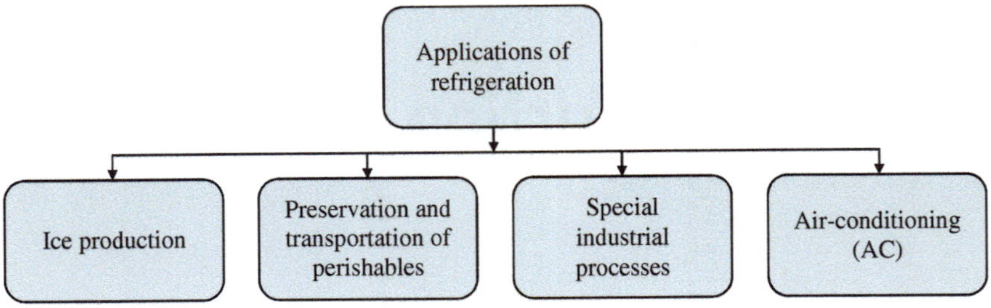

Fig. 3.1 Overall usages of refrigeration technology [1, 5].

As illustrated in Fig. 3.2, AC and refrigeration are interrelated, to the point where AC in different applications and sizes (Fig. 3.3) is currently one of the most important utilizations of refrigeration [1]. It should be noted that the AC systems used in hot and humid climates during the summer are employed for both cooling and dehumidifying. Based on a countrywide poll conducted in the USA in 2004 [15], both refrigeration and AC are considered to be among the ten most-important ideas and innovations of the last 75 years [16].

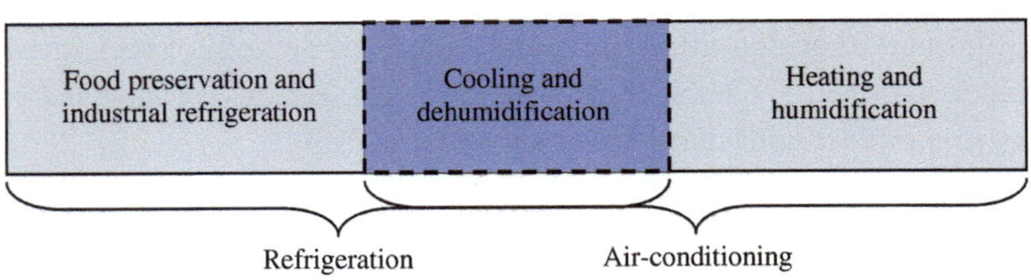

Fig. 3.2 Correlation between refrigeration and air-conditioning [1].

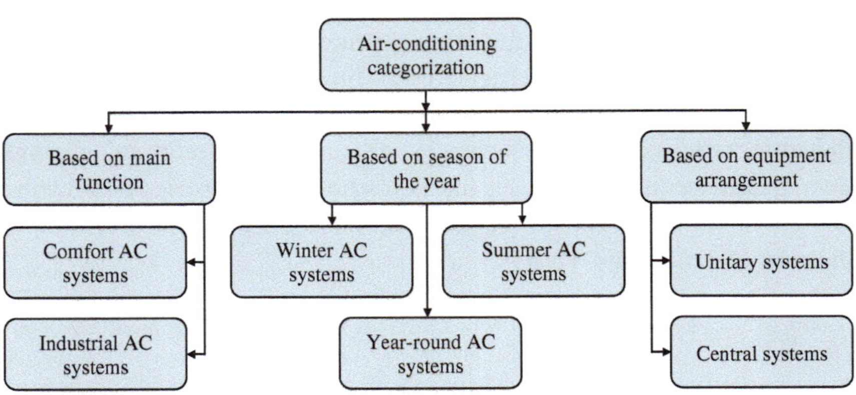

Fig. 3.3 Categorization of air-conditioning (AC) systems [1].

With artificial production of cold through refrigeration systems, as well as the ability to provide low-temperature environments by means of AC systems (especially refrigerant-based ones), humans have been able to create and even ensure the conditions of thermal comfort in hot and dry/humid regions where the summer heat is extremely intense and would without AC be almost unbearable. In addition, refrigeration systems can be able to preserve medicines, foodstuffs, beverages, agricultural products (e.g., fruits and vegetables), and so on, over both the short- and long-term, without spoiling [17]. The term "thermal comfort" is associated with physiological responses of living things [18–20], including humans, animals, and even plants. It eases, and in some cases even preserves lives, not only those of people but also of animals in such places as barns and zoos, and also plants, for instance, in greenhouses. From another perspective, thermal comfort means a satisfactory state—physiologically, psychologically, and physically—and, in fact, represents an artificial adaptation to less-than-ideal environments [21, 22]. According to the American Society of Heating, Refrigerating and Air-Conditioning Engineers (ASHRAE) [23], thermal comfort refers to "*the state of mind that expresses satisfaction with the thermal environment in which it is located,*" and it primarily depends on the following factors [24–28]:

- The environmental parameters, including the air temperature, air relative humidity, air velocity, and mean radiant temperature of each surface of a confined space;
- The thermal resistance of any clothing; and
- The metabolic rate.

A number of researchers have identified factors that affect building inhabitants' thermal comfort: gender, age, race, time of year, activity level, geographical location, weather conditions [29–32], and even culture [33] and individual expectations [34], although the perception of comfort may vary widely among humans. It is necessary to consider all these factors at the design stage of buildings [35]. Additionally, thermal comfort is directly influenced by a building's energy efficiency and vice versa [36]. Fig. 3.4 provides an illustration of the range of human comfort zones in summer and winter.

Many people spend the majority of their time (around 80%–90%) inside buildings or enclosed spaces [37]. As an example, according to a report by the US Environmental Protection Agency (EPA), Americans spend about 90% of their time indoors on average [38]. Therefore, AC systems can play an important role in providing a good indoor environmental quality (IEQ) for inhabitants. The IEQ is assessed based on four major factors of interior

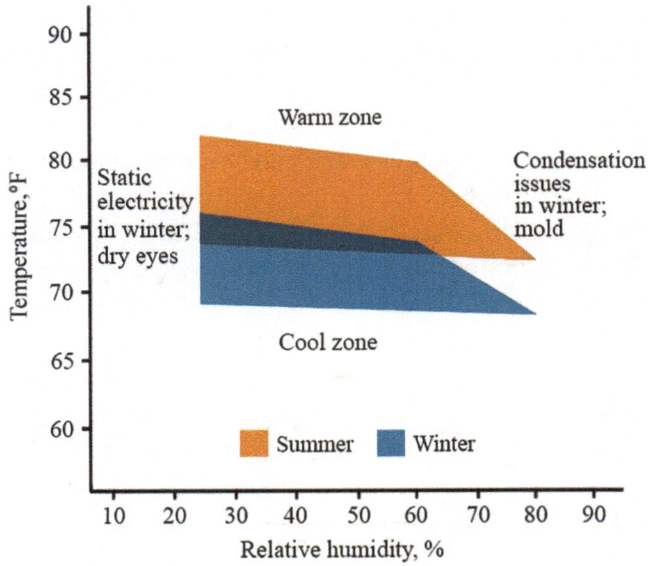

Fig. 3.4 Comparison of summer and winter comfort zones [26].

comfort—thermal, acoustic, visual, and air quality—and also has a direct impact (short- and long-term) on the health, well-being, and productivity of inhabitants [39–42]. Acoustic comfort is measured by a system's capability of protecting a building's inhabitants against various sources of noise, for example, sounds caused by transportation and industry (outdoor noise) and noises resulting from the operation of electrical appliances and mechanical systems (indoor noise). Its purpose is to provide an appropriate acoustic environment for living or any activities [36, 43]. Visual comfort means the capability of detecting, identifying, and analyzing anything in the line of vision, considering the velocity, the quality, and the accuracy of perception, and depending mainly on the lighting conditions of a confined space [36].

3.2 Methods and types of refrigeration

Refrigeration can be achieved in different ways. The commonest methods of refrigeration are [1, 5, 44–46]: ice refrigeration; evaporative refrigeration; refrigeration via dry ice; refrigeration through the expansion of gas; gas throttling refrigeration; steam jet refrigeration; liquid gas refrigeration; and vapor refrigeration.

- *Ice refrigeration*: In this method, blocks of ice are employed to produce cold naturally (through direct or indirect approaches) inside an enclosed and insulated space (e.g., a container or chamber), and maintain

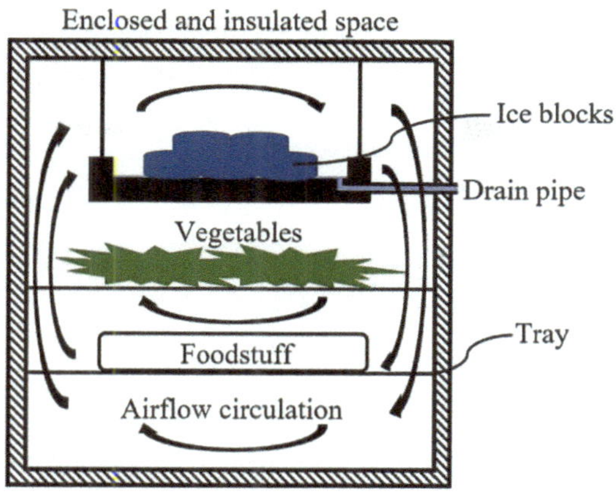

Fig. 3.5 Direct refrigeration method by ice. *(Modified from S.N. Sapali, Refrigeration and Air Conditioning, second ed., PHI Learning Pvt. Ltd., New Delhi, Delhi, India; Nov. 2013).*

its temperature at an appropriate level, less than the surrounding ambient temperature, to protect food and agricultural products, and drinks [5, 44]. Fig. 3.5 shows an example of the use of ice refrigeration for storing foodstuffs and crops via a direct approach. Fig. 3.6 indicates an indirect refrigeration approach that uses ordinary ice.

– *Evaporative refrigeration*: In this technique, which has been used since olden times, the air in a confined space is evaporatively cooled as it moves around a pot of porous pottery that is full of water. The heat needed for the evaporation process is taken in part from water inside the pot as well as from the surrounding ambient air. Consequently, both air and water cool down. The circulation of this cooled air can provide a degree of cooling in small and confined areas, perhaps even small rooms.

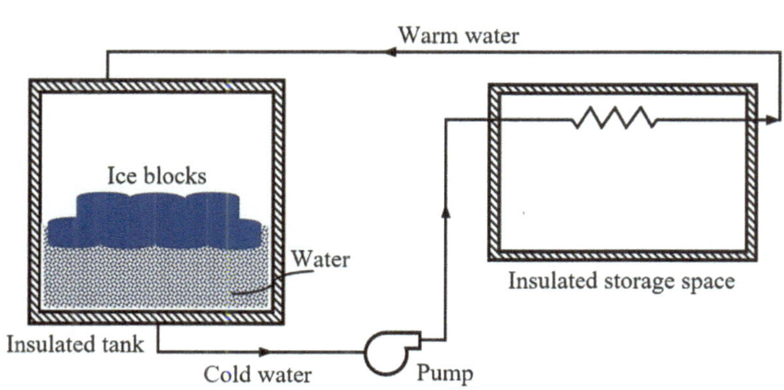

Fig. 3.6 Indirect refrigeration method by ice. *(Modified from S.N. Sapali, Refrigeration and Air Conditioning, second ed., PHI Learning Pvt. Ltd., New Delhi, Delhi, India, Nov. 2013).*

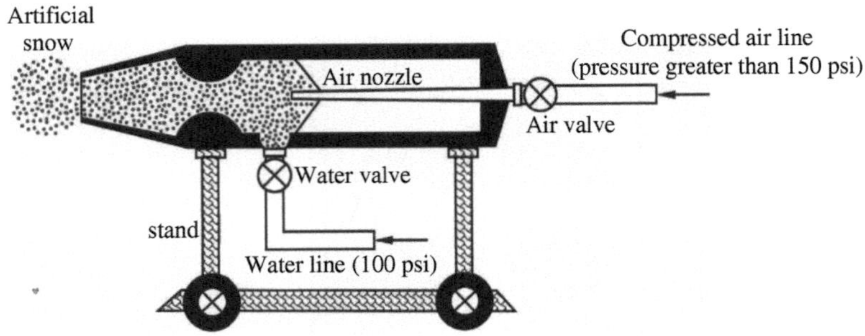

Fig. 3.7 Use of evaporative cooling for artificial snow production. *(Modified from S.N. Sapali, Refrigeration and Air Conditioning, second ed., PHI Learning Pvt. Ltd., New Delhi, Delhi, India, Nov. 2013; How It Works Team. Technology: How do snow machines work?; Dec. 2018 <https://www.howitworksdaily.com/how-do-snow-machines-works/> [Accessed 16 November 2020]).*

This air circulation occurs naturally. Evaporative or swamp coolers are modern examples of this kind of refrigeration, cool the air via the evaporation of water, and are suitable for hot, dry climates. Furthermore, as shown in Fig. 3.7, the phenomenon of evaporative cooling can be used to make artificial snow.

- *Refrigeration by dry ice*: The solid form of carbon dioxide (CO_2) is called dry ice. At atmospheric pressure, it sublimates at −78.5°C, and its sublimation heat is 571.1 kJ/kg [47]. Dry ice is mainly utilized as a cooling agent [48], typically manufactured in the forms of blocks, pellets, slabs, and nuggets [49], and has different applications, such as preserving food products at chilled or frozen temperatures during storage and transportation (the most common use) [50], as well as refrigerating laboratory biological samples [51] and carbonated beverages [52].
- *Refrigeration by expansion of gas*: In this method, the cooling effect or refrigeration is achieved by the expansion of gas in an adiabatic process (meaning a process in which no heat transfer occurs) through an expander. During this procedure, the gas pressure and subsequently its temperature intensely decrease [53]. The temperature difference in the adiabatic expansion process can be determined by [53]:

$$\Delta T = T_2 - T_1 = T_1\left[\left(\frac{P_2}{P_1}\right)^{\frac{r-1}{r}} - 1\right] \quad (3.1)$$

where T and P represent the temperature and pressure of gas, the subscripts *1* and *2*, respectively, refer to before and after expansion, and r is the adiabatic coefficient. The cooled gas (or air) can be used to withdraw heat from the intended space.

- *Gas throttling refrigeration (Joule–Thomson effect)*: Gas throttling means the pressure reduction in a gas flow when it is being forced through a throttle (such as orifice or valve), accompanied by a temperature variation but with no heat exchange with the surrounding environment. The temperature change of gas during the adiabatic throttling process can be defined by the following relationship [54]:

$$\mu_J = \left(\frac{\partial T}{\partial P}\right)_h \tag{3.2}$$

where μ_J represents the Joule–Thomson coefficient, known as the adiabatic throttling coefficient, in units of K/Pa, and h is the specific enthalpy that remains constant during the process. In Eq. (3.2), for refrigeration processes, μ_J is positive. The processes of refrigeration in air conditioners, heat pumps, and liquefiers are examples of applications of gas-cooling throttling [55, 56].

- *Steam jet refrigeration*: This system, which uses water as a refrigerant, operates based on the principle of reducing the boiling temperature of water (100°C) by decreasing the pressure of the surface of the water to below atmospheric pressure. This decrease in pressure (or increase in a vacuum) is preserved via throttling the steam produced by the boiling water by means of a nozzle or nozzles [5, 44, 57]. In this system, as represented schematically in Fig. 3.8, a heavily insulated flash chamber is filled with

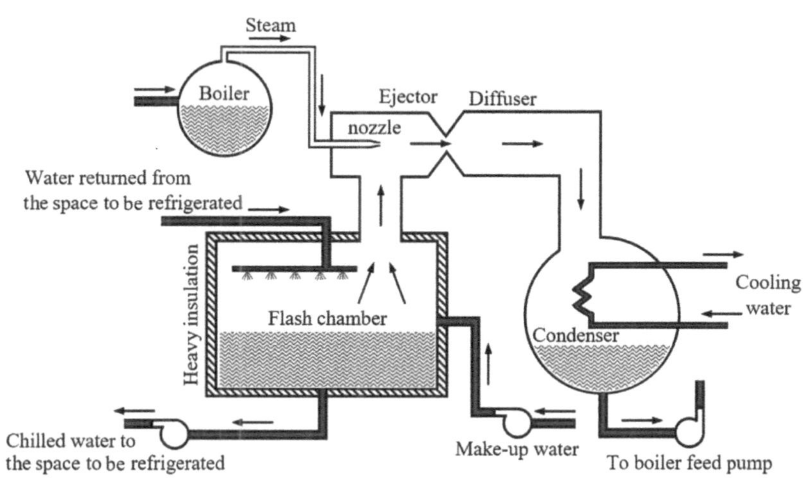

Fig. 3.8 Schematic configuration of a steam jet refrigeration system. (Modified from S.N. Sapali, Refrigeration and Air Conditioning, second ed., PHI Learning Pvt. Ltd., New Delhi, Delhi, India, Nov. 2013; V.K. Manglik, Elements of Mechanical Engineering, PHI Learning Pvt. Ltd., New Delhi, Delhi, India; Oct. 2013).

the water to be cooled, and steam produced in the boiler is expanded via the nozzle(s) and ejected at high speed. Because of this action, low pressure is generated in the chamber, and water evaporates quickly. The steam ejector entrains the evaporated water from the chamber and dispatches the water into a condenser at a high speed. The condensed water is recirculated to the boiler. The heat required for the process of evaporation is taken from the water in the flash chamber. Hence, the water temperature is decreased. The chilled water is circulated around the space to be refrigerated, absorbing heat and thereby generating the cooling effect [44, 45, 57].

- *Liquid gas refrigeration*: In producing refrigeration effect by means of liquid gases, a primary consideration must be choosing a gas that is nontoxic, as the gas may come into direct contact with the material to be frozen. Liquid nitrogen is an effective and convenient choice as a refrigerant because of its availability, low cost, and properties of interest [44, 58]. Moreover, it is the cryogen chosen for most applications related to low temperatures (for instance, in transporting of ice creams, medicines, and frozen foods [5]) due to its very low boiling temperature (−195.8°C) and high capacity for refrigeration at atmospheric pressure [58]. Several cooling methods utilize the refrigeration capabilities of liquid nitrogen in batch or continuous processes, including direct surface (semi-indirect) cooling, secondary circuit (indirect) cooling, cold gaseous nitrogen (GAN) cooling, direct liquid nitrogen injection/spray cooling, and immersion cooling [58]. As an example of this use, Fig. 3.9 illustrates the secondary circuit cooling method by means of a heat transfer fluid, in which liquid nitrogen supplies refrigeration for sample vials.
- *Vapor refrigeration*: This method is broadly used for various purposes of refrigeration, for example, for all industrial systems from small domestic refrigerators to large AC plants [59–61]. In vapor refrigeration systems, vapors (e.g., ammonia and Freon) replace air as the working fluid [62]. Vapor compression refrigeration and vapor absorption refrigeration, which are the most common approaches to this technology, will be explained in detail in the following sections.

In general, all refrigeration methods mentioned earlier can be classified into three types of refrigeration systems, as follows [5, 45, 46]:

- *Noncyclic systems*. In these types of systems, refrigeration can be achieved through processes such as melting ice, evaporating water, and sublimating dry ice [5, 63]. As mentioned previously, they are utilized for small-scale

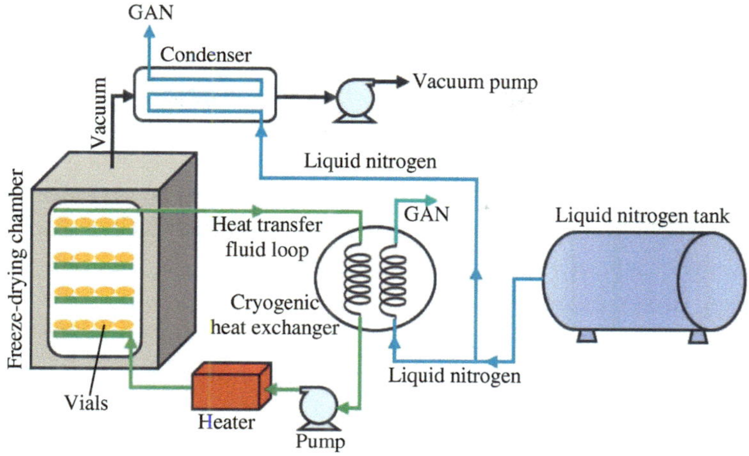

Fig. 3.9 Schematic diagram of the liquid nitrogen secondary circuit cooling by use of a heat transfer fluid to produce refrigeration. (Modified from O. Beteta, S. Ivanova, Cool down with liquid nitrogen, Chem. Eng. Prog. 111 (9) (2015) 30–35; A. Cheng, Method for Lyophilization Using Cryogenic Refrigeration System, The United States Patent No. 8938979; Jan 27, 2015).

refrigeration, for instance, in workshops and laboratories, or in portable coolers [45, 64].

- *Cyclic systems.* These systems can be divided into two major groups, based on whether they rely on a vapor cycle (including vapor compression refrigeration and vapor absorption refrigeration) or a gas cycle (mostly air as a working fluid) [45, 64, 65].
- *Other systems* include thermoelectric, steam jet, vortex tube, and magnetic refrigeration [5, 45, 66], as well as relatively new technologies such as thermoacoustic refrigeration [67], and magnetocaloric, electrocaloric, and elastocaloric refrigeration [68, 69].

3.3 History of refrigeration

As already mentioned in Chapter 2, before the invention of artificial refrigeration technologies, the ice industry and its trade formed one of the most important international activities for a fairly short period of time (particularly between 1890 and 1900 [70]). Ice was widely used to fulfill cold requirements for various purposes. However, this business gradually lost its position in the world because of contamination-related health risks, the expansion of the power grid, and the advent and prevalence of artificial methods of ice production [17, 71]. Artificial techniques have made it possible for various machines to produce ice based on demand, so we no longer have to depend on the climatic and other considerations that have made the natural methods of ice making such a struggle in the past. A number of researchers have reported on the history and evolution of artificial refrigeration technologies as well as on artificial ice production systems/machines [2, 66, 71–76]. Some of the scientists and inventors who established landmarks in the history of refrigeration technology are briefly mentioned later. Especially in the last few centuries, researches on more efficient methods of refrigeration have been driven by commercial and colonial forces.

According to historical evidence, Ibn Sina or Avicenna (980–1037), the preeminent Persian philosopher, physician, and scientist, invented the refrigerated coil capable of condensing aromatic vapors [77, 78], a major innovation in distillation technology. He used this refrigerated tubing to distill and extract essential oils [79, 80]. The initial steps needed to achieve artificial production of ice and also artificial refrigeration occurred in the 16th century with the discovery of chemical refrigeration [72, 73]. Water's temperature was found to decrease when salts (e.g., saltpeter, potassium nitrate, and sodium nitrate [2]) were dissolved in it, as they absorbed heat. This technical approach—reaching equilibrium temperature based on energy balance—was employed to achieve artificial temperature reductions and consequently produce ice. Later called "the refrigerant mixtures method," it fully described for the first time by Blasius Villafranca, a Spanish physician, around 1550 [71, 72, 75].

The earliest recognized public attempt to produce artificial refrigeration based on the principle of vacuum evaporation was by Scottish physician and scientist William Cullen (1710–1790) in 1755 [71, 72]. He was able to make a small amount of ice in the laboratory. The closed-loop refrigeration cycle for producing ice, suggested and designed by Oliver Evans (1755–1819) in 1805, can be categorized as a vapor compression refrigeration system [76, 81], although he never built a refrigerator. Elaborating on Evans's theory, Jacob Perkins (1766–1849) constructed the world's first practical ice machine based on the closed vapor compression refrigeration cycle using ethyl ether as the refrigerant and patented it in 1834 [76, 82]. American scientist John Gorrie (1803–1855) designed and built the first successful air refrigerating machine (based on a gas cycle cooling system with a compressor and expander) in 1844. It not only produced ice but also provided cold air for patients suffering from yellow fever (Gorrie's primary purpose) [72, 83]. Charles Williams Siemens (1823–1883), a British civil engineer, designed and patented the first practical ice-making apparatus in 1855, based on the refrigerant mixtures method, by means of an aqueous solution of calcium chloride as the working fluid [84, 85].

In 1850, the first commercial discontinuous absorption refrigerator was developed by Ferdinand Philippe Edouard Carré (1824–1900), a French scientist, using a working fluid of water and sulfuric acid [73]. Thereafter, in 1859, he introduced a novel continuous absorption refrigeration system (Fig. 3.10), in which water and ammonia acted as the absorbent and refrigerant, respectively [87–90]. This system was the fundamental design employed at the early age of refrigeration development [88, 89]. In 1868, the American scientist Thaddeus Sobieski Coulincourt Lowe (1832–1913) constructed a refrigerated vessel for the shipment of cooled and frozen beef, using a compression ice-making machine with carbon dioxide as the working fluid [72, 76]. In fact, this can be considered the first successful attempt to carry foodstuffs by means of a refrigerated ship [72]. In 1873, the first ammonia compression system to produce ice (indeed the earliest successful commercial liquid-vapor refrigeration) was devised and developed by David Boyle (1837–1891) in the USA [91, 92].

The Swiss physicist Raoul Pierre Pictet (1846–1929) designed and built a refrigerating machine in 1875 that utilized sulfur dioxide (SO_2), known as anhydrous sulfurous oxide, as its refrigerant agent [93, 94]. This refrigerant was employed more in Europe than in the USA [74], mostly in small- to medium-scale applications (for instance, domestic refrigerators) for a

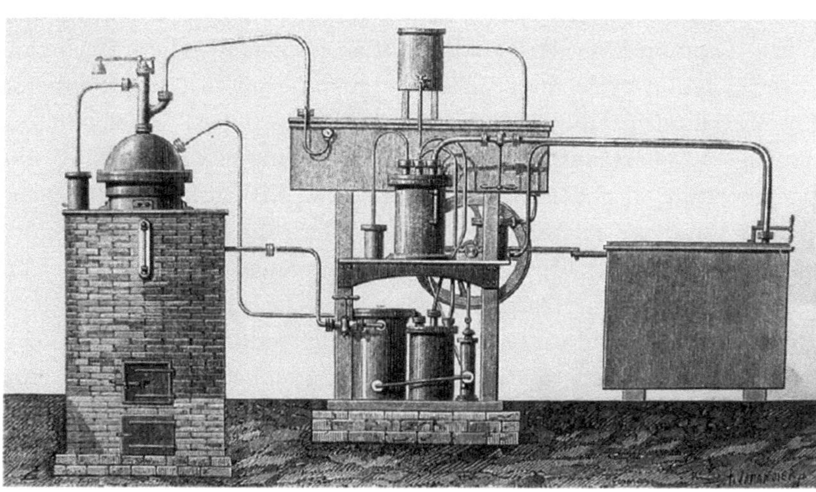

Fig. 3.10 Carré's ice production machine [86].

long period of time before being replaced by chlorofluorocarbons (CFCs). In 1876, Carl Paul Gottfried von Linde (1842–1934), a German chemist, developed a refrigeration system that incorporated the first industrial-scale air separation and gas liquefaction process [76]. In 1891, the first air-conditioning system, which was used for maintaining photographic films in Rochester, New York, was installed by the Eastman Kodak Company, and in 1894, the first domestic air-conditioning system was installed in a house in Frankfurt [66, 76]. In 1904, the American engineer Willis Carrier designed a central air-conditioning system using a very sophisticated air washer [66, 76].

In 1928, an ammonia-water absorption-type refrigeration machine was developed by Paul Crosley. It was suitable for rural regions where electricity was not available [2, 95]. In 1931, Thomas Midgley (1889–1944) and Albert Leon Henne (1901–1967) found a new refrigerant, now called "Freon-21" or "R-21," the first CFC [96, 97] in the USA. Freon's discovery prompted a revolution in the refrigeration industry, and it was thereafter used in various refrigeration applications [98] for a few decades until its negative impacts on the earth's ozone layer were discovered. Since then, its use has gradually been phased out, driven by international agreements. In 1939, the first dual-temperature (freezer-refrigerator) domestic refrigerator was introduced commercially by the General Electric Company [99]. The first commercial heat pump was installed in Switzerland in 1939 [76].

3.4 Compression refrigeration systems

The vapor compression refrigeration system is currently the most broadly used refrigeration cycle for a variety of applications in both commercial and industrial refrigeration and AC systems [100–103]. All vapor compression refrigeration systems are based on the following principle: the pressure and temperature in phase change materials (PCMs) from liquid to vapor (and vice versa) are related to each other. This dependency is indicated by the curve in Fig. 3.11, where *T* and *C* represent the triple point and the critical point of the material, respectively. The right side of the curve illustrates a superheated vapor zone, and the left side of the curve shows a compressed liquid (or a subcooled liquid) zone. The points on the curve represent the saturated vapor or liquid. Table 3.1 demonstrates the values of pressure (*P*) and temperature (*T*) for water; ammonia; and the fluids R-12 (dichlorodifluoromethane), R-22 (chlorodifluoromethane), and R-134a (1,1,1,2-tetrafluoroethane), used as refrigerants.

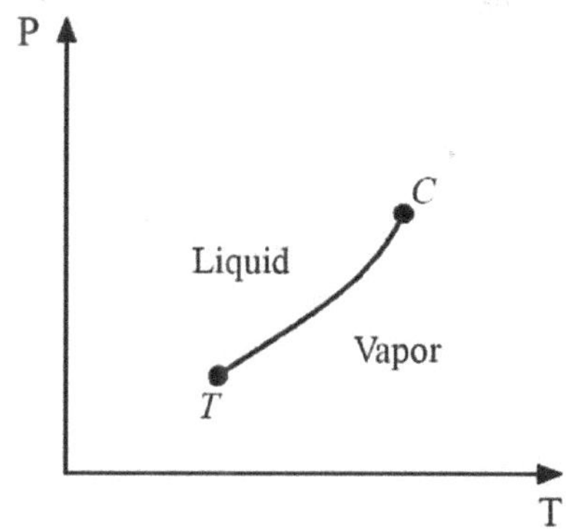

Fig. 3.11 Pressure changes versus temperature in phase change materials (PCMs) from liquid to vapor and vapor to liquid.

The following relationships can be represented by means of the information available for the refrigerants water, ammonia, R-12, R-22, and R-134a in specified temperature ranges [17]. In these relationships, the temperature and pressure are in units of °C and kPa, respectively.

- For water ($1 < T < 60°C$):

$$P = 0.6408 + 4.3 \times 10^{-2} T + 1.2 \times 10^{-3} T^2 + 3 \times 10^{-5} T^3 + 4 \times 10^{-7} T^4$$

- For ammonia ($-50 < T < 60°C$):

$$P = 429.57 + 16.075\, T + 0.2346 T^2 + 1.5 \times 10^{-3} T^3 + 3 \times 10^{-6} T^4$$

- For R-12 ($-50 < T < 60°C$):

$$P = 308.6 + 10.145 T + 0.1259 T^2 + 7 \times 10^{-4} T^3 + 9 \times 10^{-7} T^4$$

- For R-22 ($-50 < T < 60°C$):

$$P = 497.67 + 16.216 T + 0.1986 T^2 + 10^{-3} T^3 + 2 \times 10^{-6} T^4$$

- For R-134a ($-50 < T < 60°C$):

$$P = 294.05 + 10.614 T + 0.1471 T^2 + 9 \times 10^{-4} T^3 + 2 \times 10^{-6} T^4.$$

3.4.1 Main components of a compression refrigeration system

The most important part of a compression refrigeration system is the evaporator, which acts as a heat exchanger; a refrigerant, for example, ammonia, enters in it at a pressure P_1 and temperature T_1 and then evaporates. For instance, if the ammonia's pressure is 190.2 kPa, its evaporation temperature will be $-20°C$, as shown in Table 3.1. It can also be used to bring saltwater from a temperature of −5°C up to −15°C by passing the liquid over the evaporator and then recirculating it around water tanks, consequently freezing the existing water inside the tanks at 0°C or less. Ice can artificially be produced by this method.

Ammonia vapor can be reused by distilling and condensing it into liquid form. In doing so, the distillation heat has to be transferred to an available fluid such as air and water. For example, when the air temperature is around 40°C in summer, and the temperature of the available water is between 20°C and 30°C, a distillation temperature of 50°C is needed for ammonia vapor to transfer the heat to a cooling fluid (air or water) and distill the vapor. In addition, as shown in Table 3.1, the ammonia's pressure must reach 2033.1 kPa, which can be done using a compressor. Therefore, the ammonia vapor exiting the evaporator at a pressure of 190.2 kPa must be compressed by the compressor to reach the pressure of 2033.1 kPa. Then, the vapor exiting the compressor has to enter a condenser and become liquid ammonia, which is at the same pressure but at 50°C. Since the fluid pressure in the evaporator is 190.2 kPa, the pressure of fluid exiting from the condenser must be reduced to this level, in other words, from 2033.1 kPa to 190.2 kPa. This reduction is easily done with an expansion valve or a capillary tube.

Figs. 3.12 and 3.13 show the major components of two different types of compression refrigeration systems. In Fig. 3.12, the compressor is a rotary type and the condenser is cooled by air. In Fig. 3.13, the compressor is a piston type. The diagrams in Figs. 3.14–3.16 represent the theory behind compression refrigeration cycles. These figures ignore the fluid pressure drop in the evaporator, condenser, and interface tubes; the exchange of heat in the interface tubes; and the entropy production in the compressor. Fig. 3.15 shows a theoretical *T-s* diagram of the compression refrigeration cycle. This diagram in fact represents a Rankine cycle that is used to produce mechanical work, but here it flows in the opposite direction. The difference between the two processes is that instead of liquid expansion at point 3 and the production of mechanical work in process 3–4, an expansion valve or a capillary tube (to reduce a refrigeration system's cost) is used, and no work is produced. The Carnot refrigeration system is the same as the Carnot heat engine but operates in the opposite direction [104, 105]. Fig. 3.15 also provides a schematic diagram of the Carnot refrigeration cycle (processes 5-7-3-6-5).

Some differences exist between actual and theoretical compression refrigeration cycles. These differences consist of the entropy production;

Table 3.1 Pressure changes versus temperature for phase change from liquid to vapor for the refrigerants water, ammonia, R-12, R-22, and R-134a [104].

Temperature (°C)	Pressure (kPa)				
	Water	Ammonia	R-12	R-22	R-134a
−50		40.6	39.1	64.4	29.9
−40	0.0129[a]	71.7	64.2	104.9	51.8
−30	0.0381[a]	119.5	100.4	163.5	85.1
−20	0.1035[a]	190.2	150.9	244.8	133.7
−10	0.2601[a]	290.9	219.1	354.3	201.7
0	0.6108[a]	429.6	308.6	479.6	294.0
10	1.228	615.2	423.3	680.7	415.8
20	2.339	857.5	567.3	909.9	572.8
30	4.246	1167.0	744.9	1191.9	771.0
40	7.384	1554.9	960.7	1533.5	1017.0
50	12.35	2033.1	1219.3	1942.3	1318.1
60	19.94	2614.4	1525.9	2426.6	1681.8

[a] Phase change from solid to vapor.

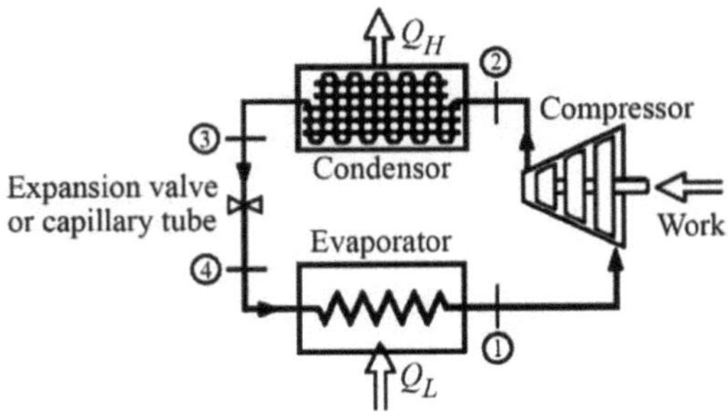

Fig. 3.12 Major components of a compression refrigeration system with a rotary compressor and an air-cooled condenser.

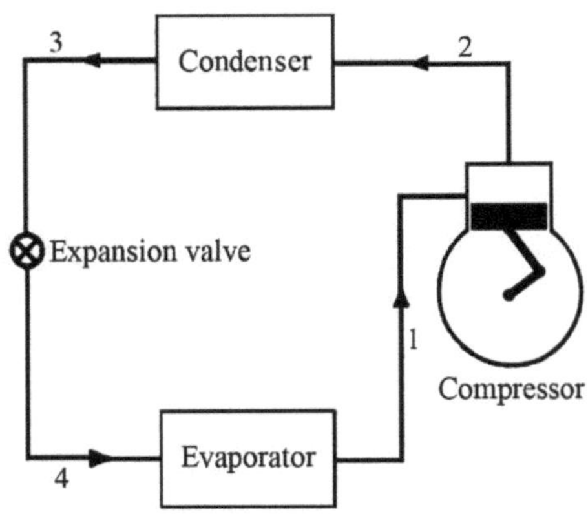

Fig. 3.13 Major components of a compression refrigeration system with a piston compressor.

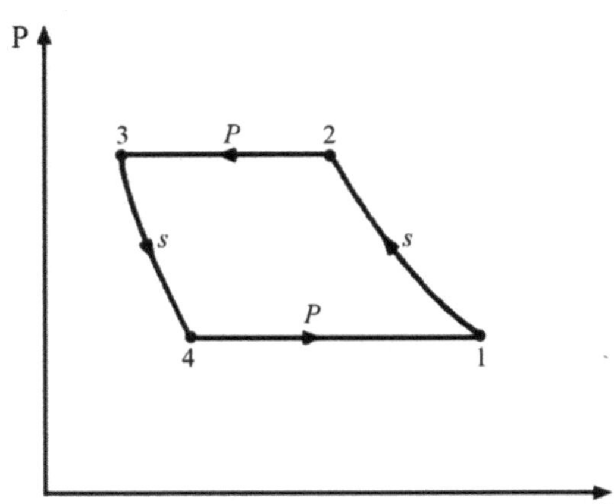

Fig. 3.14 Theoretical *P-v* diagram of compression refrigeration cycle.

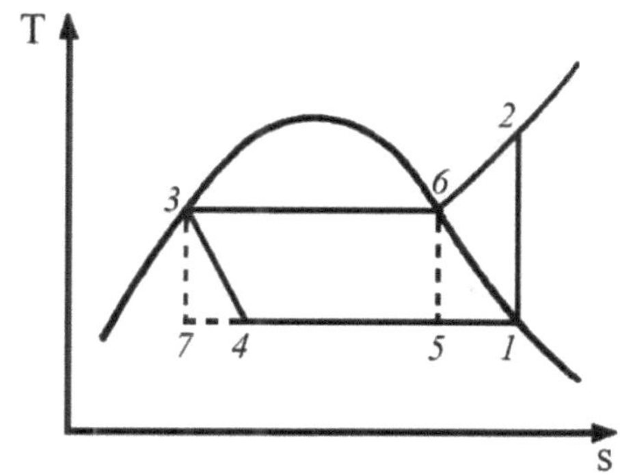

Fig. 3.15 Theoretical *T-s* diagram of compression refrigeration cycle (processes 1-2-3-4-1) and Carnot refrigeration cycle (processes 5-7-3-6-5).

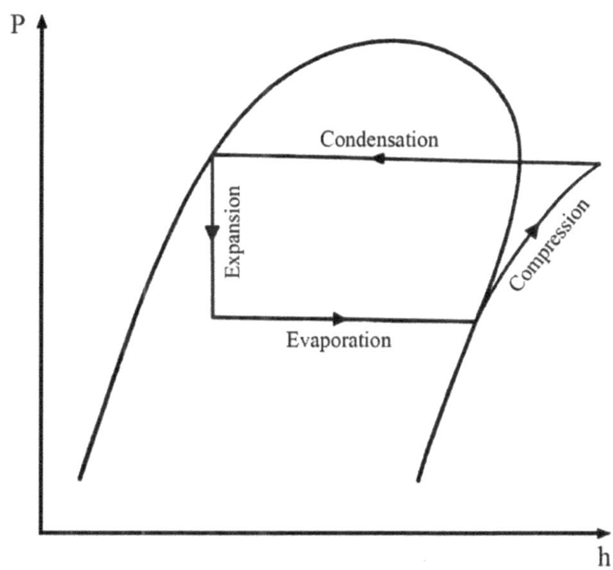

Fig. 3.16 Theoretical *P-h* diagram of compression refrigeration cycle.

the heat exchange in the vapor compression process in the compressor; the pressure drop in all directions including passing the entrance and exit valves of the compressor, interface tubes, evaporator, condenser; and finally, the heat exchange in the interface tubes. Additionally, in the actual cycle, the fluid exiting from the evaporator can be a superheated vapor, and the fluid exiting from the condenser can be a compressed liquid instead of a saturated one.

3.4.2 Analysis of theoretical compression refrigeration cycle in terms of the laws of thermodynamics

In the analysis of the theoretical cycle of compression refrigeration systems, it is assumed that the systems operate in a steady-state steady-flow (SSSF) condition and that changes in the potential and kinetic energy at the entrance and exit of each component are negligible. Therefore, the principle of mass conservation is:

$$\dot{m}_1 = \dot{m}_2 = \dot{m}_3 = \dot{m}_4 = \dot{m} \tag{3.3}$$

The first law of thermodynamics for the evaporator is given by:

$$\dot{m}_4 h_4 - \dot{m}_1 h_1 + \dot{Q}_L = 0 \tag{3.4}$$

By combining Eqs. (3.3) and (3.4), we obtain:

$$\dot{Q}_L = \dot{m}\left(h_1 - h_4\right) \tag{3.5}$$

The first law of thermodynamics for other components of the refrigeration system is as follows:

- in the compressor,

$$\dot{W} = \dot{m}\left(h_2 - h_1\right) \tag{3.6}$$

- in the condenser,

$$\dot{Q}_H = \dot{m}\left(h_2 - h_3\right) \tag{3.7}$$

- in the expansion valve,

$$h_3 = h_4 \tag{3.8}$$

In Eqs. (3.3) to (3.8), $\dot{m}$, h, $\dot{Q}_L$, $\dot{Q}_H$, and $\dot{W}$ represent the mass flow rate; enthalpy; heat absorbed from a source at low temperature by a circulating fluid; heat transferred to a source at a higher temperature; and work done on the engine by a compressor, respectively.

3.4.3 The coefficient of performance (COP) of a compression refrigeration system

The ratio of useful heating or cooling provided to the work required in refrigeration systems is called "the coefficient of performance (COP)" [104–106]. In power plants that aim to produce mechanical work or electricity using thermal energy, this ratio is called efficiency (usually indicated by η). According to the second law of thermodynamics, the amount of efficiency is always less than one. In compression refrigeration systems, the ratio is usually greater than one. With this fact in mind, instead of using the word "efficiency," the term "coefficient of performance" is used here, indicated by COP or β.

The COP in the theoretical cycle of a compression refrigeration system is characterized by:

$$\beta_{ref} = \frac{\dot{Q}_L}{\dot{W}} = \frac{h_1 - h_4}{h_2 - h_1} \tag{3.9}$$

3.4.4 COP of compression refrigeration system that works as a heat pump

In refrigeration systems, heat transferred from the condenser to the cooling fluid can be used elsewhere. Hence, this system is called a "heat pump." The heat may be used, for example, in winter for heating a building. Therefore, we have:

$$\beta_{hp} = \frac{\dot{Q}_H}{\dot{W}} \tag{3.10}$$

or

$$\beta_{hp} = 1 + \beta_{ref} \tag{3.11}$$

Note that Eq. (3.8) is used to obtain Eq. (3.11).

Example 3.1 The goal is to design a compression refrigeration ice factory to produce 1000 kg of ice per hour. The temperatures selected for the fluid phase changer in the condenser and evaporator are 60°C and − 20°C, respectively. The condenser of this refrigeration system is cooled by water. The input temperature to the condenser and the output temperature from it are considered to be 20°C and 30°C, respectively. Referring to Figs. 3.13–3.16, considering the theoretical cycle, and choosing ammonia, R-22, and R-134a as the refrigerants, calculate the refrigerants' mass flow rate, the compressor's power requirement, the transfer heat in the condenser, the COP of the refrigeration system, and the mass flow rate of the cooling water in the condenser.

Solution: The amount of $\dot{Q}_L$ can be determined by:

$$\dot{Q}_L = \dot{m}_i \left(u_f - u_i\right) \tag{3.12}$$

where $\dot{m}_i$ is the ice mass produced in kg/s, u_f is the internal energy of water poured into the ice production tanks, and u_i is the internal energy of the produced ice. The input temperature of water poured into the ice production tanks and the final temperature of the ice produced are considered to be 20°C and 0°C, respectively. Referring to any table listing the properties of water, the following numbers can be obtained [104]:

$$u_f = 83.94 \text{ kJ / kg}, u_i = -333.42 \text{ kJ / kg}$$

Thus,

$$\dot{Q}_L = 1000\frac{\text{kg}}{\text{h}} \times \frac{1\ \text{h}}{3600\text{s}}\left(83.94 + 333.42\right)\frac{\text{kJ}}{\text{kg}} = 115.93\text{kW}$$

The mass flow rate of the refrigerant can be obtained from Eq. (3.5):

$$\dot{m} = \frac{\dot{Q}_L}{h_1 - h_4} = \frac{\dot{Q}_L}{h_1 - h_3}$$

From the ammonia properties tables, the following amounts are acquired for h_1 and h_3 [104, 105]:

$$h_1 = 1418.0\text{kJ} / \text{kg}, h_4 = 472.79\text{kJ} / \text{kg}$$

Therefore,

$$\dot{m} = \frac{115.93}{1418.0 - 472.79} = 0.123\text{kg} / \text{s}$$

From Eq. (3.6), we have:

$$\dot{W} = \dot{m}\left(h_2 - h_1\right)$$

To determine h_2, given that the vapor compression process in the compressor is isentropic (a process during which the entropy remains constant), so we have:

$$s_2 = s_1 \tag{3.13}$$

Referring to the table of ammonia properties [104, 105], we obtain:

$$s_1 = 5.6155\text{kJ} / \text{kgK}$$

Therefore, we obtain:

$$s_2 = 5.6155\text{kJ} / \text{kgK}, \quad P_2 = 2614.4\text{kPa}$$

In point 2, knowing s_2 and P_2, and referring to the table of ammonia properties [104, 105] and by interpolation, we get $h_2 = 1818\,\text{kJ/kg}$. Therefore:

$$\dot{W} = 0.123\text{kg} / \text{s}\left(1818 - 1418\right)\text{kJ} / \text{kg} = 49.2\text{kW}$$

The amount of $\dot{Q}_H$ can be calculated using either of the following equations:

$$\dot{Q}_H = \dot{m}\left(h_2 - h_3\right) \Rightarrow$$
$$\dot{Q}_H = 0.123\text{kg} / \text{s}\left(1818 - 1418\right)\text{kJ} / \text{kg} = 165.46\text{kW}$$

or from:

$$\dot{Q}_H = \dot{Q}_L + \dot{W} \Rightarrow$$
$$\dot{Q}_H = 115.93 + 49.2 = 165.13\text{kW} \tag{3.14}$$

The two produce almost equal results. Eq. (3.9) is used to determine the COP:

$$\beta_{ref} = \frac{\dot{Q}_L}{\dot{W}} = \frac{115.93\text{kW}}{49.2\text{kW}} = 2.356$$

When the refrigeration system is used as a heat pump, the COP can be calculated with either Eq. (3.10) or Eq. (3.11):

$$\beta_{hp} = \frac{\dot{Q}_H}{\dot{W}} = \frac{165.13}{49.2} = 3.363$$

or

$$\beta_{hp} = 1 + \beta_{ref} = 3.356$$

The amounts obtained are approximately equal.

The mass flow rate of cooling water in the condenser can be determined by:

$$\dot{Q}_H = \dot{m}_w \left(h_6 - h_5\right) \tag{3.15}$$

where points 5 and 6 refer to the inlet and outlet water in the condenser, respectively. Referring to the water properties tables, h_5 and h_6 can be obtained. Eq. (3.15) can be written as follows:

$$\dot{Q}_H = \dot{m}_w C_w \left(T_6 - T_5\right) \tag{3.16}$$

where C_w is the specific heat capacity of water. By choosing $C_w = 4.18\,\text{kJ/kg}^\circ\text{C}$, we have:

$$\dot{m}_w = \frac{\dot{Q}_H}{C_w \left(T_6 - T_5\right)} = \frac{165.46\text{kW}}{4.18\text{kJ}/\text{kg}^\circ\text{C}\left(30 - 20\right)^\circ\text{C}} = 3.958\text{kg}/\text{s}$$

The method used to calculate the results of other refrigerants is similar to the aforementioned approach. Table 3.2 shows the results obtained for ammonia, R-12, and R-134a. A comparison of the data in this table indicates that for producing a certain amount of ice, ammonia has the highest COP and R-134a has the lowest. But the fundamental problem with using ammonia is its toxicity and the high vapor pressure at the exit of the

Table 3.2 Summary of calculations for Example 3.1 on theoretical cycle of compression refrigeration system, using three types of refrigerants (ammonia, R-12, and R-134a) to produce 1000 kg of ice per hour [17].

Conditions of various points in the refrigeration system (referring to Figs. 3.13 to 3.16)	Ammonia	R-12	R-134a
P_1 (kPa)	190.2	150.9	133.7
T_1 (°C)	−20	−20	−20
h_1 (kJ/kg)	1418.0	178.74	386.08
s_1 (kJ/kg K)	5.6155	0.7087	1.7395
P_2 (kPa)	2614.4	1525.9	1681.8
h_2 (kJ/kg)	1818.0	219	440
s_2 (kJ/kg K)	5.6155	0.7087	1.7395
P_3 (kPa)	2614.4	1525.9	1681.8
T_3 (°C)	60	60	60
h_3 (kJ/kg)	472.79	95.74	287.79
P_4 (kPa)	190.2	150.9	133.7
T_4 (°C)	−20	−20	−20
h_4 (kJ/kg)	472.79	95.74	287.79
$\dot{Q}_L$ (kW)	115.93	115.93	115.93
$\dot{m}$ (kg/s)	0.123	1.397	1.179
$\dot{W}$ (kW)	49.20	56.24	63.57
$\dot{Q}_H$ (kW)	165.46	172.19	179.46
T_5 (°C)	20	20	20
T_6 (°C)	30	30	30
$\dot{m}_W$ (kg/s)	3.958	4.119	4.293
β_{ref}	2.356	2.061	1.824
β_{hp}	3.356	3.061	2.824

compressor (P_2) and condenser (P_3). The refrigerant R-134a not only has a relatively lower refrigerating capacity compared to R-12 [107] but also has fewer harmful effects on the environment. For a long time, R-134a has been used instead of R-12 due to the former's more desirable features [108–111]. However, both have high environmental impacts (e.g., ozone layer depletion, global warming, and climate change) [1, 111–113].

Example 3.2 Energy software [114] has been used to determine the cooling load for a four-floor residential building located in the city of Tehran, Iran, that has four units per floor. There is 800-m^2 infrastructure for each floor and parking under the first floor. The load has been found to be equal to 180 kW, or 51.12 tons of refrigeration[c] [115]. To ensure residents' thermal comfort, one fan coil unit is used in each room.

[c]1 kW = 3413 Btu/h; 1 kW = 0.284345 tons of refrigeration.

In case 1, cold water at 5°C enters each unit and comes out at 10°C. The refrigerant R-134a is employed in the building's refrigeration system. The evaporation temperature of refrigerant in the evaporator is 0°C, and the distillation temperature of refrigerant in the condenser is 70°C. In case 2, because of water shortages, the water has been removed from the system, and instead, the condenser is cooled by air. The condenser's incoming air temperature is 40°C and the exiting air is 50°C. Considering the theoretical compression refrigeration cycle, calculate the mass flow rate of the refrigerant, the mass flow rate of water flowing in the fan coil units, the mass flow rate and volumetric flow rate of the cooling air in the condenser, as well as the compressor power requirement and the COP of the refrigeration system.

Solution: In case 1: The calculation method is similar to that in Example 3.1. Referring to the table on the properties of R-134a [104] and Figs. 3.12 to 3.16, the following amounts are obtained:

$$T_1 = 0^{\circ}C,\ P_1 = 294\ \text{kPa},\ h_1 = 398.36\ \text{kJ / kg}$$
$$s_1 = 1.7262\ \text{kJ / kg K},\ h_3 = 304.64,\ h_4 = 304.64\ \text{kJ / kg}$$
$$P_2 = 2117\text{kPa},\ s_2 = 1.7262\text{kJ / kg K},\ h_2 \approx 438\text{kJ / kg}$$

The mass flow rate of refrigerant can be determined from Eq. (3.5):

$$\dot{m} = \frac{\dot{Q}_L}{(h_1 - h_4)} = \frac{180\text{kW}}{(398.36 - 304.64)\text{kJ / kg}} = 1.921\text{kg / s}$$

The mass flow rate of water flowing in the fan coil units (or the water that supplies the heat necessary for evaporation of refrigerant in the evaporator) is obtained by:

$$\dot{m}_w = \frac{\dot{Q}_L}{(h_8 - h_7)} = \frac{\dot{Q}_L}{C_w (T_8 - T_7)} = \frac{180\text{kW}}{4.18\dfrac{\text{kJ}}{\text{kg}^{\circ}\text{C}}(10-5)^{\circ}\text{C}} = 8.612\text{kg / s}$$

where the subscripts 7 and 8 are related to the water entry and exit points in the fan coil units, respectively. The compressor power in the refrigeration system is given by Eq. (3.6):

$$\dot{W} = \dot{m}(h_2 - h_1) = 1.921(438 - 398.36) = 76.148\text{kW}$$

In case 2: The heat in the condenser that must be dissipated to the surrounding air is determined by Eq. (3.7):

$$\dot{Q}_H = \dot{m}(h_2 - h_3) = 1.921(438 - 304.64) = 256.185\text{kW}$$

The mass flow rate of the cooling air in condenser cooling is obtained by:

$$\dot{Q}_H = \dot{m}_a \left(h_6 - h_5\right) = \dot{m}_a C_{pa} \left(T_6 - T_5\right) \Rightarrow$$

$$\dot{m}_a = \frac{\dot{Q}_H}{C_{pa}\left(T_6 - T_5\right)} = \frac{256/185}{1.004 \frac{\text{kJ}}{\text{kg}^\circ\text{C}} \left(50 - 40\right)^\circ\text{C}} = 25.516\text{kg}/\text{s}$$

where points 5 and 6 represent the input and output air in the condenser. To determine the volumetric flow rate of the cooling air, the air can be assumed to be an ideal gas, and its density is given by:

$$\rho_5 = \frac{P_5}{RT_5} \tag{3.17}$$

The outside air pressure that is blown into the condenser is considered equal to $P_5 = 105\,\text{kPa}$, a value slightly higher than that of the atmospheric pressure in order to provide the necessary pressure drop of the airflow over the condenser. Thus, we have:

$$\rho_5 = \frac{105\text{kPa}}{0.287 \frac{\text{kJ}}{\text{kg K}} \left(273 + 40\right)\text{K}} = 1.169\ \text{kg}/\text{m}^3$$

The volumetric airflow rate is calculated by:

$$\dot{m} = \rho_5 \dot{V}_5 \tag{3.18}$$

$$\dot{V}_5 = \frac{25.516}{1.169} = 21.83\text{m}^3/\text{s}$$

The COP is given by Eq. (3.9):

$$\beta_{ref} = \frac{\dot{Q}_H}{\dot{W}} = \frac{180}{76.148} = 2.364$$

The COP is determined by Eq. (3.10) if the system is considered as a heat pump, thus:

$$\beta_{hp} = \frac{\dot{Q}_H}{\dot{W}} = \frac{256.185}{76.148} = 3.364$$

The COP can also be obtained by Eq. (3.11):

$$\beta_{hp} = 1 + \beta_{ref} = 1 + 2.364 = 3.364$$

3.4.5 Selection of refrigerants

Refrigerants play a key role in the production of cold: a refrigerant with a phase change from liquid to vapor receives the heat required from a substance at low temperature to change its phase and so produce cold. With a phase change from vapor to liquid, the refrigerant also transfers its phase-change heat to another substance at a higher temperature. This process is performed in both compression and absorption refrigeration systems. The equipment and devices that form the major and minor components of such systems are designed to work effectively in achieving the aforementioned processes, namely the evaporation of a refrigerant at low pressure and temperature and its distillation at high pressure and temperature [5, 66, 116].

Refrigerants can be classified into three groups based on their performance and properties [2, 5, 117–120]: (1) *working principle*, including primary and secondary refrigerants; (2) *safety*, including safe (or nontoxic and nonflammable) refrigerants, toxic and somewhat flammable refrigerants, and highly flammable refrigerants; and (3) *chemical composition* such as halocarbon compounds, cyclic organic compounds, azeotropes, miscellaneous, oxygen and nitrogen compounds, inorganic compounds, and unsaturated compounds. Primary refrigerants are those fluids, that are utilized directly as working fluids and that pass through the compression, cooling or condensation, expansion, and evaporation or warming stages during cyclic processes. As stated in the previous sections, these fluids such as ammonia, R-12, R-22, carbon dioxide (CO_2), and sulfur dioxide (SO_2) create refrigeration through undergoing a process of phase change in the evaporator [2, 5, 121]. Secondary refrigerants (such as water, air, brine solutions of sodium chloride and calcium chloride, and glycol solutions) are employed to transfer heat from the substance or space to be cooled to a heat exchanger (or evaporator) where the heat is absorbed via a primary refrigerant [2, 5].

Selecting a suitable refrigerant depends on many factors. For instance, to produce a certain amount of cold, the best refrigerant is that which allows us to select devices that are relatively smaller in size, cheaper in price, lower in power consumption, and higher in efficiency. Furthermore, a refrigerant must have the appropriate properties in terms of thermodynamics, heat transfer, and fluid mechanics, meaning the refrigerant requires high latent heat, a low specific heat, high vapor density, a low boiling point temperature at atmospheric pressure, high thermal conductivity, and a low viscosity coefficient [116]. On the contrary, there are more important conditions that the refrigerant must possess, the most critical being chemical stability and integrity under working conditions, low toxicity or nontoxicity (in some uses), and reduced environmental impacts [17, 122, 123]. It must also be nonflammable, inexpensive, abundant, compatible with any lubrication oil in the compressor or metals that it will come in contact with, and have low hygroscopicity [17, 105, 123].

Another important point, particularly in the last few years, is that any refrigerant leaking out of the refrigeration system must not harm the environment. Halogenated refrigerants, such as CFCs, remain stable in the atmosphere for a long time but will finally penetrate the stratosphere, where they decompose and release chlorine that can deplete the ozone layer [124, 125]. At lower atmospheric levels, CFC molecules absorb infrared radiation at long wavelengths and contribute to global warming [17, 126]. Replacing halogens with hydrogen atoms greatly reduces the molecules' lifetime in the atmosphere and consequently diminishes the harmful environmental impacts of refrigerants [17]. Refrigerants obtained in this way are known as hydrochlorofluorocarbons (HCFCs). Note that no currently available refrigerant has all the aforementioned conditions. The most appropriate refrigerant must be selected according to the intended usage, taking into account the important factors associated with that usage.

The various refrigerants are employed for different applications. As an example, Table 3.3 shows some refrigerants that were commonly utilized for a variety of applications in the past. Gradually, these refrigerants have been being replaced by new ones (ozone-friendly refrigerants). CFC refrigerants were once in use in many applications, but a few years ago, their damaging effects on the ozone layer were discovered. In 1987, the "Montreal Protocol" was signed by 46 countries to prevent further destruction of the ozone layer

Table 3.3 Different applications of several refrigerants [2, 5].

Types of applications	Refrigerant in use
Domestic refrigerators and freezers	R-12
Small retail supermarkets	R-12, R-22, R-502
Air-conditioning (AC)	R-11, R-12, R-22, R-144, R-502, R-13B1
Industrial	Ammonia (R-717),[a] R-22, R-502
Transport	R-12, R-502

[a] Ammonia is still a popular refrigerant (especially in adsorption refrigeration systems [127]), because it is very cheap and also has some interesting thermodynamics properties, including high enthalpy of evaporation, stability, wide temperature range, and low freezing temperature [2, 128, 129], despite having such undesirable features as being volatile, poisonous, and flammable [2, 130]. Fortunately, ammonia has zero ozone depletion potential when released into the atmosphere and does not directly affect global warming [2].

by, among other things, gradually phasing out the use of CFCs. Altogether, this protocol now has nearly 200 signatories. In developed countries, CFC production was first limited and eventually stopped in 1996. For the next phase of the protocol, HCFCs in these countries will no longer be in use by 2030. In developing countries, CFCs were phased out by 2010, and the use of HCFCs will be stopped by 2040 [17, 105]. Currently, there are no international agreements about the production and consumption of HCFCs such as R-32, R-134a, and R-143a, as well as for mixtures of them like R-404, R-407, and R-410a, but some countries may intend to stop their production and use [17].

3.5 Absorption refrigeration systems

Absorption refrigeration systems work based on the principle that fluid vapor (a refrigerant) can be absorbed in the liquid. The temperature decrease of the liquid is equal to the amount of heat absorption. Absorbed vapor can be removed by increasing the temperature of the solution. Fig. 3.17 outlines the method by which absorption refrigeration machines work [17, 131].

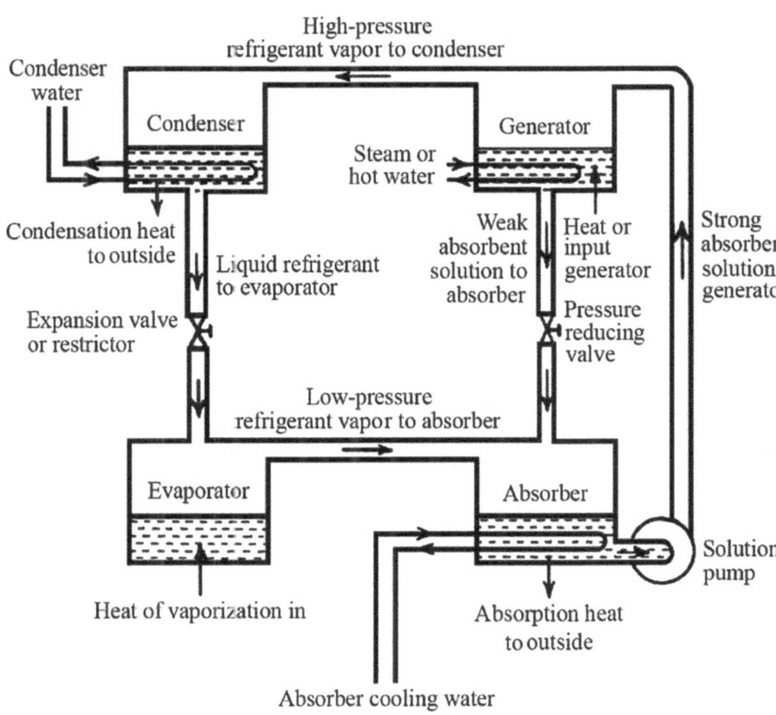

Fig. 3.17 Schematic of an absorption refrigeration system [17, 131].

The refrigerant evaporates in the evaporator, absorbing the evaporation heat from the fluid that is to be cooled. The vapor then enters the absorber, where a pump increases its pressure. It then enters the generator. Transferring the heat to the generator keeps the solution's temperature at a relatively high level and causes the refrigerant to emerge from it in the form of vapor at high pressure and temperature. The refrigerant then enters in the condenser, where it liquefies and transfers its evaporation (or distillation) heat to the water or the cooling air. The refrigerant exits the condenser in the form of liquid and, after passing through the expansion valve and pressure drop, enters the evaporator. This cycle is then repeated.

Comparing Fig. 3.17 with Figs. 3.12 and 3.13, it is observed that the evaporator, condenser, and expansion valve remain the primary components of refrigeration systems, but here, there is an absorber, a generator, and a solution flow pump instead of a compression refrigeration compressor, so heat is utilized in the compressor instead of mechanical energy. The heat required for absorption refrigeration systems may be drawn from the excess heat of factories, solar energy, geothermal energy, etc. Necessary work is usually very low in a solution flow pump (about 1% of the required thermal energy), and the system (or device) can be set up in very small units (with 40W of cooling capacity) without the use of mechanical energy [17, 105].

3.6 Comparison of compression and absorption refrigeration systems

Comparing absorption and compression refrigeration systems highlights the following advantages of the absorption systems [17]:

1. The solution pump is the system's only moving part; therefore, these systems are much smaller than compression refrigeration systems, and quieter, having no noisy compressor.
2. If the vapor that emerges from the evaporator is accompanied by some liquid, then that does not constitute a major drawback in the operation of absorption refrigeration systems, whereas it may be risky in compression refrigeration systems. To prevent this happening (vapor emerging with liquid) in compression refrigeration systems, the compressor must have special control devices. This issue can be very important in applications in which the cooling requirements vary widely.
3. Absorption refrigeration systems with high capacities take up much more room than compression refrigeration systems. However, the units

can be placed outside a building, and be designed to be taller instead of broad and deep, so they occupy less area [131].

4. Some industries produce unwanted heat through various processes, and it is just wasted by being transferred to the ambient air. However, these industries actually require cooling. Using the excess heat in absorption refrigeration systems to produce the desirable cold is a practicable solution and establishes a sound symbiotic relationship in such situations.

Comparing the two systems reveals that absorption refrigeration systems have the following disadvantages [17]:

1. The COP is very low in absorption refrigeration systems, and producing a certain amount of cold consumes about four times more energy (in the form of heat).
2. Cooling the condenser and absorber in absorption refrigeration systems consumes approximately twice the water used in the condenser of compression refrigeration systems (if these systems are cooled by water). Due to the shortage of water in countries like Iran, and the fact that the value of water is increasing daily and must be conserved, absorption refrigeration systems are not widely used.
3. Due to the development of absorption refrigeration systems using natural gas in cities, natural gas consumption has increased over the years, leading to air pollution, especially in the summer.

3.7 Thermoelectric refrigeration systems

In 1821, the Estonian-German physicist Thomas Johann Seebeck (1770–1831) noticed that an electric current flows between two dissimilar materials (conductors or semi-conductors) that are configured in a closed circuit with two junctions maintained at different temperatures, in what is now known as the "Seebeck effect" [2, 132–134]. He termed this phenomenon "the thermomagnetic current" or "thermomagnetism" [135]. Afterward, in 1834, Jean Charles Athanase Peltier (1785–1845), a French physicist, found that cooling is created at one junction of two different materials when an electric current is passed through them, and at the same time, heat is created at the other junction (Fig. 3.18), in what is called the "Peltier effect" or "Peltier cooling" [2, 66, 136]. This finding was, indeed, the reverse to that of Seebeck process. It should be noted that the direction of the current flow affects which of the junctions or ends will become hot or cold. In addition, the actual cooling created at the cold end will be decreased because of irreversible processes of Joule–Fourier heat transfers [66].

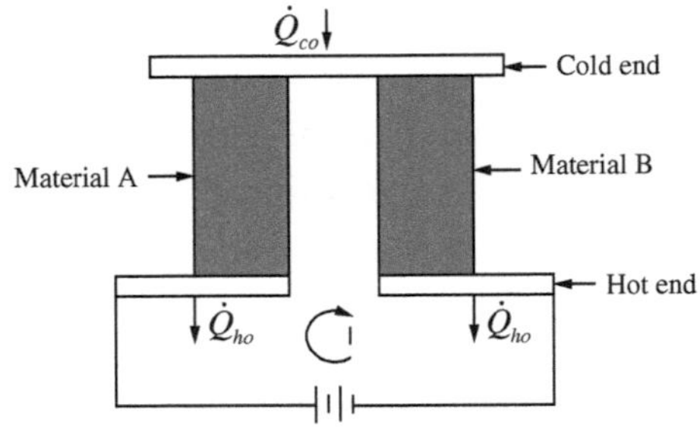

Fig. 3.18 Schematic illustration of the Peltier effect [66], in which $\dot{Q}_{co}$ and $\dot{Q}_{ho}$, respectively, refer to the actual quantities of heat flows at the cold and hot ends.

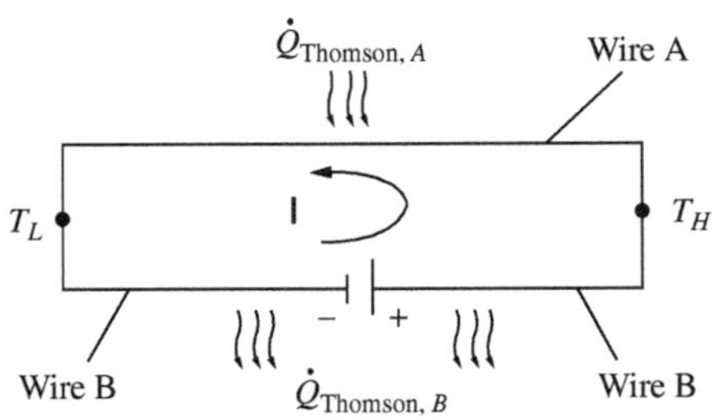

Fig. 3.19 Schematic illustration of the Thomson effect [139]. In this figure, T_L and T_H refer to a low junction temperature and high junction temperature, respectively.

In 1838, the Russian-German physicist Heinrich Friedrich Emil Lenz (1804–1865) employed a direct current (DC) source and two materials (bismuth and antimony) joined together to freeze a droplet of water and make ice. When the current was reversed, the ice melted [2, 66]. In fact, he applied the principle of the Peltier thermoelectric effect to produce refrigeration [137, 138]. Lenz's finding resulted in the concept of thermoelectric refrigeration [2, 138], meaning the process of transferring thermal energy from a material, medium, or one side of a device to another material, medium, or side using electrical energy. In 1854, the Irish-Scottish scientist William Thomson (1824–1907), later Lord Kelvin, found that heat energy is either absorbed or released if there is a temperature difference between any two points of a current-carrying conductor (Fig. 3.19). This effect depends on the material used and the direction of current flow [140] and is known as "Thomson effect" or "Thomson heat" [141].

The three effects (Seebeck, Peltier, and Thomson) are jointly called the "thermoelectric effects" [141]. Additionally, two important relationships between the aforementioned effects were developed by Thomson using the first and second laws of thermodynamics, plus assuming that the reversible and irreversible processes in thermoelectricity are separable [139, 141]. They were named the Thomson (or Kelvin) relationships. Later, these relationships were totally confirmed via experiments [139, 141].

Nowadays, thermoelectric cooling is utilized in a variety of applications, such as in electronic systems and computers to cool sensitive components (e.g., power amplifiers and microprocessors) [142], portable refrigerators, water coolers [143–145], scientific machines used in space exploration and aircraft [106, 145], and medical instruments [139, 146]. Generally, the efficiency of thermoelectric cooling systems is lower than that of compressor-based refrigeration systems [147, 148]. Further information about thermoelectric refrigeration can be found in many sources, such as Refs. [2, 66, 139, 143].

3.8 Vortex tube refrigeration systems

A vortex tube is a mechanical-thermal device that separates a compressed flow of air (or any inert gas) into hot and cold streams and has no mechanical moving parts [149–152]. In 1928, the vortex tube was devised by Georges Joseph Ranque (1898–1973), a French physicist, after he had noticed the temperature splitting effect in a vortex-type pump [153, 154]. A few years later, in 1947, the German physicist Rudolf Hilsch (1903–1972) conducted a systematic study of the device and also improved its design [155]. Hence, this device is now known as the Ranque-Hilsch vortex tube. As shown in Fig. 3.20, the working principle of the vortex tube is as follows: when a flow of highly compressed air is injected into the device, it first passes through a nozzle that is designed to move the air tangentially within the tube in order to set up a high-speed vortex movement. Thereafter, the whirling airflow is forced by the nozzle to alter its direction and go through

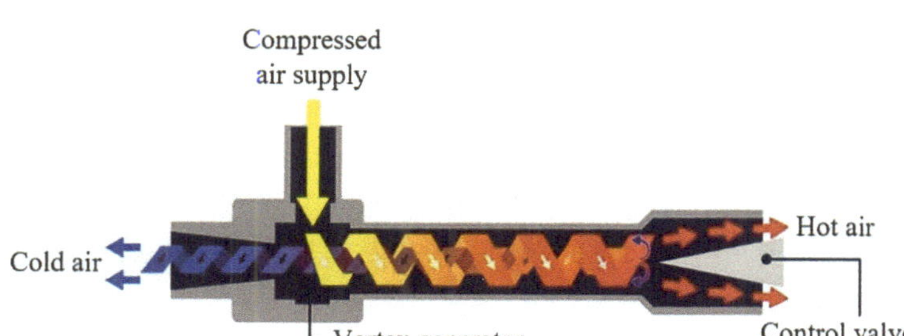

Fig. 3.20 Schematic illustration of working mechanism of a vortex tube [156, 157].

the length of the tube in the form of a rotating shell. A special conical valve at the end of one side of the tube (the hot end) allows a part of the air to escape, whereas the rest of the air turns back in the tube and moves as a second vortex inside the larger outer vortex. The inner vortex gives off its kinetic energy in the form of heat (through viscous dissipation) to the outer vortex, and finally, it exits at another end of the tube (the cold end) [149–152, 158]. Note that by controlling the conical valve, the amount of the cold air and its temperature can be changed.

The vortex tube can be applied in a variety of applications, such as instrument cooling, gas species separation, dehydration of natural gas, fire-fighting equipment cooling [159, 160], spot cooling (e.g., for electronic components), and spot heating [161, 162]. It has several advantages compared to other refrigeration systems, being simple in construction, having no moving parts, experiencing no leakage problems because air (or an inert gas) is used as the working fluid, needs no electricity, is light in weight, costs little, requires little space, allows adjustable instant cold and hot flows, and so on [158, 161–164]. Nonetheless, the vortex tube has a low COP [163], limited capacity, and is not appropriate for large-capacity refrigeration units [164].

3.9 Magnetic refrigeration systems

Magnetic refrigeration is an emerging, ecofriendly, and promising cooling technology developed on the basis of a magnetic solid that behaves as a refrigerant (e.g., gadolinium and some of its alloys) through magnetocaloric effect (MCE)[d] [167–171]. In 1881, the German physicist Warburg (1846–1931) initially discovered this effect in an iron sample [172]. Thereafter, in 1917, the French physicist Weiss (1865–1940) and the Swiss physicist Piccard (1884–1962) described the MCE [173, 174]. MCEs were increasingly being considered when Brown [175], in 1976, indicated that magnetic cooling was possible at about room temperature by means of gadolinium. Following Brown's initial work, in 1978, Steyert [176] introduced and explained the concept of using an active magnetic regenerator in a cooling machine to facilitate heat transfer via assessment of the Stirling cycle for magnetic heat engines and refrigerators [171]. A few years later, Barclay and Steyert [177] and Barclay [178] designed and developed an active magnetic regenerative (AMR) cycle, which is a Brayton-like cycle.

[d]MCE is an intrinsic magneto-thermodynamic property of magnetic materials [165, 166].

Today, the AMR refrigeration is the reference cycle for magnetic refrigeration [167, 168, 179]. According to the research conducted by Chen et al. [180], Yang et al. [181], and Yu et al. [182], for near-room temperature in magnetic refrigerators, a regenerative cycle is more efficient than the Carnot, Stirling, and Ericsson cycles. Supplementary information on magnetic refrigeration can be achieved by different sources, such as Refs. [106, 168–171, 180–182]. Currently, studies related to magnetic refrigeration are concentrated on improvements of magnetocaloric materials' properties, magnets, and an optimal design for magnetic refrigerator devices for room temperature applications [183, 184] in order to commercialize this technology widely.

3.10 Conclusion

Today, refrigeration and AC systems consume huge amounts of energy [185, 186], and the electrical power used by these systems is mainly provided by fossil fuel power plants (approximately 80% globally [187]). According to a report provided in 2019 by the International Institute of Refrigeration (IIR) [188], around 20% of the total worldwide electricity production (e.g., ~20% in the USA and ~25% in Japan [189]) is utilized in various types of refrigeration, AC, and heat pump systems. On the contrary, for a variety of reasons (e.g., the world's population growth, rising living standards, human comfort needs, and expanding industries [190–192]), the demand for these systems is growing day-to-day on a global level. Hence, they make a significant contribution to the emissions of greenhouse gases (GHGs). In addition, the leakage of those refrigerants that are harmful to the environment and public health (such as CFCs and HCFCs) continues to cause damage to the ozone layer.

Refrigeration and AC technology is now regarded as being essential to modern life; however, it also clearly influences global warming and associated climate change. In fact, paradoxically, the greater the global warming, the greater the need and desire for this technology. Therefore, the efficiency of conventional refrigeration and AC systems must be both optimized and improved in terms of energy consumption and performance. In addition, sustainable and environmentally friendly alternatives must be sought. The use of low-energy approaches such as making ice semi-naturally to cool confined spaces would be an effective, economical, and practical solution to supply thermal energy needs and, simultaneously, reduce emissions of GHGs and harmful gases associated with ozone layer depletion.

References

[1] A. Ameen, Refrigeration and Air Conditioning, Prentice-Hall of India Pvt. Ltd., New Delhi, Delhi, India, Jan. 2006.
[2] M. Prasad, Refrigeration and Air Conditioning, New Age International (P) Ltd., New Delhi, Delhi, India, 2011.
[3] M. Gaikwad, D. Shevade, A. Kadam, B. Shubham, Review on thermoelectric refrigeration: materials and technology, Int. J. Curr. Eng. Technol. 4 (Special Issue–4) (March 2016) 67–71.
[4] S. Chamarthi, A. Saxena, S. Ratna, Solar refrigeration system, Int. J. Sci. Res. 6 (5) (May 2017) 483–486.
[5] S.N. Sapali, Refrigeration and Air Conditioning, second ed., PHI Learning Pvt. Ltd., New Delhi, Delhi, India, Nov. 2013.
[6] ASHRAE, Terminology of Heating, Ventilation, Air Conditioning, and Refrigeration, second ed., The American Society of Heating, Refrigerating and Air-Conditioning Engineers (ASHRAE), Inc., Atlanta, GA, 1991.
[7] The American Society of Refrigerating Engineers (ASRE), Air Conditioning Refrigerating Data Book, Design Volume, ninth ed., The American Society of Refrigerating Engineers (ASRE), New York, NY, 1955.
[8] C. Xu, J. Zhang, H. Dinh, Q. Xu, Process synthesis of mixed refrigerant system for ethylene plants, Ind. Eng. Chem. Res. 56 (28) (2017) 7984–7999.
[9] J. Wagner, M. Schäfer, A. Schlüter, L. Harsch, J. Hesselbach, M. Rosano, C.-X. Lin, Reducing energy demand in production environment requiring refrigeration—a localized climatization approach, HVAC&R Res. 20 (6) (2014) 628–642.
[10] A. Dehghani, H. Bridjanian, A. Farshi, Selection separation method, for hydrogen and light hydrocarbons gases from waste-gas streams of a petroleum refinery, Pet. Sci. Technol. 30 (21) (2012) 2196–2207.
[11] C.W. Gellings, New electrification technology, Strateg. Plan. Energy Environ. 31 (2) (2011) 69–78.
[12] N.M. Shah, A.F.A. Hoadley, G.P. Rangaiah, Inherent safety analysis of a propane precooled gas-phase liquified natural gas process, Ind. Eng. Chem. Res. 48 (10) (2009) 4917–4927.
[13] H.-M. Chang, A thermodynamic review of cryogenic refrigeration cycles for liquefaction of natural gas, Cryogenics 72 (Part 2) (Dec. 2015) 127–147.
[14] G. De Guido, S. Langè, L.A. Pellegrini, Refrigeration cycles in low-temperature distillation processes for the purification of natural gas, J. Nat. Gas Sci. Eng. 27 (2015) 887–900.
[15] The Henry Ford, Annual Report, The Henry Ford Organization, Dearborn, MI, 2004.
[16] Y. Hwang, Editorial: expansion work recovery machine: worth using or not? HVAC & R Res. 15 (4) (July 2009) 671–672.
[17] M.N. Bahadori, M. Yaghoubi, Ventilation and Natural Cooling System in Traditional Buildings of Iran, Markaz-e Nashr-e Daneshgahi, Tehran, Iran, 2006 (in Farsi).
[18] O. Kaynakli, M. Kilic, An investigation of thermal comfort inside an automobile during the heating period, Appl. Ergon. 36 (3) (May 2005) 301–312.
[19] T.G. Cengiz, F.C. Babalik, An on-the-road experiment into the thermal comfort of car seats, Appl. Ergon. 38 (2007) 337–347.
[20] O. Kaynakli, U. Unver, M. Kilic, Evaluating thermal environments for sitting and standing posture, Int. Commun. Heat Mass Transfer 30 (8) (Nov. 2003) 1179–1188.
[21] ISO 7730, Moderate Thermal Environments—Determination of the PMV and PPD Indices and Specification of the Conditions for Thermal Comfort, International Organization for Standardization, Geneva, Switzerland, 1994.
[22] J. Li, The science of clothing comfort, Text. Prog. 31 (1–2) (2001) 1–135.

[23] ASHRAE, ASHRAE standard 55-1992 thermal environmental conditions for human occupancy, The American Society of Heating, Refrigerating, and Air-Conditioning Engineers (ASHRAE), Inc., Atlanta, GA, 1992.
[24] J. Huang, W. Xu, A new practical unit for the assessment of the heat exchange of human body with the environment, J. Therm. Biol. 31 (4) (May 2006) 318–322.
[25] P.O. Fanger, Thermal Comfort: Analysis and Applications in Environmental Engineering, McGraw-Hill, New York, NY, 1972.
[26] M. Boduch, W. Fincher, Standards of Human Comfort: Relative and Absolute, Seminar in Sustainable Architecture, The University of Texas, Austin, TX, 2009. https://soa.utexas.edu/sites/default/disk/preliminary/preliminary/1-Boduch_Fincher-Standards_of_Human_Comfort.pdf. (Accessed 21 September 2020).
[27] Y. Epstein, D.S. Moran, Thermal comfort and the heat stress Indices, Ind. Health 44 (3) (2006) 388–398.
[28] M.N. Bahadori, Natural air-conditioning systems, Chapter 5, in: K.W. Boer (Ed.), Advances in Solar Energy: An Annual Review of Research and Development, vol. 3, Plenum Press, New York, NY, 1986, pp. 283–356.
[29] T.N. Quang, C. He, L.D. Knibbs, R. Dear, L. Morawska, Co-optimisation of indoor environmental quality and energy consumption within urban office buildings, Energ. Buildings 85 (Dec. 2014) 225–234.
[30] J.F. Nicol, M.A. Humphreys, Adaptive thermal comfort and sustainable thermal standards for buildings, Energ. Buildings 34 (6) (July 2002) 563–572.
[31] J. Smolander, Effect of cold exposure on older humans, Int. J. Sports Med. 23 (2) (2002) 86–92.
[32] M. Frontczak, P. Wargocki, Literature survey on how different factors influence human comfort in indoor environments, Build. Environ. 46 (4) (2011) 922–937.
[33] A. Lovins, Air-Conditioning Comfort: Behavioral and Cultural Issues, U.S. Department of Energy, Office of Scientific and Technical Information, Oak Ridge, TN, 1992.
[34] R.J. de Dear, G.S. Brager, Developing an adaptive model of thermal comfort and preference, ASHRAE Trans. 104 (1) (1998) 145–167.
[35] T. Steemers, J.O. Lewis, J.R. Goulding, Energy in Architecture: The European Passive Solar Handbook, BT Batsford for the Commission of the European Communities, Directorate General XII for Science, Research and Development, London, UK, 1992.
[36] M.C. Katafygiotou, D.K. Serghides, Indoor comfort and energy performance of buildings in relation to occupants' satisfaction: investigation in secondary schools of Cyprus, Adv. Buil. Energy Res. 8 (2) (2014) 216–240.
[37] ASHRAE, ASHRAE Guideline 10P, Interactions Affecting the Achievement of Acceptable Indoor Environments, Second Public Review, The American Society of Heating, Refrigerating and Air-Conditioning Engineers (ASHRAE), Inc., Atlanta, GA, 2010.
[38] U.S. Environmental Protection Agency (EPA), Report to congress on indoor air quality: vol. II, EPA/400/1-89/001C, Washington, DC, USA, Aug. 1989.
[39] P.S. Nimlyat, M.Z. Kandar, E. Sediadi, Multitrait-multimethod analysis of subjective and objective methods of indoor environmental quality assessment in buildings, Build. Simul. 11 (2) (April 2018) 347–358.
[40] Y. AlHorr, M. Arif, M. Katafygiotou, A. Mazroei, A. Kaushik, E. Elsarrag, Impact of indoor environmental quality on occupant well-being and comfort: A review of the literature, Int. J. Sustain. Built Environ. 5 (1) (June 2016) 1–11.
[41] V. De Giuli, O. Da Pos, M. De Carli, Indoor environmental quality and pupil perception in Italian primary schools, Build. Environ. 56 (Oct. 2012) 335–345.
[42] A. Poje, I. Potocnik, Legal basis, standards and criteria for evaluating thermal conditions in forest work, Zbornik Gozdarstva in Lesarstva 86 (2008). pp. s.13–s.20.
[43] S. Korur, D. Aydın, S.Z. Korkmaz, Assessing a sample of mass housing produced with the industrialized building system in terms of inner space comfort conditions, Int. J. Arts Sci. 4 (19) (2011) 321–328.

[44] Mechanical Engineering Technician, Unit 1: Fundamentals of Refrigeration, Board of Intermediate Education, Andhra Pradesh (BIEAP), India, 2020. http://bieap.gov.in/Pdf/METPaperIIIYR2.pdf. (Accessed 11 April 2020).
[45] P. Dhankhar, A study on refrigeration, Int. J. Sci. Res. 3 (5) (May 2014) 1212–1220.
[46] P.G. Gajare, S. Mitkari, A. Nighot, Design and analysis of vapour absorption refrigeration system using R134a and Dimethylformamide (DMF), Int. Res. J. Eng. Technol. 6 (4) (April 2019) 4167–4170.
[47] J. Gallant, Doing Physics with Scientific Notebook: A Problem Solving Approach, John Wiley & Sons, Inc., New York, NY, May 2012.
[48] N. Gokoglu, P. Yerlikaya, Seafood Chilling, Refrigeration and Freezing: Science and Technology, Wiley-Blackwell, Hoboken, NJ, July 2015.
[49] J. Vansant, C. Rogiers, CO_2 cleaning and pH control in the food industry, Chapter 10, in: Gases in Agro-Food Processes, Academic Press, Cambridge, MA, 2019, pp. 571–581.
[50] C. Yaws, Matheson Gas Data Book, seventh ed., McGraw-Hill Professional, New York, NY, 2001.
[51] T. Sung, K.C. Kim, LNG cold energy utilization technology, in: X. Zhang, I. Dincer (Eds.), Energy Solutions to Combat Global Warming, Lecture Notes in Energy, vol. 33, Springer, Cham, 2017, pp. 47–66.
[52] H. Herzog, D. Golomb, Carbon capture and storage from fossil fuel use, Enc. Energy 1 (6562) (2004) 277–287.
[53] H. Li, K. Li, B. Liu, J. Li, Q. Meng, C. Yang, Study on the passive refrigeration for main control room of nuclear power plant in power outage accident, Nucl. Eng. Des. 326 (Jan. 2018) 183–189.
[54] Y. Demirel, Nonequilibrium Thermodynamics: Transport and Rate Processes in Physical, Chemical and Biological Systems, third ed., Elsevier Science, Amsterdam, The Netherlands, 2014.
[55] M.J. Moran, H.N. Shapiro, Fundamentals of Engineering Thermodynamics, fifth ed., John Wiley & Sons, Inc., New York, NY, 2006.
[56] A.T.A.M. De Waele, Basics of Joule–Thomson liquefaction and JT cooling, J. Low Temp. Phys. 186 (5–6) (2017) 385–403.
[57] V.K. Manglik, Elements of Mechanical Engineering, PHI Learning Pvt. Ltd., New Delhi, Delhi, India, Oct. 2013.
[58] O. Beteta, S. Ivanova, Cool down with liquid nitrogen, Chem. Eng. Prog. 111 (9) (2015) 30–35.
[59] A.K. Rai, S.A. Mohamed, Study of performance evaluation of domestic refrigerator working with mixture of propane, butane and isobutene refrigerant (LPG), Int. J. Mech. Eng. Technol. 7 (3) (2016) 161–169.
[60] M. Ouali, M.A. Djebiret, R. Ouali, M. Mokrane, N.K. Merzouk, A. Bouabdallah, Thermal control influence on energy efficiency in domestic refrigerator powered by photovoltaic, Int. J. Hydrogen Energy 42 (13) (2017) 8955–8961.
[61] S.N.K. Reddy, M.L. Kumar, Performance of domestic refrigeration system with sub cooling by use of eco-friendly refrigerants, Int. J. Eng. Res. Sci. Technol. 4 (3) (Aug. 2015) 185–191.
[62] R. Kumar, R. Shekhar, Performance analysis of vapour compression refrigeration system by using Azeotrops, Int. J. Sci. Res. Dev. 3 (4) (2015) 255–267.
[63] K.S. Moorthy, S.R. Kumar, S.P.S. Singh Sivam, K. Saravanan, N. Pradeep, Design and fabrication of vapour absorption generator to increase the effectiveness, J. Ind. Pollut. Control. 33 (2) (2017) 1726–1731.
[64] N. Satwik, B.S. Kumar, T.G. Krishna, M. Ravinder, B.K. Kumar, Refrigeration system by using LPG (liquefied petroleum gas), Int. J. Mag. Eng. Technol. Manag. Res. 3 (5) (May 2016) 226–235.

[65] X. Zhang, Organic molecules for application of engineering thermodynamics: refrigeration and organic Rankine cycle, in: Encyclopedia of Physical Organic Chemistry, John Wiley & Sons, Inc., Hoboken, NJ, Oct. 2016, pp. 1–45.
[66] R.C. Arora, Refrigeration and Air Conditioning, PHI Learning Pvt. Ltd., New Delhi, Delhi, India, 2012.
[67] M.E.H. Tijani, J.C.H. Zeegers, A.T.A.M. de Waele, Construction and performance of a thermoacoustic refrigerator, Cryogenics 42 (2002) 59–66.
[68] A. Kitanovski, U. Plaznik, U. Tomc, A. Poredoš, Present and future caloric refrigeration and heat-pump technologies, Int. J. Refrig. 57 (Sep. 2015) 288–298.
[69] C. Aprea, A. Greco, A. Maiorino, C. Masselli, A comparison between different materials in an active electrocaloric regenerative cycle with a 2D numerical model, Int. J. Refrig. 69 (Sep. 2016) 369–382.
[70] P. Jahren, T. Sui, Snow and ice, in: How Water Influences Our Lives, Chemical Industry Press & Springer, Singapore, 2016, pp. 39–57, Chapter 3.
[71] S. Reif-Acherman, Domestic ice-making machines 1830–1930, in: K. Gavroglu (Ed.), History of artificial cold, scientific, technological and cultural issues, Boston Studies in the Philosophy and History of Science, vol. 299, Springer, Dordrecht, The Netherlands, 2014, pp. 135–170, Chapter 8.
[72] S. Reif-Acherman, The early ice making systems in the nineteenth century, Int. J. Refrig. 35 (2012) 1224–1252.
[73] M.C. Elvas, I.M. Peres, S. Carvalho, Making science cooler: Carré's apparatus, in: Proceedings of the 4th International Conference of the ESHS, Barcelona, Spain, 2010.
[74] C. Gantz, Refrigeration: A History, McFarland & Company, Inc., Jefferson, NC, 2015.
[75] J. Farndon, The World's Greatest Idea, Icon Books Ltd., London, UK, July 2011.
[76] S. Karellas, T.C. Roumpedakis, N. Tzouganatos, K. Braimakis, Solar Cooling Technologies, CRC Press, Oct. 2018.
[77] V. Pitman, Aromatherapy: A Practical Approach, New edition, Nelson Thornes, UK, 2004.
[78] R.L. Myers, The Basics of Chemistry, Greenwood Publishing Group, Westport, CT, 2003.
[79] M. Ericksen, Healing with Aromatherapy, McGraw-Hill, New York, NY, 2000.
[80] G.E. Kondhalkar, V.N. Kapatkat, Performance analysis of spiral tube heat exchanger used in oil extraction system, Int. J. Mod. Eng. Res. 2 (3) (2012) 930–936.
[81] E.S. Ferguson, O. Evans, Inventive Genius of the American Industrial Revolution, Hagley Museum and Library, Greenville, DE, 1980.
[82] G. Bathe, D. Bathe, Jacob Perkins: His Inventions, his Times and his Contemporaries, The Historical Society of Pennsylvania, Philadelphia, PA, 1943.
[83] M.S. Bhatti, Open air cycle air conditioning system for motor vehicles, SAE technical paper 980289, Society of Automotive Engineers (SAE), Warrendale, PA, 1998, pp. 579–599.
[84] C.W. Siemens, Improvements in cooling and in freezing water and other bodies, British Patent No. 1105, Oct. 30, 1855.
[85] J. Gamgee, On artificial refrigeration, in: United States Fish Commission, 1877, pp. 903–940. Rep. 5.
[86] G. Tissandier, Les Merveilles de L'Industrie, L'eau, Furne, Jouvet et Cie, Paris, France, 1867 (in French).
[87] Z.Y. Xu, R.Z. Wang, Absorption refrigeration cycles: categorized based on the cycle construction, Int. J. Refrig. 62 (2016) 114–136.
[88] P. Srikhirin, S. Aphornratana, S. Chungpaibulpatana, A review of absorption refrigeration technologies, Renew. Sustain. Energy Rev. 5 (2001) 343–372.
[89] S. Aphornratana, T. Sriveerakul, Experimental studies of a single-effect absorption refrigerator using aqueous lithium–bromide: effect of operating condition to system performance, Exp. Therm. Fluid Sci. 32 (2007) 658–669.

[90] Y. Chen, W. Han, H. Jin, Thermodynamic performance optimization of the absorption-generation process in an absorption refrigeration cycle, Energ. Conver. Manage. 126 (2016) 290–301.
[91] D. Boyle, Improvement in ice machines, The United States Patent No. 128448, June 25, 1872.
[92] H.-L. Dienel, Ingenieure zwischen hochschule und industrie: kältetechnik in Deutschland und Amerika, 1870–1930, Vandenhoeck und Ruprecht, Göttingen, Germany, 1995 (in German).
[93] R. Pictet, Improvements in Producing Cold, and in the Manufacture of Sulphurous Acid, and in Apparatus for these Purposes, British Patent No. 2727, Aug. 3, 1875.
[94] R. Pictet, Improvements in the Manufacture of Ice, for Domestic Use, British Patent No. 1420, March 24, 1882.
[95] J. Tomczyk, E. Silberstein, B. Whitman, B. Johnson, Refrigeration and Air Conditioning Technology, eighth ed., Nelson Education, Scarborough, ON, Canada, 2016.
[96] T. Midgley, A.L. Henne, Organic fluorides as refrigerants, Ind. Eng. Chem. 22 (5) (1930) 542–545.
[97] R.J. Thompson, Freon, a refrigerant, Ind. Eng. Chem. 24 (6) (1932) 620–623.
[98] B.A. Nagengast, How HVAC&R changed the world, ASHRAE J. 61 (9) (2019) 34–49.
[99] A.F. Smith (Ed.), The Oxford Companion to American Food and Drink, Oxford University Press, Oxford, UK, 2007.
[100] S. Aphornratana, T. Sriveerakul, Analysis of a combined Rankine–vapour–compression refrigeration cycle, Energ. Conver. Manage. 51 (12) (2010) 2557–2564.
[101] Y.A. Çengal, M.A. Boles, Refrigeration cycles, in: Thermodynamics an Engineering Approach, eighth ed., McGraw-Hill Education, New York, NY, 2014, Chapter 11.
[102] M.M. Rathore, Thermal Engineering, McGraw-Hill Education, New York, NY, April 2010.
[103] S. Kubba, Impact of energy and atmosphere, in: Handbook of Green Building Design and Construction, Butterworth-Heinemann, Oxford, UK, 2012, pp. 385–492, Chapter 9.
[104] R.E. Sonntag, C. Borgnakke, G. Van Wylen, Fundamentals of Thermodynamics, sixth ed., John Wiley & Sons, Inc., New York, NY, 2003.
[105] ASHRAE Handbook, Fundamentals, The American Society of Heating, Refrigerating and Air-Conditioning Engineers (ASHRAE), Inc., Atlanta, GA, 2017.
[106] I. Dincer, Refrigeration Systems and Applications, John Wiley & Sons, Inc., New York, NY, 2017.
[107] K. Mani, V. Selladurai, Experimental analysis of a new refrigerant mixture as drop-in replacement for CFC12 and HFC134a, Int. J. Therm. Sci. 47 (2008) 1490–1495.
[108] UNEP, Montreal Protocol on Substances that Deplete the Ozone Layer—Final Act, United Nations Environmental Program (UNEP), Nairobi, Kenya, 1987.
[109] D. Butler, Life after CFCs and HCFCs, in: CIBSE National Conference, Regents College, London, UK, 2001.
[110] S. Devotta, S. Gopichand, Comparative assessment of HFC 134a and some refrigerants as alternatives to CFC 12, Int. J. Refrig. 15 (1992) 112–118.
[111] P. Thangavel, P. Somasundaram, T. Sivakumar, C. Selva Kumar, G. Vetriselvan, Simulation analysis of compression refrigeration cycle with different refrigerants, Int. J. Eng. Innov. Technol. 2 (10) (2013) 127–131.
[112] B. Cyriac, A.G. Shaligram, M.D. Nadar, CFD modelling of finned tube gas cooler for transcritical operation of CO_2 systems, Int. J. Mech. Eng. Robot. Res. 4 (1) (2015) 467–472.
[113] K. Mani, V. Selladurai, N. Murugan, Development of statistical models for predicting performance of R12, R134a and R290/R600 mixture refrigerants using design of experiments, Int. J. Thermodyn. 16 (1) (2013) 43–53.

[114] A. Ahmadi-Nodoushan, Cooling and Thermal Energy Reduction Potential in Residential Complex in Tehran, Using Ceiling Fans, M.Sc. Thesis, School of Mechanical Engineering, Sharif University of Technology, Tehran, Iran, 1997 (in Farsi).
[115] M.N. Bahadori, Reducing Heating and Cooling Energy Consumption in Buildings and How Subsidies to Save These Energies, Research Report Conducted for the National Iranian Oil Company (NOIC), Tehran, Iran, 2008 (in Farsi).
[116] A. Ghimire, Basic Principles of Engineering, Morrisville, NC, Oct. 2017. www.lulu.com.
[117] Addenda to ANSI/ASHRAE Standard-1999, Addenda to Designation and Safety Classifications of Refrigerants, American Society of Heating, Refrigerating and Air-Conditioning Engineers (ASHRAE), Inc., Atlanta, GA, 2000.
[118] ANSI/ASHRAE Standard 34-2013, Designation and Safety Classification of Refrigerants, American Society of Heating, Refrigerating and Air-Conditioning Engineers (ASHRAE), Inc., Atlanta, GA, 2016.
[119] ISO 817, Refrigerants—Designation and safety classification, third ed., 2014, Switzerland.
[120] N.L. Ryder, P.S.J. Jordan, P.B. Sunderland, Flammable refrigerants firefighter training: Hazard assessment and demonstrative testing, Report No. FPRF-2019-04, Fire Protection Research Foundation, Quincy, MA, May 2019.
[121] S.K. Bandey, A study on refrigerant, CSVTU Res. J. Eng. Technol. 8 (1) (2019) 21–30.
[122] L.W. Wang, R.Z. Wang, R.G. Oliveira, A review on adsorption working pairs for refrigeration, Renew. Sustain. Energy Rev. 13 (3) (2009) 518–534.
[123] C. Bogatu, A. Duţă, T.W. de Loos, D. Geană, Modelling fluid phase equilibria in the binary system trifluoromethane + 1-phenylpropane, Fluid Phase Equilib. 428 (2016) 190–202.
[124] A. Seikya, S. Misaki, A continuing search for new refrigerants, Chem. Technol. 26 (12) (Dec. 1996) 44–48.
[125] J.A. Mader, J. Staehelin, T. Peter, D. Brunner, H.E. Rieder, W.A. Stahel, Evidence for the effectiveness of the montreal protocol to protect the ozone layer, Atmos. Chem. Phys. 10 (2010) 12161–12171.
[126] N.J. Mason, J.A. Davies, J.M. Gingell, The spectroscopy of atmospheric molecules probed by synchrotron radiation and electron impact, in: D.M. Campbell, H. Kleinpoppen (Eds.), Selected Topics on Electron Physics, Physics of Atoms and Molecules, Springer, Boston, MA, 1996.
[127] E.E. Anvanwu, N.V. Ogueke, Thermodynamic design procedure for solid adsorption solar refrigerator, Renew. Energy 30 (2005) 81–96.
[128] L. Wang, L. Chen, H.L. Wang, D.L. Liao, The adsorption refrigeration characteristics of alkaline-earth metal chlorides and its composite adsorbents, Renew. Energy 34 (4) (2009) 1016–1023.
[129] C.Z. Zhuo, C.H.M. Machielsen, Performance of high-temperature absorption heat transformers using Alkitrate as the working pair, Appl. Therm. Eng. 16 (3) (1996) 255–262.
[130] R. Zehioua, C. Coquelet, A.H. Meniai, D. Richon, p−T−x Measurements for some working fluids for an absorption heat transformer: 1,1,1,2-tetrafluoroethane (R134a)+ dimethylether diethylene glycol (DMEDEG) and dimethylether triethylene glycol (DMETrEG), J. Chem. Eng. Data 55 (8) (2010) 2769–2775.
[131] R.J. Dossat, Principles of Refrigeration, third ed., Prentice-Hall, Inc., Englewood Cliffs, NJ, 1991.
[132] E. Velmre, Thomas Johann Seebeck (1770–1831), Est. J. Eng. 13 (4) (2007) 276–282.
[133] A. Molki, Simple demonstration of the seebeck effect, Sci. Educ. Rev. 9 (3) (2010) 103–107.
[134] N. Rahbar, J.A. Esfahani, Experimental study of a novel portable solar still by utilizing the heatpipe and thermoelectric module, Desalination 284 (2012) 55–61.

[135] C. Goupil, H. Ouerdane, K. Zabrocki, W. Seifert, N.F. Hinsche, E. Müller, Thermodynamics and thermoelectricity, in: C. Goupil (Ed.), Continuum Theory and Modeling of Thermoelectric Elements, Wiley, West Sussex, UK, 2015, pp. 1–74, Chapter 1.
[136] P.R. Barros, G.A. Júnior, J.B. dos Santos, Two-input two-output laboratory-scale temperature system based on Peltier modules, IFAC Proc. vol. 41 (2) (2008) 9767–9772.
[137] T. Shachtman, Absolute Zero and the Conquest of Cold, HMH Books, Boston, NY, 2000.
[138] R.K. Rajput, Non-conventional refrigeration systems, in: Refrigeration and Air Conditioning, S.K. Kataria & Sons, New Delhi, Delhi, India, 2005, pp. 328–331.
[139] H. Lee, Thermoelectrics: Design and Materials, John Wiley & Sons, Inc., New York, NY, 2016.
[140] W. Thomson, Account of researchers in thermo-electricity, Philos. Mag., Ser. 4 (8) (1854) 62–69.
[141] H. Lee, The Thomson effect and the ideal equation on thermoelectric coolers, Energy 56 (2013) 61–69.
[142] S.D. Patil, K.D. Devade, Review on thermoelectric refrigeration: applications and technology, Int. J. Mod. Trends Eng. Res. 2 (7) (2015) 2349–9745.
[143] W.L. Wright, Thermoelectric refrigeration, Electr. Eng. 79 (5) (1960) 380–384.
[144] R. Harshitha, V.B. Aaditya, B.M. Bharathesh, B.V. Chaluvaraju, U.P. Raghavendra, M.V. Murugendrappa, Studies of thermo-electric power and dielectric modulus of polypyrrole/zirconium oxide-molybdenum trioxide (PZM) composites, J. Mater. Sci. Mater. Electron. 29 (8) (2018) 6564–6578.
[145] C.F. Afonso, Recent advances in building air conditioning systems, Appl. Therm. Eng. 26 (16) (2006) 1961–1971.
[146] Y.G. Gurevich, J.E. Velazquez-Perez, Peltier effect in semiconductors, in: Wiley Encyclopedia of Electrical and Electronics Engineering, John Wiley & Sons, Inc., New York, NY, 1999, pp. 1–21.
[147] B.C. Sales, Thermoelectric devices: refrigeration and power generations with no moving parts, in: Encyclopedia of Materials: Science and Technology, second ed., Elsevier, 2001, pp. 9179–9185.
[148] H. Kuscu, K. Kahveci, B.T. Tanju, Natural convection in a square enclosure cooled by Peltier effect, in: Proceedings of the World Congress on Mechanical, Chemical, and Material Engineering (MCM 2015), Paper No. 315, Barcelona, Spain, July 20–21, 2015.
[149] T. Bornare, A. Badgujar, P. Natu, Vortex tube refrigeration system based on compressed air, Int. J. Mech. Eng. Technol. 6 (7) (2015) 97–102.
[150] N. Pourmahmoud, A.H. Zadeh, O. Moutaby, A. Bramo, Computational fluid dynamics analysis of helical nozzles effects on the energy separation in a vortex tube, Therm. Sci. 16 (1) (2012) 151–166.
[151] M. Mia, G. Singh, M.K. Gupta, V.S. Sharma, Influence of Ranque-Hilsch vortex tube and nitrogen gas assisted MQL in precision turning of Al 6061-T6, Precis. Eng. 53 (2018) 289–299.
[152] M.H. Bin Yusof, H. Katanoda, H. Morita, Temperature and pressure measurements at cold exit of counter-flow vortex tube with flow visualization of reversed flow, J. Therm. Sci. 24 (1) (2015) 67–72.
[153] G.J. Ranque, Method and apparatus for obtaining from a fluid under pressure two currents of fluids at different temperatures, The United States Patent No. 1952281, March 27, 1934.
[154] G.J. Ranque, Experiments on expansion in a vortex with simultaneous exhaust of hot air and cold air, J. Phys. Rad. 4 (1933) 112–114.
[155] R. Hilsch, The use of expansion of gases in a centrifugal field as a cooling process, Rev. Sci. Instrum. 18 (2) (1947) 108–113.

[156] Airtx.net, AiRTX Vortex Tubes (review), 2020. https://www.airtx.net/airtx-vortex-tubes-review. (Accessed 21 September 2020).

[157] AiRTX International, 2020. https://airtx.com/search?type=product&q=Vortex-+Tube. (Accessed 23 September 2020).

[158] Y. Xue, M. Arjomandi, R. Kelso, The working principle of a vortex tube, Int. J. Refrig. 36 (6) (2013) 1730–1740.

[159] A.D. Gutak, Experimental investigation and industrial application of Ranque-Hilsch vortex tube, Int. J. Refrig. 49 (2015) 93–98.

[160] A.D. Gutak, Numerical simulation of temperature separation in methane stream in Ranque-Hilsch vortex tube, Холодильна техніка та технологія 51 (вип. 3) (2015) 76–83.

[161] S. Nagane, S. Vanage, A. Anpat, S. Hole, A review on vortex tube refrigeration system, Int. J. Innov. Sci. Eng. Technol. 2 (9) (Sep. 2015) 317–318.

[162] J. Prabakaran, S. Vaidyanathan, Effect of orifice and pressure of counter flow vortex tube, Indian J. Sci. Technol. 3 (4) (2010) 374–376.

[163] A.N. Jejurkar, A.N. Shukla, An overview on vortex tube applications, Int. J. Earth Sci. Eng. 5 (1) (2012) 378–384.

[164] T.K. Sharma, G.A.P. Rao, K.M. Murthy, Numerical analysis of a vortex tube: a review, Arch. Comput. Meth. Eng. 24 (2) (2017) 251–280.

[165] Y. Li, H. Zhang, K. Tao, Y. Wang, M. Wu, Y. Long, Giant magnetocaloric effect induced by reemergence of magnetostructural coupling in Si-doped Mn0.95CoGe compounds, Mater. Des. 114 (2017) 410–415.

[166] A. Mleiki, S. Othmani, W. Cheikhrouhou-Koubaa, A. Cheikhrouhou, E.K. Hlil, Enhanced relative cooling power in Ga-doped La 0.7 (Sr, Ca) 0.3 MnO_3 with ferromagnetic-like canted state, RSC Adv. 6 (59) (2016) 54299–54309.

[167] N.A. Mezaal, K.V. Osintsev, T.B. Zhirgalova, Review of magnetic refrigeration system as alternative to conventional refrigeration system, in: IOP Conference Series: Earth and Environmental Science, vol. 87, IOP Publishing, Bristol, UK, Oct. 2017, 032024.

[168] C. Aprea, G. Cardillo, A. Greco, A. Maiorino, C. Masselli, A rotary permanent magnet magnetic refrigerator based on AMR cycle, Appl. Therm. Eng. 101 (2016) 699–703.

[169] E. Brück, Developments in magnetocaloric refrigeration, J. Phys. D Appl. Phys. 38 (23) (2005) R381–R391.

[170] E.L.T. França, A.O. dos Santos, A.A. Coelho, L.M. da Silva, Magnetocaloric effect of the ternary Dy, Ho and Er platinum gallides, J. Magn. Magn. Mater. 401 (2016) 1088–1092.

[171] K.A. Gschneidner, V.K. Pecharsky, Thirty years of near room temperature magnetic cooling: where we are today and future prospects, Int. J. Refrig. 31 (6) (2008) 945–961.

[172] E.G. Warburg, Magnetische Untersuchungen, Annalen der Physik (Leipzig) 249 (5) (1881) 141–164 (in German).

[173] P. Weiss, A. Piccard, Le phénomène magnétocalorique, J. Phys. Théor. Appl. 7 (1917) 103–109 (in French).

[174] A. Smith, Who discovered the magnetocaloric effect? Eur. Phys. J. 38 (4) (2013) 507–517.

[175] G.V. Brown, Magnetic heat pumping near room temperature, J. Appl. Phys. 47 (8) (1976) 3673–3680.

[176] W.A. Steyert, Stirling-cycle rotating magnetic refrigerators and heat engines for use near room temperature, J. Appl. Phys. 49 (3) (1978) 1216–1226.

[177] J.A. Barclay, W.A. Steyert, Active magnetic regenerator, The United States Patent No. 4332135, 1982.

[178] J.A. Barclay, Wheel-type magnetic refrigerator, The United States Patent No. 4408463, 1983.

[179] C. Aprea, A. Greco, A. Maiorino, Magnetic refrigeration: a promising new technology for energy saving, Int. J. Ambient Energy 37 (3) (2016) 294–313.
[180] F.C. Chen, R.W. Murphy, V.C. Mei, G.L. Chen, Thermodynamic analysis of four magnetic heat-pump cycles, J. Eng. Gas Turbines Power 114 (1992) 715–720.
[181] Y. Yang, J. Chen, J. He, E. Brück, Parametric optimum analysis of an irreversible regenerative magnetic Brayton refrigeration cycle, Phys. B Condens. Matter 364 (1–4) (2005) 33–42.
[182] B. Yu, M. Liu, P.W. Egolf, A. Kitanovski, A review of magnetic refrigerator and heat pump prototypes built before the year 2010, Int. J. Refrig. 33 (6) (2010) 1029–1060.
[183] A. Kitanovski, P.W. Egolf, Application of magnetic refrigeration and its assessment, J. Magn. Magn. Mater. 321 (7) (2009) 777–781.
[184] E. Talik, A. Guzik, M. Oboz, P. Zajdel, Magnetocaloric effect of the multiphase $Gd_{4-x}Tb_xCO_3$ system, Philos. Mag. 98 (36) (2018) 3300–3314.
[185] L.O.S. Buzelin, S.C. Amicoa, J.V.C. Vargasa, J.A.R. Parise, Experimental development of an intelligent refrigeration system, Int. J. Refrig. 28 (2005) 165–175.
[186] S.M. Jadhav, M. Arulprakasajothi, U. Chandrasekhar, D. Yuvarajan, Experimental investigation of vapour absorption refrigeration cycle for automobile cabin cooling, in: U. Chandrasekhar, L.J. Yang, S. Gowthaman (Eds.), Innovative Design, Analysis and Development Practices in Aerospace and Automotive Engineering (I-DAD 2018), Springer, Singapore, 2019, pp. 41–51.
[187] K.F. Fong, T.T. Chow, C.K. Lee, Z. Lin, L.S. Chan, Comparative study of different solar cooling systems for buildings in subtropical city, Sol. Energy 84 (2) (2010) 227–244.
[188] J.L. Dupont, P. Domanski, P. Lebrun, F. Ziegler, 38th Note on Refrigeration Technologies: The Role of Refrigeration in the Global Economy: Introduction, The International Institute of Refrigeration (IIR), Paris, France, June 2019.
[189] M. Almanza, A. Kedous-Lebouc, J.P. Yonnet, U. Legait, J. Roudaut, Magnetic refrigeration: recent developments and alternative configurations, Eur. Phys. J. Appl. Phys. 71 (2015) 10903.
[190] N. Kalkan, E.A. Young, A. Celiktas, Solar thermal air conditioning technology reducing the footprint of solar thermal air conditioning, Renew. Sustain. Energy Rev. 16 (2012) 6352–6383.
[191] J. Aman, D.K. Ting, P. Henshaw, Residential solar air conditioning: energy and exergy analyses of an ammonia–water absorption cooling system, Appl. Therm. Eng. 62 (2) (2014) 424–432.
[192] A.R. Dehghani-Sanij, E. Tharumalingam, M.B. Dusseault, R. Fraser, Study of energy storage systems and environmental challenges of batteries, Renew. Sustain. Energy Rev. 104 (2019) 192–208.

CHAPTER 4

The history of traditional Iranian ice-houses

4.1 Introduction

In some parts of the vast land of Iran, particularly in its arid and semi-arid areas, huge and beautiful ice-houses remain, having withstood the passage of time as well as natural disasters such as hurricanes, floods, earthquakes, and even the negative consequences of precipitation. Nowadays, despite generally being disregarded and almost forgotten, they are drawing the admiration of civil, mechanical, and energy engineers as well as architects. These natural ice-making structures, or "*Yakhchals*,[a]" represent one of the most important and sophisticated technologies of traditional Iranian architecture.

No accurate information had been preserved about the background of Iranian ice-houses before the Safavid era (1501–1722). According to Mokhlesi [1] and Hosseini and Namazian [2], and based on available evidence (including historical pictures and texts, as well as the remnants of old ice-houses), the history of traditional ice-houses in Iran most likely goes back to the 1600s, nearly 400 years ago. However, several researchers have claimed that Iranian ice-houses—in different forms and structures—date back many centuries (to approximately 400 BC and even before) [3–5], since ancient Persians invented, developed, and used various technologies/systems/structures related to water (for instance, Qanats[b]). Jorgensen [3] in his book entitled "Ice Houses of Iran, Where, How, Why" referred to the long history of Persian ice-houses several times, in statements such as "*The origins of the Iranian ice houses may be traced back to that of the Qanat, …*" and "*Ice houses have existed in Iran since c. 400 BC, or even earlier.*" Gantz [4] noted

[a]In some parts of Iran (especially in Meybod City, and Kerman, Razavi Khorasan, and South Khorasan provinces, Iran), local people also call them *Yakhdan*. *Yakh* means *ice* and *dan* means *container*; together they mean an underground place to keep/store ice.
[b]The history of the Qanat, which originated in Persia, goes back to approximately the early part of the 1st millennium *BC* [6–9].

Ice-Houses
https://doi.org/10.1016/B978-0-12-822275-1.00002-3
Copyright © 2021 Elsevier Inc.
All rights reserved.

that "*In the Persian Empire in 400 BC, an above ground, several-story-tall ice-house of adobe, shaped like a pointed dome, was called a yakhchal.*" Moreover, Patil and Ranadive [5] have also reported that by 400 BC, Iranians had learned a technique of using specially constructed buildings to store ice for summer use in hot and arid desert regions.

Historical evidence reveals that the early ice-houses in Iran were simple ice reservoirs/trenches (most likely without a ceiling), in which Persians stored the compressed snow and lumps of ice that formed naturally in winter. Their frozen state was prolonged into the summer months by coverings of straw or other insulating materials (e.g., reeds and dry twigs) and soil for protection against the Sun's rays and heat. Such reservoirs and trenches appear to have been used since the first millennium BC, if not earlier [10]. Unfortunately, no archeological proof or textual documents have been found to show the gradual transformation of these simple structures into the traditional Persian ice-houses and when exactly their development took place. What is clear is that from very early on, snow was stored in trenches at altitudes of 2600–2700 m on Mount Derak [11], located alongside the city of Shiraz, Fars, Iran (at 2870 m above sea). Local people called them "snow pits," and the remnants of these pits are still visible. In general, they were roughly 8–10 m in diameter and about 1.5–2 m deep [11].

This chapter describes the history of traditional ice-houses in Iran according to the pieces of evidence available, including literary texts, accounts (usually diary entries) by foreign travelers who visited Persia in past times, and historical photographs/paintings. In addition, it covers the distribution and history of a number of ice-houses in Iran that are still visible in diverse forms—anything from traces, to severely wind- and weather-eroded remains, to seemingly intact structures due to recent restoration efforts by the government. Maps showing the distribution of Iran's old ice-houses between the Safavid and Qajar periods, and also throughout the city of Isfahan around a century ago, are provided at the end of this chapter.

4.2 The history of traditional ice-houses in literature

In ancient and historical texts, and even poems before the Safavid period, the use of ice for a variety of purposes (e.g., cooling drinks) was reported abundantly, although no documentation describes how it was produced, stored, and transported, or, indeed, who initiated and was responsible for this activity. It is generally assumed to have been a commercial undertaking by private individuals, as evidenced from the names of some old ice-houses (e.g., "Haji Agha Ali Yakhdan" in Rafsanjan City, Kerman or "Haj

Mohammad Rahim Yakhchal" in Estahban City, Fars), perhaps with some sponsorship or encouragement from rich benefactors or investors.

The word *ice* can be seen in the poems of many Iranian poets. For instance, Ferdowsi (940–1020), the most famous epic poet of Iran, referred to ice both real and metaphorical. Two such examples exist in the following lines translated from different poems [12, 13]:

> *"When he wakes up, take him some beer and some ice, and see that someone is there to serve Goshasp ...[c]"*

> *"Now when the king's wife, Sudabeh, saw Seyavash, she grew strangely pensive and her heart beat faster; she began to waste away like ice before fire, worn thin as a silken thread ...[d]"*

Saadi (1213–1291), also known as Saadi of Shiraz, a poet and one of the greatest figures in classical Persian literature, also used *ice* in striking imagery and referred to the actual substance in both his prose and verses; for example, he says [12, 14]:

> *"... As adverse as the wind, as unpleasant as the frost; sitting as cold as snow, frozen as solid as ice ...[e]"*

> *"Thus, how long the ice stays in the water/how old can the snow be in the sunlight.[f]"*

Furthermore, in the works of other Iranian poets such as Manuchehri (d. 1040), Khaqani (1120–1199), Suzani Samarqandi (d. 1166), Nizami Ganjavi (1141–1209), Attar Neishabouri (d. 1220), Jalal ad-Din Muhammad Rumi (1207–1273), etc., the word *ice* is frequently used. For instance, Suzani Samarqandi refers to *ice* and *Yakhdan* in his poems, as in the following verses, which also supports the idea of private ownership of ice-houses [12]:

> *"I do not own a Yakhdan/as my lyric is not meaning less ...[g]"*

> *"It is not easy to harvest sweet fruit and ice/from any grape tree and Yakhdan.[h]"*

In addition, Rumi, known as Mawlana or Mawlawi, points out to both *ice* and *Yakhdan* in his poems and sonnets, such as in the following extended metaphor [15]:

> *"The ice that remains in the shade is frozen/It saw not the brilliance of my glowing Sun.[i]"*

[c] همی باش پیش گشسب سوار / چو بیدار گردد فقاع و یخ‌آر ...

[d] چنان شد که گفتی طراز نخ است / و یا پیش آتش نهاده یخ است ...

[e] چون برف نشسته و چو یخ بربسته ...

[f] این یخ در آب چند بتواند بود / وین برف در آفتاب تا کی باشد

[g] نه به مرد یک اندرم یخدان / نه سخن چون فقاع یخدانی ...

[h] کس از محلت مرد یک از رز و یخدان / نه میوه آرد و نه یخ نماند پندارم

[i] فسرده ماند یخی که به زیر سایه بود / ندیده شعشعه آفتاب تابانم

"All ice that has seen the smile of the Sun's face/Grows itself again, and says, 'I am the Water of Life.[j]*"*

"Have you or anyone else ever seen/the Yakhdan give a sign of fire?[k]*"*

"Dear, a Yakhdan has no knowledge of the sun and its glitter/so how can the soul have knowledge of its creator.[l]*"*

In some historical texts, such as the book "Anis al-talibin va'uddat al-salikin," written by Bukhari between 1389 and 1391 [12, 16], the word Yakhdan occurs, but unfortunately, no details explain the speaker's unhappiness. Doing the work "while hungry," however, is understood to mean to work quickly: "*A vintner said it is important for me to clean the dust from the Yakhdan ... You both should clean the Faqai*[m] *Yakhdan, and we started to work. Khajeh*[n] *said the work should be done while hungry ... I went to the Yakhdan with fear and sorrow ... I bought bread from the bakery and went to the Yakhdan in a hurry from Charsou.*[o]"

4.3 The history of traditional ice-houses in foreign travelers' accounts

A number of travelers who visited Iran (or Persia) in the past confirm the importance of ice in local culture, writing extensively in their travel diaries about the use of ice and snow for cooling water, wine, and other beverages; preserving food products and fruits; and making ice cream. They have also left us detailed descriptions of the natural ice production methods they witnessed, and also of the structure and components of traditional ice-houses. Ruy González de Clavijo (d. 1412), who visited various cities between 1403 and 1406 as an ambassador to Timur (or Tamerlane) from King Henry III of Castile and Leon, noted the widespread availability of cold water in the city of Tabriz during the summer [17]: "*In many of the roads and open places of the city are seen fountains and troughs for drinking from, and in these last in summer time they [local people] are wont to cool the water with pieces of ice, and they stand therein their brass and copper pots from which folk drink.*"

[j] تبسم رخ [خوش] خورشید هر یخی که بدید / بخویش [سبال] مالد و گوید که آب حیوانم

[k] هرگز دیدی تو یا کسی دید / یخدان ز آتش دهد نشانه

[l] یخدان چه داند ای جان خورشید و تابشش را / کی داند آفرین را این جان آفریده

[m] A seller of liquor such as wine and beer.

[n] "Khajeh" is a term of honor once applied to persons who were eminent.

[o] "Charsou" refers to the intersection of two main and important rows in a bazaar.

In 1620, the Italian traveler and musician Pietro della Valle (1586–1652) described local people manufacturing natural ice on winter nights and storing it within large and deep pits around the city of Isfahan, which was the capital of Persia at that time [18]:

"Outside the city [Isfahan] were built enormous south-facing walls, as high as fifty feet[p] [~ 15 m], which provided permanent shade throughout the day without blocking the cooling north wind. Beneath the walls, large deep trenches [holes] were dug. These trenches were filled with ice collected from shallow, man-made troughs that were lined with tile, filled with water, and allowed to freeze overnight during the winter. The trenches were then covered with straw, and the ice would be collected when desired. Wealthy Persians would use the ice to chill fruit and ice drinks."

Thomas Herbert (1606–1682), an English traveler and historian, who visited Iran from 1627 to 1629, mentions ice-houses outside of Isfahan City in 1628 in his itinerary [19]: "*towards the outside of the city, [are] a large castle; unflanked but moated about, and several houses within; which guard the treasure, arms and ice there stored.*" Of interest is the coupling of treasure and ice as things that are worthy of being guarded.

The German traveler and scholar Adam Olearius (1599–1671), who journeyed to Iran in 1637, has left a detailed description (included here in the form of an early translation) of how ice was naturally produced in the city of Isfahan [20]:

"And though it freezes there so little, that in a night's time it does not make an ice as thick as a man's finger, which thaws as soon as the sun appears over the horizon, yet have they a way to make the ice above two foot [more than 60 cm] thick and to keep it to be used to cool their drinks in summer. To do this, they make choice of a commodious place, that is cool, and towards the north, paved with free-stone or marble, but between, and with little descent, upon which they pour the water and as soon as that is congealed, they pour on more, and by this means in one night they make ice a foot thick, which in the daytime they cover, that the sun may not shine upon it: and so continuing the exercise for two or three nights together, they provide ice enough to serve them all summer. Having made as much as they desire, they break it in pieces and put it up into snow-houses [ice-houses], whereof there are so many at Isfahan that for two of three 'kasbekis'[q] [a Persian coin], a man may have as much as will suffice him all summer."

[p] 1 ft is equal to 0.3048 m or 30.48 cm.

[q] Olearius mentions in his book [20]: "At the time of our travels, the kasbeki were marked at Isfahan with a lion, ..."

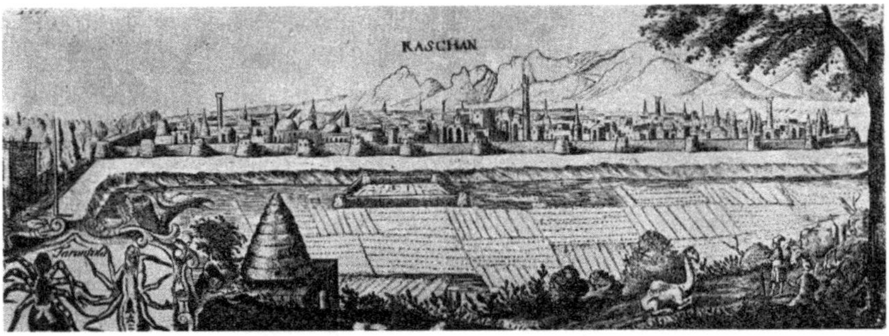

Fig. 4.1 An old picture of Kashan City with an ice-house dome, drawn by Olearius [20].

Fig. 4.1 displays an old picture of the city of Kashan, drawn by Olearius, clearly showing the dome of an ice-house (at the bottom left of the picture).

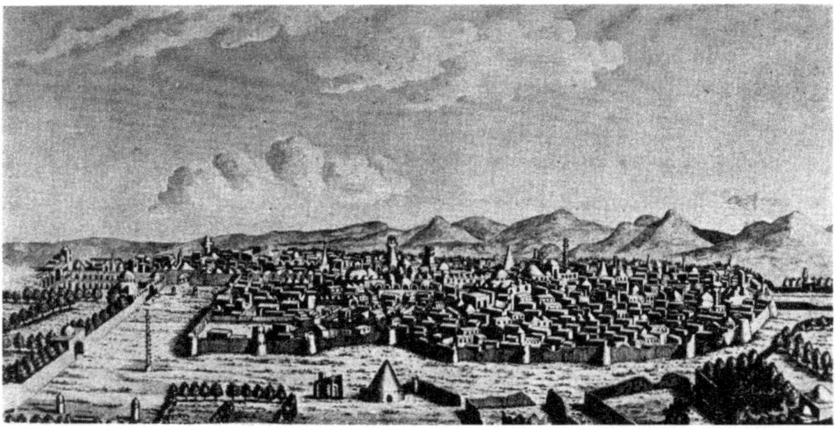

Fig. 4.2 An old picture of Kashan City and its ice-house (at the bottom middle of the picture), drawn by Chardin [21].

Fig. 4.2 shows another old picture of the city of Kashan and its ice-house, depicted by Jean Chardin (1643–1713), a French traveler and jeweler. Some scholars believe that Kashan's old ice-houses, which were attached to the city walls along with the "Jalali Castle," belong to the period of Malik-Shah I[r] (1055–1092) [1]. On the contrary, based on a number of historical texts, all structures, buildings, and edifices belonging to the Castle were demolished in the repulse of a Turkmen attack, except the thick walls of the castle and its moat [22]. It seems unlikely that the ice-houses drawn by Olearius and Chardin in the pictures of Kashan City (Figs. 4.1 and 4.2) belong to the period before the attack. Given that the ice-house's dome and its shadowing walls are intact in these pictures, they were most likely built

[r]The Seljuq Sultan Jalal-al-Din Malik Shah, the third and most famous of the Seljuq Sultans.

(or reconstructed) during the rule of Shah Abbas the Great (1571–1629) after the defeat of the Turkmen and before the visits of Olearius (in 1637) and Chardin (in 1666) to the city of Kashan [23].

Chardin's book describes (included here in the form of an early translation) in detail how ice was naturally manufactured during the winter and stored in a vast, long, and deep pit near the city of Isfahan [21]:

> *"Their way of making it is thus; they make a deep hole, at the farther end of the cellar towards the north, and before it they dig deep squares of sixteen or twenty inches[s] [around 40 or 50 cm], like so many little basins; they fill 'em with water over night, when it begins to freeze, and in the morning, when all is frozen, they break it to bits, and put all these pieces together in the hole (the ice-pit), where they break 'em again into little bits, as well as they can; for the more ice is broken, the better it is; then they fill the square holes with fresh water, as they had done the day before, and at night, they go and water with your ground bottles, with handles to them, these pieces of ice, which are broke in the hole, to the end, that they may hold the better together. In less than eight days working after this manner, they have pieces of ice five or six foot [about 1.5 to 1.8 m] thick; and then they gather the people of the quarter together, who with loud shouts of joy, and fires lighted upon the edges of the hole, and with the sound of instruments to animate them, go down into it, and lay these lumps of ice one upon the other, which they call Codrouc, [meaning] basis or foundation, as much as to say, and throw water between, to make them hold together the better. Thus, in six weeks, an ice-pit of 6 ft [~ 1.8 m] or more in depth, long and wide as they choose, is filled with ice right to the top."*

A notable point in Chardin's descriptions is the enjoyment of the locals, who lived around the ice-house, during the task of storing ice pieces inside an underground reservoir to the accompaniment of instruments and happy music making along with loud shouts of joy. This practice of communal toil driven by merrymaking was probably a tradition during that time. Chardin also mentions the clarity of the ice produced in this way [21]: "*What is very remarkable, as well as agreeable in their ice, is its beauty and clearness; you can't see the least dirt, nor gloominess; rock-water is not clear, nor more transparent than it is.*"

Another traveler from France, Jean de Thévenot (1633–1667), explains about ice storage and insulation in Isfahan's ice-houses as follows [24]:

> *"The ice was stored in a long trench [ice-pit] along the north side of a [shadowing] wall; the trench was three fathoms[t] [~ 5.5 m] deep and three to four feet wide [~ 91 to 122 cm], and the trench did not seem to be roofed with mud brick. The ice in storage, a fathom and a half deep, was covered with two to three feet [~ 61 to 91 cm] of straw and reeds, and probably earth on the top. When with time [came a period that] the ice trench was no longer used, it was filled up with earth, and the only visible remains would be the tall wall."*

[s] 1 in. is equal to 2.54 cm.

[t] A unit of length equal to six feet (~ 1.8 m), meaning "the arms extended," which is mainly used to measure the depth of water [25].

Other travelers have also commented on Iranians' use of ice for different purposes. One, John Fryer (circa 1650–1733), a British traveler and physician who traveled to Iran and India between 1672 and 1681, refers to the use of ice and snow for cooling beverages, and ice storage as well [26]: "*They mightily covet cool things to the Palate. Wherefore they mix snow, or dissolve ice in their water, wine or Sherbets [beverages]...*" and "*At this time, ice storage is considered a conventional and old tradition and possibly it entered Iran with the Mongols...*"

However, this idea that ice production and storage entered Iran with the Mongolians is totally wrong, because for a few centuries, cold drink consumption remained "unknown" in most parts of the world, especially the Far East, such as Japan and Mongolia [27]. On the contrary, the Mongols were a nomadic people, so they had no need for permanent structures such as ice-houses.

Ice-house remnants alongside the urban and interurban caravansaries in Iran, and the mention of ice-houses by travelers around different cities such as Isfahan and Kashan, confirm the construction of such structures in a particular style during the Safavid period. Paintings and photographs left by foreign travelers are added evidence. Those drawn by Olarius (Fig. 4.1) and Chardin (Fig. 4.2) are the earliest visual evidence of the existence of ice-houses in that period. A Dutch artist and traveler to Iran, Cornelis de Bruijn (1652–1727), drew a picture of an ice-house in Qom (Fig. 4.3) in 1703, which is one of the most important historical documents from the Safavid era [1, 28]. It shows the ice-house's cone-shaped dome, with its sharp tip and a shadowing wall. Fig. 4.4 reproduces an old picture of the Safavid caravansary located in Deh-Namak (near the Garmsar), done by

Fig. 4.3 An old picture of an ice-house in Qom City, drawn by de Bruijn [28].

Fig. 4.4 An old picture of an ice-house and a caravansary located in Deh-Namak, drawn by Laurens [29].

Jules Laurens (1825–1901), a French artist and lithographer, who came to Iran in 1847. The huge ice-house illustrated, next to the Tehran-Khorasan caravan route, has a cone-shaped cover over its reservoir, with shadowing walls next to it [29]. Laurens drew another picture that demonstrates the view from outside one of the old entrances to the city of Tehran from the south (Shah Abdol-Azim) during winter (Fig. 4.5). In this figure, the cone-shaped dome of a traditional ice-house is observed.

Gaspard Drouville (1783–1856), a French cavalry officer serving in the Russian army, who visiting Iran from 1812 to 1813 during the Fa'th-Ali Shah Qajar era, described the importance of ice and drinking cold water to Persians [31]: "*There is perhaps no country in the world where one drinks beverages with ice as much as in Persia; this luxury seems even indispensable, as the people bring it to their camps from more than sixty leagues [Farsang*[u]*] away; ... The suffocating heat of certain parts of Persia sufficiently justifies this luxury: it is necessary to drink cold water, or experience serious diseases, which can be cured only by drinking ice water. Moreover, this remedy is common in hot countries.*[v]"

[u]~374 km (1 Farsang or parasang is approximately equal to 6.23 km [32]).

[v]"*Il n'est peut—être pas de pays au monde où l'on boive à la glace autant qu'en Perse, ce luxe y semble même indispensable, car les Grands en font venir aux camps de plus de soixante lieues. ... Les chaleurs étouffantes de certaines parties de la Perse, justifient suffisamment ce luxe: il faut y boire frais, ou éprouver des maladies graves, dont on ne guérit qu'en buvant à la glace. Au reste, ce remède est commun dans les pays chauds.*"

Fig. 4.5 An exterior view of an old entrance of Tehran City (Shah Abdol-Azim) in winter, drawn by Laurens [29, 30].

The physician Jakob Eduard Polak (1818–1891), from Austria, who lived in the country from 1851 to 1860, refers in detail to the role of ice in the lives of Iranians, how they manufacture ice in winter and store it by means of ice-houses in the city of Isfahan for summer use, and also how ice is distributed at a low price as follows [33][W]:

"Having ice in Iranian beverages is unavoidable; even the most indigent of the poor of the city cannot easily do without ice, insofar as it is used for any beverage during the winter. Much effort is made to provide and store the ice and offer it at a low price, and a policeman [an official] determines a specific tariff for the ice. To provide ice in cities and around them, ice-houses are built, and they contain parallel holes, which are enclosed within long cob walls on three sides, with an area of 350 to 550 m^2. The fourth side, located to the north, is open so that the sun's rays do not fall on it [the ice]. In winter, these holes are filled with water, and as soon as a thin layer of ice is created during the night, they pour water constantly on it until the ice is thick enough. The ice is then crush[ed] into small pieces and taken underground. Due to the dry weather in Isfahan, this ice remains throughout summer and fall, even for two years. In Isfahan, I saw an ice-house whose ceiling had been demolished, and despite that, the ice, which was covered with some straw and dry twigs, lasted all summer. The ice produced inside the city is contaminated because the holes are used for storing garbage and carcasses during the summer; however, they clean the holes in fall ..."

Charles James Wills (1842–1912), an English physician who lived in Persia between 1866 and 1880, describes the use of ice for making a variety of cold beverages and ice creams as an art for Iranians during the warm months [34]:

[W]Translated from the original Persian by the authors.

"A great thing in such a warm place as Shiraz is the cheapness of ice; for about fifteen shillings[X] in dear years and five in cheap ones, ice can be obtained all through the warm weather, and in fact is used from May to October, as no one would think of drinking anything uncooled. A huge block is thrown down in one's doorway each morning by the ice-seller; it is supposed to weight two Tabriz maunds, or fourteen pounds. The Persians well understand the art of making water-ices and ice-creams, and various ices unknown to us are made by them, as tamarind juice, pomegranate and cherry-water ices; iced 'mast,' or curdled milk, and various ices of pounded fruits, as apricots and cherries, which are very good."

Wills also details the characteristics of Mushir ice-house located in the city of Shiraz (10 km from north of Shiraz to Isfahan, Akbarabad), of which almost nothing is left today, and how ice was naturally produced on winter nights as follows [34]:

"When I accompanied Captain St. J— in a call he made upon the Muschir, four conical ices, the size and shape of an ordinary sugar-loaf, were placed in handsome Chinese porcelain basins before each of us. In fact, the cheapness of provisions generally causes among the well-to-do a lavishness and profusion (not to call it waste) unknown in Europe.

The Muschir has a 'yakhjal' [Yakhchal] or place for the making and collecting and storing ice, in an open plain some six miles from the town, at the side of the Ispahan [Isfahan] road. The earth is dug out to a depth of two feet [~ 61 cm]; with this earth a mud wall some twenty feet [around 6 m] high is constructed of sun-dried bricks. The excavation is some ten yards[Y] by one hundred, and the wall is ten yards high by a hundred long; this wall runs in such a way as to protect the open pond thus excavated from the hottest rays of the sun.

The delicious Ab-i-Rookhni ('stream of Rookhnabad') is diverted from its course during the first cold night. A few inches of still clear water is collected in the pond; by morning it is frozen; at night the water is again admitted, and another inch or two of ice made. When three or six inches [~ 7 or 15 cm] thick, the ice is broken and collected for storage in a deep well on the spot: and so, day by day the process goes on during the short winter until the storehouses are full. Should the supplies from these be exhausted by a very large demand, ice, or rather blocks of snow, are brought from the mountains; but as these are some distance, and as snow melts faster than ice, the weights being equal, the price rises.

An order is generally issued when the ice is running short that each house is to be on half allowance – a wise measure, as it makes the cooks careful, and everybody gets some. So common is the use of ice that the poorest are enabled to have it, a big bit

[X]A former British coin and monetary unit equal to one twentieth of a pound or twelve pence. The name shilling derives from the old English "scilling" or "scillinga," meaning "cutting" or "slice" [35].

[Y]1 yard is equal to 3 ft. or 91.44 cm.

being sold for a farthing,[z] and even the bowls of water for gratuitous drinking at the shop doors are cooled by it. Ripe fruit is generally also cooled prior to being eaten."

The telegraph engineer and photographer, Ernst Hoeltzer (1835–1911) from Germany, traveled to Iran during the period of Naser al-Din Shah Qajar (the King of Persia from 1848 to 1896) and lived in Isfahan for about 20 years. He left a description of the manufacture and storage of natural ice in Isfahan's ice-houses, as well as how to transport, distribute, and sell the ice produced [37][aa]:

"Iranians are doomed to decline without ice. Especially in the summer, the poor suffer because they cannot afford cold-water without bearing the costs. Therefore, they have made very large Yakhchals, so that they can meet the people's needs even for two years. There is rarely a winter without frost, and there is enough ice in January. They dig several holes 5 to 6 m deep next to a tall wall, so that the holes are shaded all summer. The ground has a slope, so water-ices can flow down from one side; hence, a deep pit is dug at the end of the water path. They fill a large area [the holes] with water. Then, they break up the thick layers of ice that form there in winter, and store the pieces of ice in a storage pit. Pieces are stacked on top of each other regularly with a rake, as uniformly as possible. Afterwards, they pour water on the stack until all the pieces of ice stick together. During night and day, this action is rapidly repeated until the storage pit is filled – one to two meters at the top of the storage pit usually remains empty. They then put rice straw on the ice to a height of 1 to 2 m to insulate the ice from warm air. To remove ice, either early in the morning or during the night, they break off the pieces of ice from the corner nearest a staircase, so that they can remember the place and cover up the remaining ice easily.

There are plenty of Yakhchals in Isfahan; some of them are for private use. Nevertheless, the poor can also use the Yakhchals to cool fruit and water. Basically, beverages and fruit are preserved with ice in all shops. Huge chunks of ice are carried by donkeys and sold all over the province. In Isfahan, people can buy ice either in the bazaar or straight from a Yakhchal building."

Fig. 4.6 shows an ice seller in the city of Isfahan using a donkey to carry big pieces of ice from the ice-house during the hot/warm months in 1880 [38].

Another traveler from France, the archaeologist and explorer Jane Paule Henriette Dieulafoy, known as Madame Dieulafoy (1851–1916), visited Iran in 1881 and noted Yakhchals in the city of Tabriz [39]: "*Returning from the consulate, we passed by a Yakhchal, constructions meant to store ice in winters. People bring ice out in summers to sell. It is easy to provide ice with little labor.*

[z]A farthing was a coin of the Kingdom of England worth one quarter of a penny, $^1/_{960}$ of a pound sterling. The word "farthing" is derived from the Anglo-Saxon feorthing, which means a "fourthling" or "fourth part" [36].

[aa]Translated from the original Persian by the authors.

Fig. 4.6 Photograph of an ice seller, carrying blocks of ice from the ice-house on a donkey, taken by Hoeltzer [38].

People fill pitfalls adjacent to high walls in winters, and the water freezes during the night. In the morning, workers break the ice into pieces and store them underground for summer." Dieulafoy also refers to the price of ice produced via Yakhchals in Tabriz City in her itinerary [39, 40]: "*Though the cost of ice is very reasonable, each Yakhchal earns its owner between one hundred and two hundred tomans.*[ab]"

Al-Isfahani, in his book entitled "Nesf-e Jahan Fi Tarif-e Esfahan" (written in 1882), mentions Isfahan's ice-houses in the Qajar period as an industry at that time [41]: "*Another component of Isfahan industries is the Yakhchal(s), which have been built inside and in the suburbs of the city. They are used to make ice during the winter and collect it in the same place, then [local people] bring it to town and the bazaar in the summer, and sell it at a low price.*"

George Nathaniel Curzon (1859–1925), known as Lord Curzon, traveled in Iran for around four months, entering in late September 1889 [42]. He depicted an ice-house around the village of Mazinan, near the city of Sabzevar, Razavi Khorasan in his book (Fig. 4.7). A British artist Frederick Charles Richards (1878–1932) drew a picture of Kerman's ice-houses (Fig. 4.8) during his journey to Iran in 1930 [43]. Maxime Siroux (1907–1975), a French architect and archaeologist who collaborated with André Godard (1881–1965) in Iran for many years, described how to produce, collect, and store ice, as well as the characteristics of traditional ice-houses, briefly but in detail, by drawing a cross-section of a typical Iranian ice-house (Fig. 4.9) [44]. A notable point in Siroux's descriptions is his mention of the

[ab]The toman is worth 9 Fr. [French francs] 60 of our currency at the moment; its rate is variable [40].

Fig. 4.7 Photograph of an ice-house around Mazinan village, near Sabzevar City, Razavi Khorasan, taken by Curzon [42].

Fig. 4.8 Ice-houses in Kerman, drawn by Richards [43, 44].

substantial temperature difference between sunny and shady places during ice making, which was reported to range from 15°C to 20°C.

4.4 Distribution with limited timeframe of traditional ice-houses in Iran

Fig. 4.10 provides a map showing the distribution of traditional ice-houses in various Iranian cities with different climatic conditions. In addition, the names, locations, and periods of a number of Iran's old ice-houses are listed in Table 4.1. Alaie et al. [46] reported that the number and application of

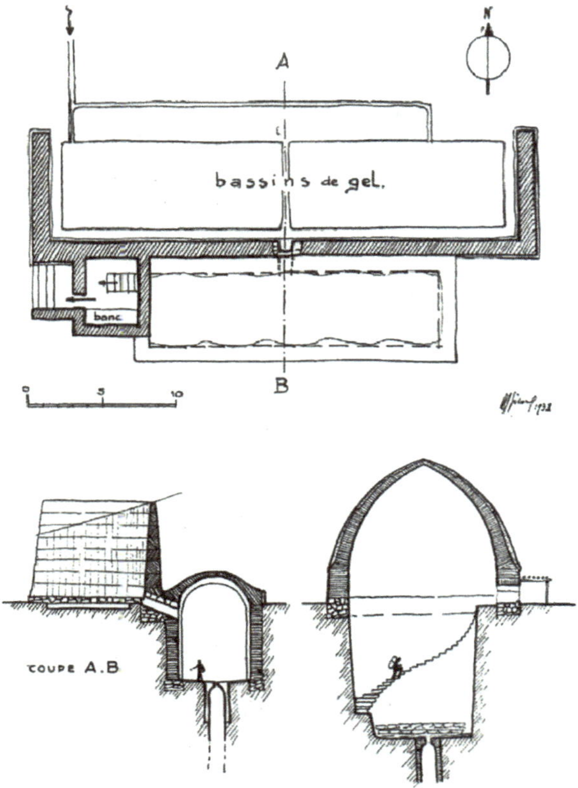

Fig. 4.9 Schematic illustration of traditional ice-houses in Iran, drawn by Siroux [45].

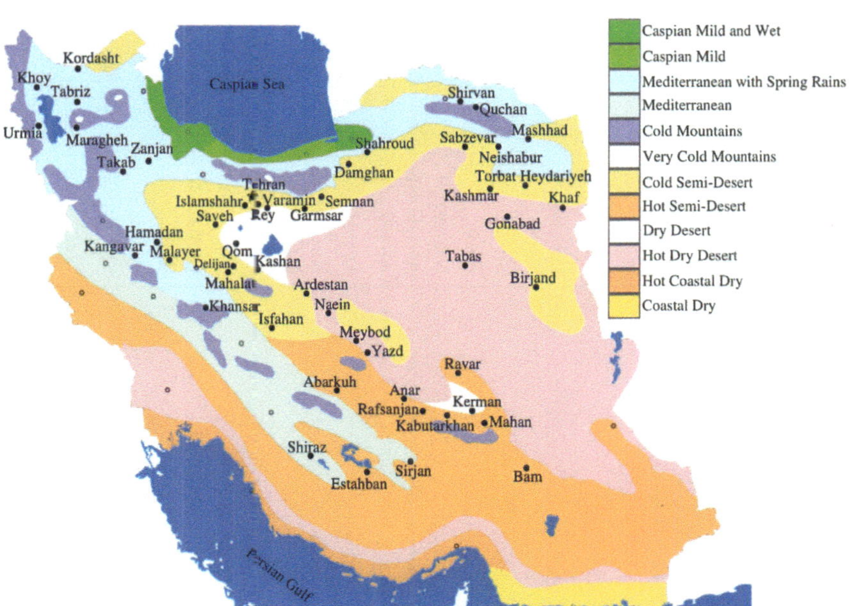

Fig. 4.10 Map showing cities that have or once had traditional ice-houses and the range of climate conditions.

Table 4.1 The name, location, and period of a number of old Iranian ice-houses.

Name	Location	Period[a]	Name	Location	Period
Moayedi	Kerman	Late Safavid	Rahim Abad	Birjand	Late Qajar
Rig Abad	Kerman	Qajar	Shokat Abad	Birjand	Qajar
Haji Agha Ali	Rafsanjan	Qajar	Kalateh Nayeb	Birjand	Qajar
Kabutarkhan	Rafsanjan	Qajar	Tarshanbeh	Islamshahr	Safavid
Anar	Anar	Qajar	Ali Abad	Islamshahr	Qajar
Ravar	Ravar	Late Qajar	Khalili	Tehran	Late Qajar
Haj Rashid	Sirjan	Qajar	Kan	Tehran	Late Qajar
Mahmoud Abad	Sirjan	Qajar	Abbas Abad	Shahroud	Qajar
Arg-e Bam	Bam	Safavid	Bastam	Shahroud	Qajar
Langar	Mahan	Qajar	Dowlat Abad	Shahroud	Qajar
Aghazadeh	Abarkouh	Qajar	Deh-Namak	Aradan	Safavid
Hek	Abarkouh	Qajar	Qatul	Garmsar	Qajar
Meybod	Meybod	Safavid	Kardovan	Garmsar	Qajar
Jalali	Kashan	Safavid	Taghrud	Qom	Safavid
Zavareh	Ardestan	Late Safavid	Mahdi Abad	Saveh	Qajar
Twin	Khor, Naein	Safavid	Nimehvar	Mahalat	Qajar
Triplets	Sabzevar	Qajar	Mirfattah	Malayer	Qajar
Kusar Bidokht	Gonabad	Late Qajar	Alagh Beyk	Takab	Qajar
Khaf	Khaf	Qajar	Noh Pelleh	Urmia	Qajar
Namadmal	Quchan	Late Qajar	Churs	Khoy	Safavid
Dowlat Abad	Neishabur	Qajar	Moradian	Maragheh	Qajar
Ordoghsh	Neishabur	Qajar	Sadeghieh	Tabriz	Qajar

[a] The age of the traditional ice-houses is not exactly known; hence, their time periods are estimated.

ice-houses grew during the Qajar dynasty (particularly in the Naser al-Din Shah period), insofar as they formed an important part of the routine lives and culture of Persians. The enhanced economic power of society led to an increased number of ice-houses. On the contrary, an excessive supply of ice precipitated a drop in the price of ice, and so the poor could consume it, too [47]. Jorgensen [3], who conducted extensive research on Iranian ice-houses located in the country's hot and arid/semi-arid areas, detected, and recorded 129 ice-house sites, but was able to find the remnants of only 104. Obviously, there were originally, and still are, the ruins of many more traditional ice-houses in Iran. Jorgensen [3] did not investigate other ice-houses that were situated outside the study areas, for example, cold regions such as the cities of Tabriz, Urmia, Khoy, Maragheh, Takab, Hamadan, Zanjan, and Malayer (see Fig. 4.10).

Shafaghi [48] reported that, based on the map drawn by Seyyed Reza Khan in 1923 (Fig. 4.11), the number of old ice-houses in the city of Isfahan was approximately 20, including Mushir, Mirza Ismail, Mirza Mohammad, Agha Seyyed Bagher, and Khorram Yakhchals. The latest traditional Iranian ice-houses were used up to 1970 in the cities of Tehran, Isfahan, Kashan, Qom, and Kerman [48, 50]. Some of the old ice-houses have been repaired, such as the Meybod and Arg-e Bam ice-houses, and some have been completely destroyed, such as the Mushir ice-house in the city of Shiraz and all ice-houses inside Isfahan City, but most have just been abandoned to fall into ruin over time.

As illustrated in Fig. 4.10, there were no ice-houses in hot and humid regions of Iran, such as next to the Persian Gulf, Oman Sea, and Caspian Sea, because the production of ice naturally was not possible in these areas during the winter because of weather conditions. In addition, to take advantage of their economic prosperity and bustling trade, the majority of ice-houses were built in cities located on traditional economic routes, for instance, the "Silk Road" and "Spice Road" (international routes) as well as the "Shahi Road" or the "Royal Road" (on the national scale), such as Isfahan, Kerman, Yazd, Sabzevar, Shahroud, Gramsar, and Shahr-e-Rey (in modern-day Tehran).

4.5 Conclusion

The historical studies and documentation provided in this chapter confirm that Persians clearly long understood the art of producing ice in winter by means of natural methods and indigenous knowledge and

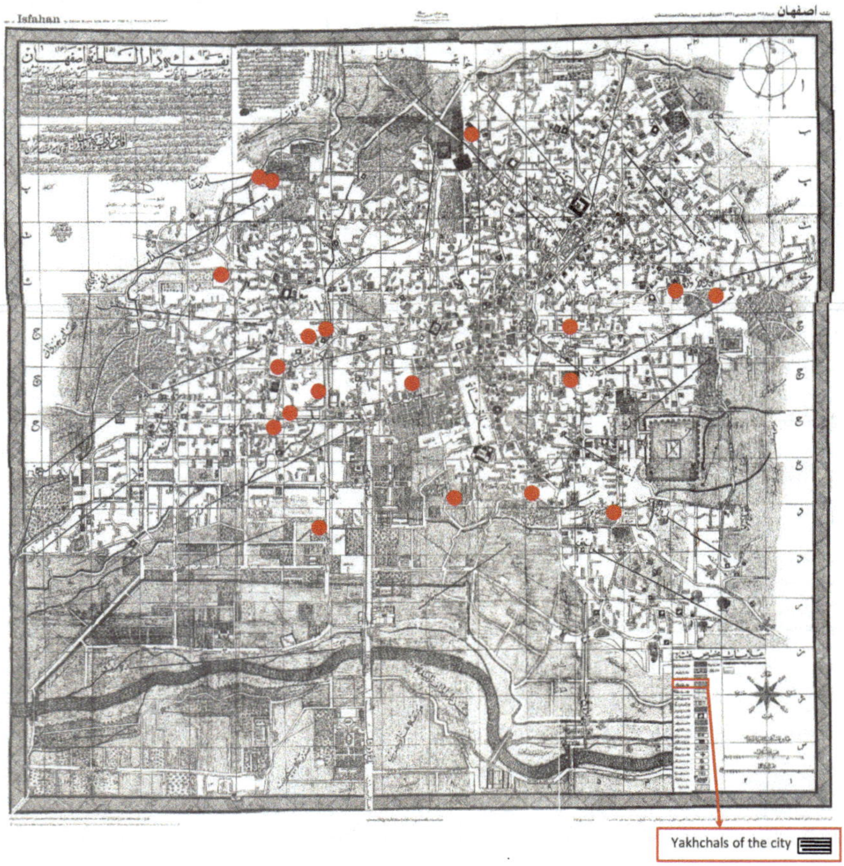

Fig. 4.11 Seyyed Reza Khan's map of the city of Isfahan in 1923 [49], showing the distribution of old ice-houses inside the city.

technologies,[ac] as well as how to store it for use during hot/warm summer months in arid and semi-arid regions. Through the use of ice-houses that were designed to be environmentally friendly, they achieved a sustainable technology that benefitted both the comfort and the health of their communities for centuries. Updating and using these kinds of sustainable and not polluting indigenous technologies can help overcome some of the ongoing and emerging challenges and problems that today's world faces.

[ac]Indigenous techniques are based on indigenous knowledge [9], that is, the sum of local or traditional knowledge that has been generated, evolved, and transferred, over time, by those who reside in a specific geographical area, to deal with their sociocultural and agroecological environments [51, 52]. This knowledge has been obtained via experiments and trial and error procedures by local people, and passed down from generation to generation. Based on that knowledge, technologies have been developed to meet the needs of the community [9].

References

[1] M.A. Mokhlesi, The Traditional Ice-houses, the Forgotten Architectural Masterpieces, the Second Congress on Architecture and Urban Planning History, vol. II, Cultural Heritage Organization of Iran, 1995, pp. 690–695 (in Farsi).
[2] B. Hosseini, A. Namazian, An overview of Iranian ice repositories, an example of traditional indigenous architecture, METU J. Fac. Archit. 29 (2) (2012) 223–234.
[3] H. Jorgensen, Ice Houses of Iran: Where, How, Why, Mazda Publishers, Costa Mesa, CA, 2012.
[4] C. Gantz, Refrigeration: A History, McFarland Publishing, Jefferson, NC, 2015.
[5] A.G.S. Patil, A.S.A. Ranadive, Role of traditional wisdom in design education for global product development, in: A. Chakrabarti, R. Prakash (Eds.), ICoRD'13, Lecture Notes in Mechanical Engineering, Springer, India, pp. 1357–1367, 2013.
[6] E. Goldsmith, The Qanats of Iran, Sci. Am. 218 (4) (1968) 94–105.
[7] H. Ahmadi, A. Nazari Samani, A. Malekian, The Qanat: a living history in Iran, in: G. Schneier-Madanesand, M.F. Courel (Eds.), Water and Sustainability in Arid Regions, Springer, Dordrecht, The Netherlands, 2010.
[8] A.A. Semsar Yazdi, M. Labbaf Khaneiki, Qanat Knowledge: Construction and Maintenance, Springer, Dordrecht, The Netherlands, 2017.
[9] A.R. Dehghani-Sanij, in: A. Sayigh (Ed.), Cisterns: Sustainable Development, Architecture and Energy, River Publishers, Denmark, 2016.
[10] H. Jorgensen, "YAḴČĀL," Encyclopædia Iranica, 2012, online ed. http://www.iranicaonline.org/articles/yakcal. (Accessed 21 July 2020).
[11] P. Javaheri, M. Javaheri, Challenge for Water in Ancient Fars, vol. 2, Iranian National Committee on Irrigation & Drainage (IRNCID) and Water National Museum of Iran (WNMI), Tehran, Iran, 2000 (in Farsi).
[12] A.A. Dehkhoda, Dictionary, second ed., vol. II, University of Tehran Press, Tehran, Iran, 1998 (in Farsi).
[13] A. Ferdowsi, Shahnameh: The Persian Book of Kings, Translated by D. Davis, Penguin Classics, Deluxe ed., London, UK, 2007.
[14] S.M. Sa'di of Shiraz, The Gulistan (Rose Garden) of Sa'di: Bilingual English and Persian, Translated by W.M. Thackston, Bilingual ed., Ibex Publishers, Inc., Bethesda, MD, 2008.
[15] J.M. Rumi, Selected Poems From the Divan-e Shams-e Tabrizi: Along With the Original Persian, Edited and translated by R.A. Nicholson, Kitab Bhavan, New Delhi, Delhi, India, March 2001.
[16] S. ibn Mubarak Bukhari, Anis al-talibin va'uddat al-salikin, Translated by T. Hashem Poursobhani, Association of Cultural Works and Scholars, Tehran, Iran, 2004 (in Farsi).
[17] R.G. de Clavijo, Embassy to Tamerlane, 1403–1406, Translated by Guy Le Strange, G. Routledge & Sons, Harper, New York and London, 1928.
[18] H. Chisholm (Ed.), Valle, Pietro della, Encyclopædia Britannica, eleventh ed., vol. 27, Cambridge University Press, Cambridge, UK, 1911.
[19] T. Herbert, in: S.W. Foster (Ed.), Travels in Persia 1627–9, Hakluyt Society, London, UK, 1928, p. 120.
[20] A. Olearius, Vermehrte Newe Beschreibung Der Muscowitischen und Persischen Reyse, Holwein, Schleswig, Schleswig-Holstein, Germany, 1656 (in German).
[21] J. Chardin, The Travels of Sir John Chardin into Persia and the East Indies, The Coronation of Solyman III, King of Persia, Printed for Moses Pitt, London, UK, 1686.
[22] A.A. Shatri, Kashan Historical Monuments, Hangam ba Hasti Publications, Kashan, Iran, 2008 (in Farsi).
[23] J. Chardin, Chardin's Travelogue, Translated by Iqbal Yaghmaei, Toos Publications, Tehran, Iran, 1993 (in Farsi).
[24] J. de Thévenot, The Travels of Monsieur de Thevenot into the Levant, In Three Parts Viz. Into, I. Turkey. II. Persia. III. The East-Indies, Printed by H. Clark, for H. Faithorne, J. Adamson, C. Skegnes, and T. Newborough, London, UK, 1687.

[25] H. Chisholm (Ed.), Fathom, Encyclopædia Britannica, eleventh ed., vol. 10, Cambridge University Press, Cambridge, UK, 1911, p. 201.
[26] J. Fryer, A New Account of East-India and Persia, in Eight Letters, Being Nine Years Travels, Begun in 1672, and Finished in 1681, R. Chiswell, London, UK, 1698.
[27] M.H. Papoli Yazdi, M. Labaf Khaniki, Ice-house and artificial ice production, in: Second Regional Climate Change Conference, Iran National Meteorological Organization, Zanjan, Iran, Nov. 13–14, 1999 (in Farsi).
[28] G. Homayoun, The Illustrated European Documents From Iran, University of Tehran Press, Tehran, Iran, 1969 (in Farsi).
[29] The Collection of Traditional Iranian Architecture (Islamic Era), second ed., Iranian National Geographical Organization of Armed Forces, Tehran, Iran, 1997 (in Farsi).
[30] J. Laurens, 'Nazar-endaz', Episode d'un voyage en Perse, Revue orientale et algérienne, 1853–54, in: repr. in J. Laurens, La légende des ateliers, Paris, France, 1901, pp. 342–351 (in French).
[31] G. Drouville, Voyage en Perse: pendant les années 1812 et 1813, Masson & Yonet, Paris, France, 1828 (in French).
[32] H.-A. Mustawfī of Qazwīn, The Geographical Part of the Nuzhat-al-Qulūb, Translated by G. Le Strange, E.J. Brill, Imprimerie Orientale, Leiden, The Netherlands, 1919.
[33] J.E. Polak, Iran and Iranians, Translated by K. Jahandari, Kharazmi Publications, Tehran, Iran, 1982 (in Farsi).
[34] C.J. Wills, In the Land of the Lion and Sun or Modern Persia, Being Experiences of Life in Persia From 1842 to 1912, Ward, Lock and Co., London, New York, Melbourne, 1891.
[35] Pounds, shillings & pence, Royal Mint Museum, 2020. http://www.royalmintmuseum.org.uk/coins/british-coinage/pounds-pence-shillings/index.html. (Accessed 11 April 2020).
[36] Spink Numismatic Circular, vol. XCI, Spink & Son, Ltd., London, UK, 1983, p. 39. No. 2.
[37] E. Hoeltzer, Iran by One Hundred and Thirteen Years Ago (Part I: Isfahan), Translated by M. Asami, Ministry of Culture and Art and Anthropological Center of Iran, Tehran, Iran, 1976 (in Farsi).
[38] C.J.M. Achour-Vuurman, Nineteenth-Century Persia in the Photographs of Albert Hotz, Images From the Hotz Photograph Collection of Leiden University Library, Barjesteh van Waalwijk van Doorn & Co., Rotterdam/Gronsveld, The Netherlands, 2011.
[39] J.P.H. Dieulafoy, La Perse, La Chaldee et la Susiane, Hachette, Paris, France, 1887 (in French).
[40] Y. Porter, A. Thevenart, Palaces and Gardens of Persia, Flammarion, Paris, France, 2004 (in French).
[41] M.M. Al-Isfahani, in: M. Sotudeh (Ed.), Nesf-e Jahan Fi Tarif-e Esfahan, Amir Kabir Publications, Tehran, Iran, 1989 (in Farsi).
[42] G.N. Curzon, Persia and the Persian Question, 2 vols., Longmans, Green and Co, London, UK, 1892.
[43] F. Richards, A Persian Journey: Being an etcher's Impressions of the Middle East, with Forty-Eight Drawings, Jonathan Cape, London, UK, 1931.
[44] F. Richards, Travelogue of Fred Richards, Translated by M. Dokht Saba, Scientific & Cultural Publications, Tehran, Iran, 2000.
[45] M. Siroux, Caravansérails d'Iran et petites constructions routières, de l'Institut Français d'Archéologie Orientale, Série t. 81, Le Caire, Egypt, 1949 (in French).
[46] A. Alaei, S. Ranjbar, Z. Sabzi, Guar water in the desert, the study of two examples of masterpieces of desert architecture: cistern and traditional ice-house, in: The First National Conference on Architecture, Civil, and Urban Environment, Hamadan, Iran, 2014 (in Farsi).

[47] T. Tavousi, Effective climatic conditions on the evolution of the ice-house architecture, J. Iran. Stud. Shahid Bahonar Univ. Kerman 20 (Fall 2011) 169–192 (in Farsi).
[48] S. Shafaghi, Traditional ice-houses of Central Iran, an example of architectural genius, J. Isfahan Cult. 35 (2007) 42–49 (in Farsi).
[49] M. Azizkhani, Interactions and Contradictions of Preservation and Modernization: The Case of Isfahan, Iran, M.Sc. Thesis in Historic Preservation, School of Architecture, The University of Texas at Austin, Austin, TX, USA, 2014.
[50] B. Hourcade, M.P. Berthou, Les glaciéres du plateau iranien, Luqman, Numéro 20 (1994) 87–98 (in French).
[51] R.S. Mosha, The Heartbeat of Indigenous Africa: A Study of the Chagga Educational System (Indigenous Knowledge and Schooling), Garland Publishing, Inc., New York and London, 2000.
[52] V. D'Errico, M. Fico, G. Franco, Knowledge based business model to manage opportunities in a distributed network of partners, in: Proceedings of the 12th International Conference on Product Focused Software Development and Process Improvement, Torre Canne, Brindisi, Italy, June 20–22, 2011, pp. 135–140.

CHAPTER 5

The architecture of traditional Iranian ice-houses

Traditional Iranian architecture successfully exemplifies designing for climate [1, 2] and is thus considered to follow sustainable patterns of architecture [3–6]. Such architecture endeavors to meet the requirements and desires of people living in a specific period and area [6], based primarily on local weather and geographical conditions. Other factors such as material availability and local cultural and religious beliefs also contribute to the design, and are certainly evident in Iran's traditional architecture [1, 7]. Structures such as porches, central courtyards, wind towers (or wind-catchers), cisterns, domed roofs, and ice-houses exemplify the country's early sustainable architecture. Ice-houses are of particular interest because they combine traditional structure with indigenous technology in a particularly successful model of climate-influenced design. This chapter comprehensively explains the specifications of traditional Iranian ice-houses and the associated details, such as the structural components, typologies, building methods, construction materials, decorations, vulnerabilities, and socioeconomic features.

In certain areas, such as at high altitudes, where climatic conditions allow the winter's snow and ice to remain, people have felt no need to develop sophisticated structures and complex techniques for making ice. Instead, they have used the available natural ice resources. For example, making use of a mountain near the city of Kazerun, Fars, local people filled a shaded hillside cave with snow during the winter. They then compressed the snow until it became like ice [8]. Additionally, they dug holes in the shadowy places of the mountain as well as at the foot of high cliffs where the sunlight never reached. They then lined the holes' walls and floors with stones and a vernacular mortar called "Sarooj[a]" in Persian, and other materials. This step both provided reinforcement against cave-ins and created smooth and clean surfaces on which to deposit the snow, thus reducing contamination from loose debris. Once there was sufficient snow in the winter and especially

[a]It consists of clay, water, lime, ash, and straw fibers (or rice stems) [9–12]. In some texts, it is reported that egg white was added for its sticky qualities [9, 10].

Ice-Houses
https://doi.org/10.1016/B978-0-12-822275-1.00003-5
Copyright © 2021 Elsevier Inc.
All rights reserved.

when snow had been converted into ice, local workers filled the holes with it and covering of wood and tree branches. They then insulated the mass by applying sand mixed with a particular type of soil that was in gray color like cement, called "Shushk" locally [8]. In this way, the people of the area stored and kept winter's frozen snow for use in summer.

In fact, these holes were not true ice-houses and were just called "snow pits" [13]. Their creators made use of naturally occurring features of the land. During the hot/warm months, their owners opened them and brought the frozen snow to cities and villages for various purposes, for example, cooling water and various beverages (mostly), preserving food, or even making ice cream. However, in most parts of Iran such as the desert and semi-desert areas (see Fig. 4.10), Iranians manufactured ice by means of skillfully designed structures and methods developed by trial and error over many generations. These developments allowed them to both produce ice and store it in specially constructed underground reservoirs. These structures, known as ice-houses and called "Yakhchals" or "Yakhdans" in Persian, were built using traditional architectural techniques to provide ice naturally and store it from winter to late summer according to the climatic conditions of the region [14].

5.1 The components of ice-houses

The construction technique and architectural style of the ice-houses in different parts of Iran show that their designers and builders carefully considered many important aspects, such as insulation of the structure, the creation of suitable conditions for storing ice, continuous removal of melt water from the ice stored inside the reservoir, the use of appropriate construction materials, and the means of producing and keeping ice under the specific local weather and geographical conditions [15]. In general, traditional Iranian ice-houses consist of three main components [15–18]: (1) shadowing walls (omitted from some ice-houses for reasons explained later), (2) shallow ice-making ponds, and (3) ice pits/reservoirs. It should be noted that the main and secondary components of ice-houses differed from region to region in response to their geographical and climatic conditions.

5.1.1 Shadowing walls

The main shadowing wall was usually very long and tall, and extended from east to west. This wall, which sometimes reached above 15 m in height [19], prevented solar radiation from hitting the surface of frozen water in the ice-making ponds during winter days. If sunlight would hit ponds' surfaces from the east and the west, respectively, in the early morning and late afternoon, an appropriate configuration of side shadowing walls would be attached to the main shadowing wall, as displayed in Figs. 5.1 and 5.2. Depending on the architect's experience, the junction between the main wall and side walls would either be curved or be set at an appropriate angle, for instance, with the two walls perpendicular to each other [23]. For stability, the shadowing walls (both main and side) were thick at the bottom and became thinner as they rose in height. Table 5.1 indicates the dimensions of the shadowing walls of some ice-houses in Kerman Province.

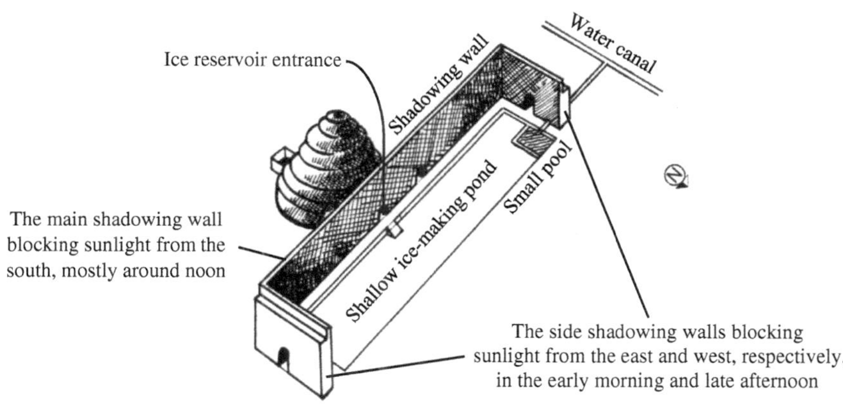

Fig. 5.1 Schematic illustration of the shadowing walls of an ice-house in Yazd, Iran [15, 20].

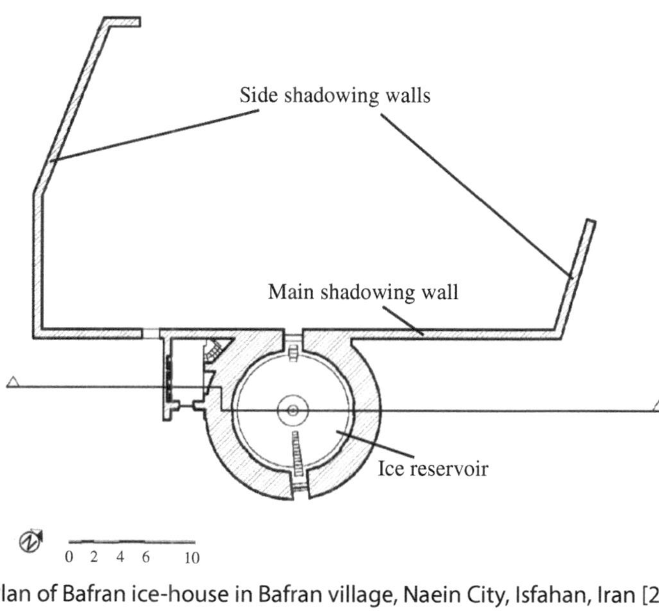

Fig. 5.2 Plan of Bafran ice-house in Bafran village, Naein City, Isfahan, Iran [21, 22].

Table 5.1 The dimensions of the shadowing walls of several ice-houses located in Kerman Province, Iran [23].

	Ice-house name and its location	Height of shadowing walls (m)	Thickness of shadowing walls at the bottom (m)	Thickness of shadowing walls in the middle (m)	Thickness of shadowing walls at the top (m)
1	Haj Rashid in the city of Sirjan	11.5	2.7	0.98	0.35
2	Ravar in the city of Ravar	6.7	2.3	1.07	0.1
3	Rig Abad in the city of Kerman	8.2	2.2	1.3	0.3
4	Haji Agha Ali in Abbas Abad Haji village, Rafsanjan City	13.65	2.4	1.45	0.22
5	Kabutarkhan in Kabutarkhan village, Rafsanjan City	12	2	0.8	0.2
6	Zarisef in the city of Kerman	15.4	3	2	1
7	Langar in Langar village, Mahan City	7.4	2.6	1.1	0.15
8	Moayedi in the city of Kerman	9	3.5	2.3	0.05

Shadowing walls were sometimes part of the city's defensive walls, as are, for instance, the wall of Jalali ice-house (Fig. 5.3) in the city of Kashan, Isfahan. In Haji Agha Ali ice-house situated in Abbas Abad Haji village, Rafsanjan City, Kerman, the architect has built a small space inside the width of a side shadowing wall for guards. It can be accessed near the bottom part of the shadowing wall's width by a narrow staircase, as indicated in Fig. 5.4. It should be noted that the height of the shadowing walls had a direct relationship with the angle of sunlight and solar noon in the winter in the different regions, as explained in the following subsection.

- *The influence of the angle of sunlight on formation of shadowing walls*
  Ice-houses located in hot and dry desert areas of Iran were built with the most-complete range of components/architectural features. As mentioned earlier, among the major features were the shadowing walls. Some ice-houses, as a result of local conditions, could function fully

Fig. 5.3 View of the Jalali ice-house's shadowing wall that was part of the city's defensive walls, Kashan City, Isfahan, Iran.

Fig. 5.4 View of a guard room and its entrance within the width of side shadowing wall of Haji Agha Ali ice-house in Abbas Abad Haji village, Rafsanjan City, Kerman, Iran.

when less well equipped with walls, and others needed especially long and tall walls. Differences in the existence or absence of shadowing walls were due to the angle of solar radiation in a location. In areas at a greater latitude such as cities located in the northern parts of Iran, the more-oblique angle of the sun's rays meant that ice-houses needed no shadowing walls. However, approaching the hot and dry semi-desert regions, the height of the shadowing walls had to increase. In hot and dry desert climates such as those of Kerman and Yazd, ice-houses had higher walls because the angle of the sun's radiation was closer to vertical [8, 24]. Table 5.2 indicates the relation of the angle of sunlight to the formation of shadowing walls in different regions of Iran.

Table 5.2 Effect of angle of solar radiation on formation of shadowing walls in different areas of Iran [8].

Climate location of ice-house		Cities studied[a]	Status of shadowing wall	Examples
Cold climate		Hamadan, Tabriz, Takab, Kangavar, Malayer, and Urmia	No shadowing wall	Alagh Beyk, Mirfattah, and Noh Pelleh ice-houses
Hot and dry semi-desert climate	North	Semnan, Sabzevar, Aradan, and Quchan	No shadowing wall, or low shadowing wall	Zavaghan, Shahjouq, and Deh-Namak ice-houses
	South	Isfahan, Kashan, and Naein	High shadowing wall	Mohammadieh, Zavareh, and Hassan Abad ice-houses
Hot and dry desert climate		Kerman, Yazd, and Meybod	Very high shadowing wall	Moayedi, Rig Abad, Anar, and Meybod ice-houses

[a] Exceptions may be found in the cities located in the three considered climates. For instance, in the cities where ice-houses generally do not have shadowing wall(s), it is possible to find an ice-house with one.

Architects were also known to have constructed shadowing walls and ice-making ponds based on the number of frost days in each area [8]. In some regions, due to the high number of frost days plus extreme cold, there was no need to build shadowing walls, so ice was just acquired naturally during cold winter days and stored in pits. In hot and dry desert areas, the ice-houses were made with all components, including shadowing walls, ice-making ponds, an underground storage reservoir, and subcomponents, despite the fact that the number of frost days in these areas was not great. What mattered was that the temperature difference between night and day was high. Table 5.3 shows the impact of the number of frost days on the two main components of ice-houses (the shadowing wall and the ice-making pond) for various cities located in different climates of Iran.

– *The stability of shadowing walls*

To decrease the effect of various forces (particularly horizontal and/or thrust forces caused by wind and earthquakes) and torques on the

Table 5.3 Effect of the number of frost days on shadowing walls and ice-making ponds under different Iranian climates [8].

Climate location of ice-house		Cities studied	Existence of shadowing wall	Existence of ice-making pond
Cold climate		Hamadan, Tabriz, Takab, Kangavar, Malayer, and Urmia	×	×
Hot and dry semi-desert climate	North	Semnan, Sabzevar, Aradan, and Quchan	×	√
	South	Isfahan, Kashan, and Naein	√	√
Hot and dry desert climate		Kerman, Yazd, and Meybod	√	√

Note: The symbol √ represents that elements are present, and × shows that they are not.

shadowing wall, architects often incorporated a series of blind arcades[b] running along its length. Besides boosting wall stability, these additions created a light and shade pattern on the wall, reducing its appearance of blank uniformity and providing a decorative element. Examples of the use of blind arcades can be found in the Meybod and Haji Agha Ali ice-houses (Fig. 5.5). In other ice-houses, shadowing wall stability, especially against the wind, was promoted by the construction of large buttresses, mostly, along both sides of the wall, as shown in Fig. 5.6. However, these structures in certain ice-houses were only built along the wall's southern-facing side (Fig. 5.7). Where buttresses were not used, architects were forced to enhance the thickness of the wall, although the tradition of having thinner upper walls was maintained. When a shadowing wall

[b]A blind arcade, a small vault with low cover and depth that was created to beautify the structure, has been described thus: "A blind arcade is a representation of the vault," and "the purpose of making the blind arcade is for the sake of beauty." The top of blind arcades generally formed an arch. Although the arcades played a beautifying role in structures, they also boosted solidity and stability. In ice-houses, the blind arcades played both roles [25, 26].

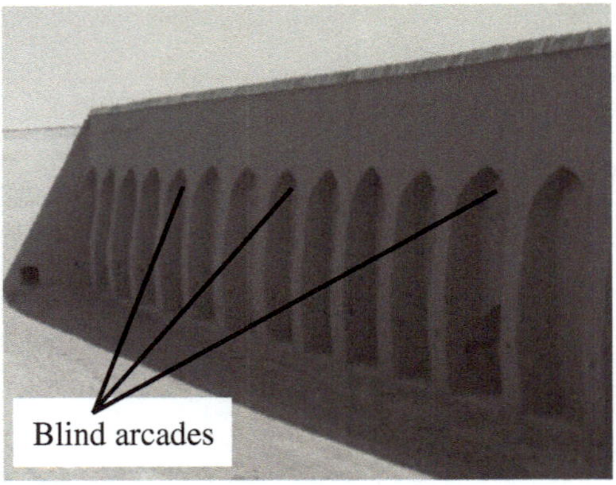

Fig. 5.5 Blind arcades created in the shadowing wall of Haji Agha Ali ice-house in Abbas Abad Haji village, Rafsanjan City, Kerman, Iran.

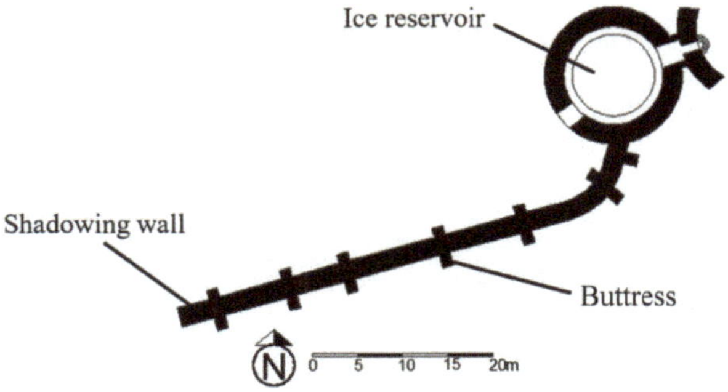

Fig. 5.6 Plan of Rig Abad ice-house (current situation) in Kerman City, Kerman, Iran [8, 24].

Fig. 5.7 Buttresses made in the southern part of the shadowing wall of Rig Abad ice-house in Kerman City, Kerman, Iran.

formed part of a city's defensive walls, stability was increased in two ways: (1) adding its width to the width of the city walls (in other words, creating thicker (double) walls) and (2) constructing a number of buttresses.

– *The number of shadowing walls*

In some ice-houses located in desert areas, such as the Barzan ice-house in the city of Abarkouh, Yazd, several other shadowing walls were constructed parallel to the main one. Consequently, the number of ice-making ponds was increased. In general, each desert area ice-house would have one, two, three, or even more shadowing walls if its surrounding space was large enough. These walls might run parallel (Fig. 5.8), perpendicular (Fig. 5.9), or even nonperpendicular to one another (Fig. 5.2), depending on factors such as the direction of the prevailing wind, the angle of solar radiation, and the nature of the topography.

In addition to blocking sunlight from hitting the surfaces of ice-making ponds, shadowing walls also deflected wind away from the ponds. When James Justinian Morier (1782–1849) visited the city of Shiraz early in the nineteenth century, he noted in his diary the importance of protecting the ponds against wind, stating that: "*A wall is built the whole length of the reservoir to screen the ice from the south wind, which is here the hottest*" [28]. However, he did not mention the shadowing role of the wall. Maxime Siroux (1907–1975) in his book entitled "Caravansarais of Iran and small buildings along the roads"[c] also referred to the ice-house

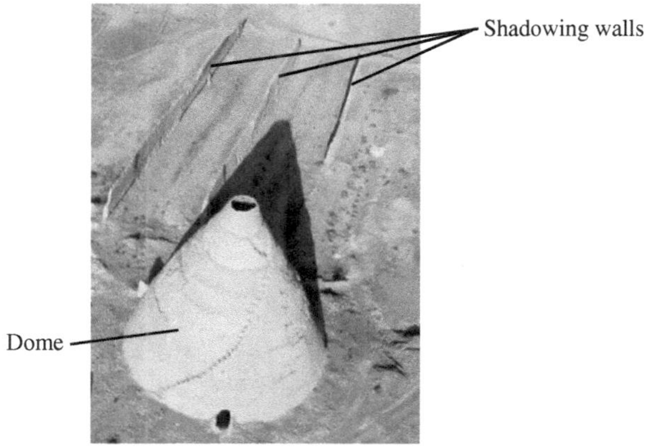

Fig. 5.8 View from above of the shadowing walls of Barzan ice-house in Abarkouh City, Yazd, Iran. The walls run parallel to one another.

[c]"Caravanserais d'Iran et Petites Constructions Routieres" (the original French title of the book).

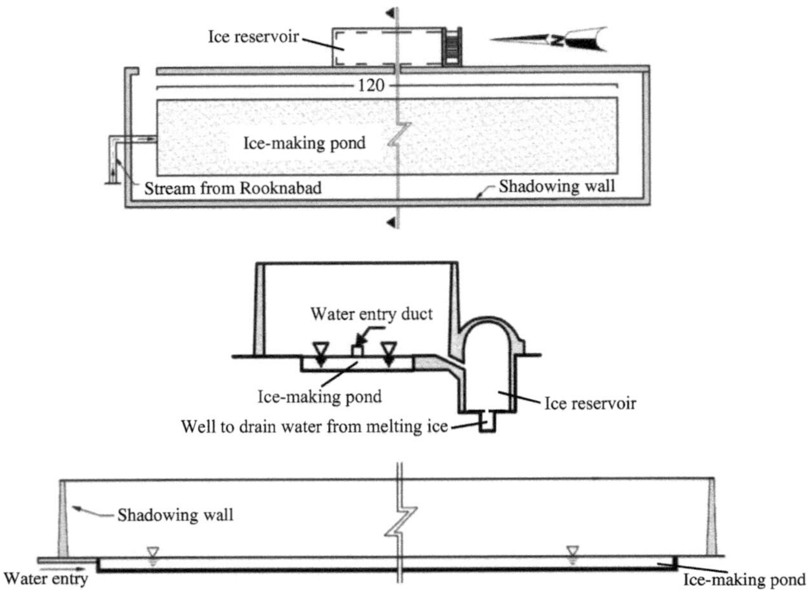

Fig. 5.9 Plan, side view, and cross-section of Mushir ice-house in Shiraz City, Fars, Iran [27]. The shadowing walls run perpendicular to each other. It seems that architect employed this technique to protect the ice-making pond and consequently its ice from any contamination.

shadowing walls, as follows [29]: "*An ice-house is made of a shallow basin [shallow ice-making pond] (thirty to forty centimeters [in depth]), dug in the ground, and protected by a wall 7 to 8 m high to the south, west and east. … The height of the walls blocks the wind, promoting rapid and regular ice production.*"[d]

The shadowing walls of the ice-houses located in Kerman Province trapped the north and northwest winds because of their U-shape and rotation to the northwest, the direction of prevailing cold wind in winter, thereby reducing the temperature in the northern part of the structures [24]. In some cities of Iran, these walls were the tallest local ones. For example, in the city of Yazd, the local people called the main one "the High Wall," and the neighborhood in which the ice-house was built was known as "the High Wall Neighborhood" [30].

As already mentioned, some ice-houses, in addition to the main shadowing wall, had side shadowing walls. Usually, the two ends of the main

[d]"*La glacière est composée d'un bassin peu profond (trente à quarante centimètres), creusé en pleine terre, protégé au Sud, à l'Ouest et à l'Est par une muraille de 7 a 8 métrés de hauteur… Les hautes murailles arrêtent le vent ce qui provoque une prise rapide et régulière de la glace.*"

shadowing wall were extended to the north for protection of the eastern and western parts of the ice-making pond(s). The additional walls, on the one hand, prevented early morning and late afternoon sunshine from reaching the eastern and western sides of the ice-making ponds, respectively. On the other hand, the colder air that formed in the space around ice-making ponds was trapped by the shadowing walls on three sides, consequently creating a temperature inversion above the ponds. Cold winter air entering from the north would have completed the blockade. This mechanism helped reduce heat exchange between the air confined inside the shadowing walls and the surrounding environment, and further chilled the cool air over ice-making ponds. Thus, the duration and intensity of frost in the space around these ponds were greater than those of the surrounding environment, conditions that accelerated the nighttime freezing and facilitated this process on future nights [30].

The length of side shadowing walls was less than that of the main shadowing wall, although their length was more than the maximum length of shadows created by the main wall on winter days. Considering the north orientation, in some areas, the main shadowing wall was located in front of the ice reservoir and its domed roof, and in other areas, it was located behind them [31]. The side shadowing walls contributed to the stability of the main shadowing wall in addition to adding shade and reducing wind impacts (Fig. 5.10).

5.1.2 Shallow ice-making ponds

The ice-making pond was a shallow rectangular pit that was dug beside and parallel to the main shadowing wall in the northern part of the structure. Its length was slightly less than the length of the main wall, and its depth was

Fig. 5.10 View of main and side shadowing walls of the Moayedi ice-house in Kerman City, Kerman, Iran.

usually between 30 and 60 cm. In general, the shallow pond was made in the form of a large, flat rectangular terrace, and sometimes two or more ponds were created in a space surrounded by main and side shadowing walls. These ponds were a place for producing ice naturally on cold winter nights, and in many cases, no materials were used in their construction; they were simply cut into the ground. However, in some cities like Yazd, they were built very well, with their floor and walls lined with Sarooj. The quality of the earth, especially its salinity level, was likely one of the most important considerations for the type of construction chosen. In sandy soils where water penetrates easily, the ponds' walls and floor had to be lined. In slow-draining clay and marl[e] soils, the floor and walls of the ice-making ponds could be left natural. For instance, in the Bafran ice-house, located in Bafran village, the city of Naein, Isfahan, the walls and floor of the ponds were natural and had simply been dug into the stiff clay [21].

At ice-making sites, the shadows created by the main and side walls kept the area relatively cool during the day, ready for ice-producing operations during the night. On cold winter nights, by filling the shallow ponds with water, local workers could naturally manufacture ice. In these ponds, water was protected from direct exposure to airflow. To improve the process of freezing under particular climatic conditions and increase the rate of freezing, the workers gradually filled the ponds with water and waited. When the ambient temperature was not cold enough to produce ice in this way, they sprayed water into the air in different ways to help make ice. In areas such as Isfahan and Kerman, after ending their ice-making operations when spring warmed the air, locals cultivated the pond basins (often by planting vegetables), thus taking advantage of their irrigation feature [33]. Fig. 5.11 illustrates this work done in the past inside Haj Rashid ice-house ponds in the city of Sirjan, Kerman, as an example.

5.1.3 Ice pits/reservoirs

The main pit or reservoir was a place for ice storage dug into the flat ground. The pit's walls were usually covered with Sarooj or cob, called "Kahgel" in Persian, a mixture of mud (mainly) and chopped straw [34, 35]. Water from ice melting in this pit drained into a well situated at the bottom of the pit to keep it separate from the ice stored in the reservoir [36]. The structure built

[e]Marl is a type of calcium carbonate soil that contains several kinds of minerals and aragonite. In scientific terms, marl refers to soil that includes 35% to 65% clay and 65% to 35% carbonate [32].

Fig. 5.11 Crops growing in Haj Rashid ice-house basins during summer in Sirjan City, Kerman, Iran.

Fig. 5.12 View of a stepped domed roof over ice reservoir of an ice-house near Shahroud City, Semnan, Iran.

over the main pit was generally a dome- or tunnel-shaped cover [15]. For example, Fig. 5.12 shows an ice-house with a dome-shaped roof, located near the city of Shahroud, Semnan. A small hole was usually created at the tip of the dome to pull warm air from inside into the surrounding environment during the hot/warm months when the ice-house was utilized for ice sales. This hole was closed for the rest of the year.

In most of the hot desert and semi-desert areas of Iran, the ice storage reservoirs were covered with a domed roof in the form of an oval (prolate shape) with the ratio of the distance between the foci and the length of the minor axis equal to $\frac{4}{3}$ (Fig. 5.13), which is known as a "Bastoo,[f]"

[f]This dome is obtained through the rotation of the Bastoo arch around the vertical axis that crosses its vertex (Fig. 5.13), and has a higher rise (called "Khiz" in Persian) compared with other oval domes used in Iranian architecture [38]. The literal meaning of Bastoo in Persian is "jug."

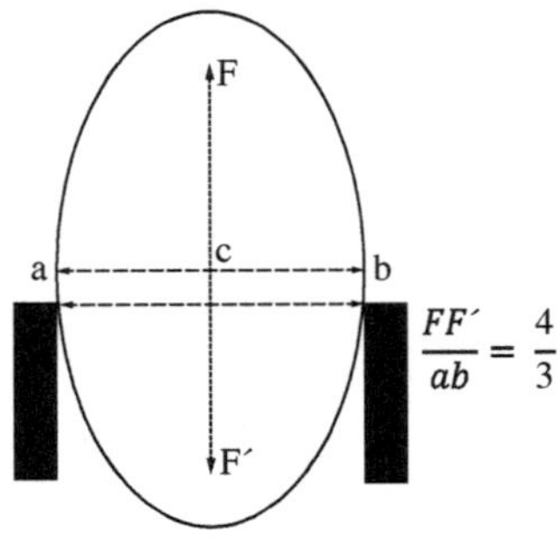

Fig. 5.13 Method of drawing a Bastoo arch [37].

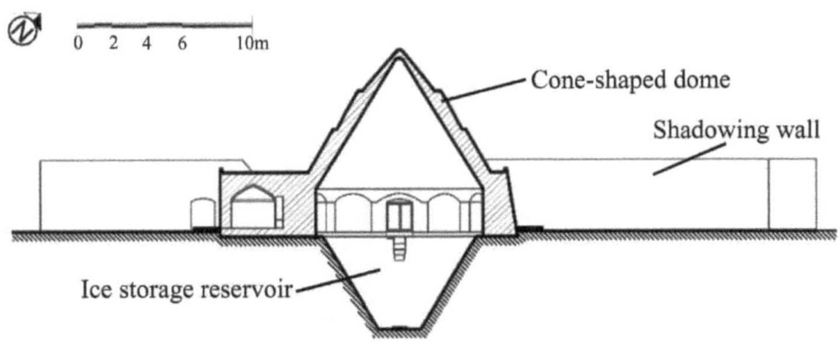

Fig. 5.14 Cross-section of the Bafran ice-house in Bafran village, Naein City, Isfahan, Iran [21, 22].

or a sharp cone-shape (Fig. 5.14), called a "Rok,[g]" in Iranian architecture [37, 38]. These types of roofs were built on wide openings[h] so that they could withstand the stress of different loads (e.g., pressure from weight and stretching) without the use of techniques such as adding extra weight to either the base or walls [37]. The diameter and thickness of these domed roofs, where the base of the dome arch begins (called the "impost[i]" [35]), were high, and approaching the vertex, their diameter and thickness diminished. Sometimes, to decrease the pressure from the dome on the base, architects constructed the outer surface of the dome in the form of a staircase, otherwise known as a stepped pattern.

The ice reservoirs in hot desert and semi-desert climates were in the form of either an inverted circular truncated cone (mostly) or cylinder (with a radius of about 4 m or more). For instance, the shape of the ice storage reservoir in the Bafran ice-house was an inverted circular truncated cone, with a

[g]Rok is a type of cone-shaped (conical) dome with a sharp tip, so that it looks like an inverted funnel [37, 38].

[h]The opening or mouth is the interval between walls that are covered with the vault or dome.

[i]It is known as "Pakar" in Iranian architecture [25].

Table 5.4 Type of staircase inside some ice-houses' reservoirs in Kerman Province, Iran [24].

Ice-house name	Type of staircase		Size of the dome opening (m)
	Straight	Circular	
Anar	√		7.2
Arg-e Bam		√	12
Haj Rashid		√	14.3–15.5
Ravar	√		11.5
Rig Abad	√		11
Haji Agha Ali		√	15
Kabutarkhan		√	16
Zarisef		√	15.2
Langar	Unknown		
Mahmoud Abad	√		11
Moayedi[a]	√		15

[a] The reservoir staircase was initially circular, but after repair and use change, to facilitate access, was converted to a straight up-down form.

diameter of 8.8m at the top and 2.4m at the bottom [21] (Fig. 5.14). Ice-house reservoirs had a staircase for accessing the ice stored inside them, constructed either to run straight down, along the reservoir wall, or circling the wall to the bottom of the reservoir. These staircases were made based on the type of internal "ambulatory," a place by which laborers could walk inside the ice-house, around the upper edge of the ice reservoir. If the dimensions of the ambulatory were large enough, the staircase was circular; if small, it ran straight up and down. Domes with wide openings and sufficient space almost always had circular staircases [24]. For example, Table 5.4 displays the type of staircase within several ice-houses' reservoirs located in Kerman Province.

In some Iranian ice-houses, particularly in cold regions, the ice reservoirs were constructed in the form of a long, rectangular atrium similar to a tunnel. The width of these atriums was around 5m, and their length reached up to 10m. Their vaults were generally in the form of a barrel vault or ribbed vault[j] (Fig. 5.15). A small room was often constructed next to the atrium, half of it located underground, housing the stairs that allowed workers to reach the bottom of the pit. Some ice-houses such as Khalili in Tehran, Noh Pelleh in Urmia, and Sadeghieh in Tabriz were of this type. The depths of these kinds of reservoirs differed. Depending on the amount of ice produced and the ground properties, they might reach up to 10m.

[j] This vault is formed by supportive elements called "ribs." The ribs were prefabricated arches installed in desired locations, and then covered with bricks. The shape obtained is known as a "strip of vault."

Fig. 5.15 Ribbed vault made of brick in an ice-house in Saveh City, Markazi, Iran [20]. The air vents of these types of ice-houses were opened during cold winter days and then closed in warmer weather.

Sometimes, the reservoirs of ice-houses with shadowing walls were located on one side of the main wall, with a short space between the two structures, as is the case at Zarisef ice-house in Kerman City, Kerman (Fig. 5.16). In other cases, they were built into the middle of the main shadowing wall, interrupting its straight run either slightly (such as in Meybod ice-house, in the city of Meybod, Yazd) or markedly. In addition, ice reservoirs might be placed anywhere along the length of the main shadowing wall; for instance, at Kashmar ice-house in the city of Kashmar, Razavi Khorasan, it is located between the middle and the end of one side of the main wall (Fig. 5.17). In Haji Agha Ali ice-house in Abbas Abad Haji village, Rafsanjan City, Kerman, it stands at the end of one side of the main wall. In some parts of Iran, such as the city of Sabzevar, Razavi Khorasan, three to six ice-houses were usually built next to one another or close together to facilitate ice production and storage as well as general operations. Their strategic location on the Silk Road meant that demand was heavy, so sales and delivery had to be efficient [39].

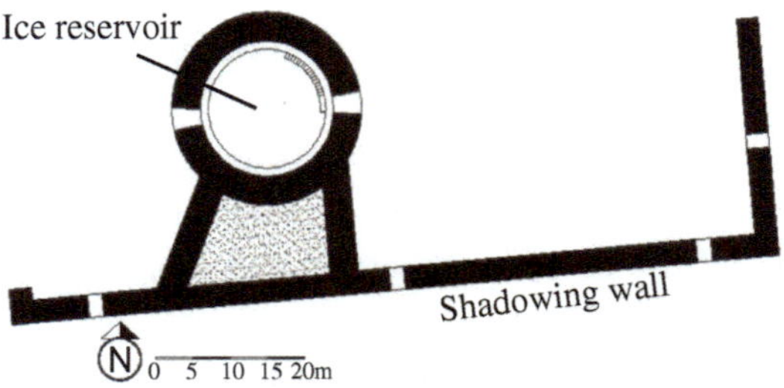

Fig. 5.16 Plan of Zarisef ice-house, Kerman City, Kerman, Iran [8, 24].

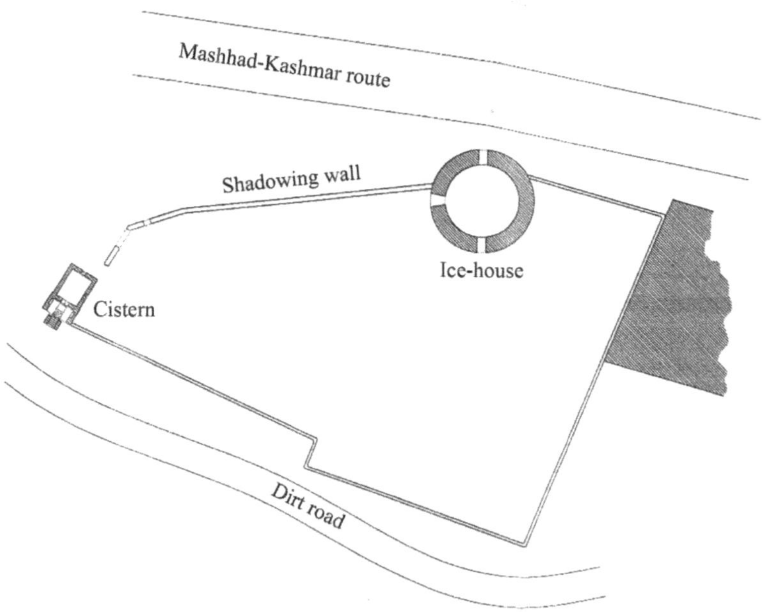

Fig. 5.17 Plan of an ice-house and a cistern in Kashmar City, Razavi Khorasan, Iran.

The ice-houses' domed roofs and their reservoirs' walls were made of rubble, lime mortar, mud-brick (or sun-dried mud-brick, called "Khesht" in Persian), brick, baked brick, or stratum (or mud layers, called "Chineh" in Persian) to reduce heat exchange with their surroundings[k] [40, 41]. The exterior surfaces of these domes were sometimes smooth and polished, but most were constructed as "stepped" levels to make their maintenance and repair easier. The ice blocks produced in shallow ponds during cold winter nights were divided into portable pieces and then stored in the ice pits (Fig. 5.18). The successive layers of ice blocks were kept apart with a thin layer of straw for insulation, as shown in Fig. 5.19. These layers of straw also made ice separation and removal easier during the hot/warm months. In some cities such as Isfahan and Naein, a type of plant or reed that grew alongside rivers and creeks was utilized to cover on the top level of ice stored in reservoirs. In certain desert areas, the uppermost layer of insulating material (straw, wheat stems, dry twigs, etc.) was covered with a layer of soil, mud, or Kahgel to provide extra insulation.

[k]Sometimes, most parts of some ice-houses located in hot and dry desert and semi-desert areas, excluding parts in contact with ice or water and places with high erosion (for instance, the lower parts of the dome's wall), were exclusively made of clay and Kahgel [29], because they had suitable thermal properties that reduced the penetration of heat into the place of ice storage.

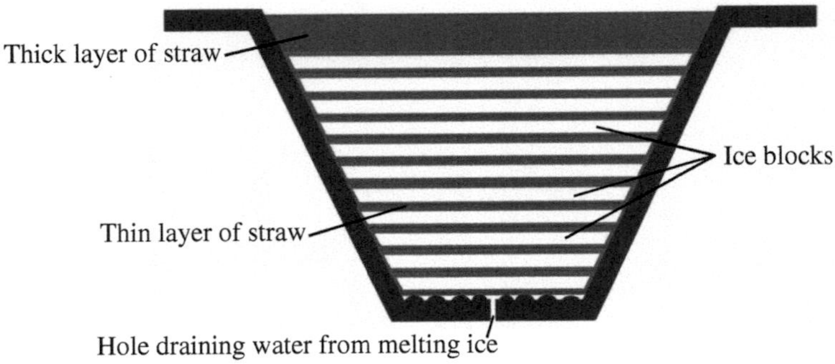

Fig. 5.18 Overview of the arrangement of ice blocks in ice-house reservoir.

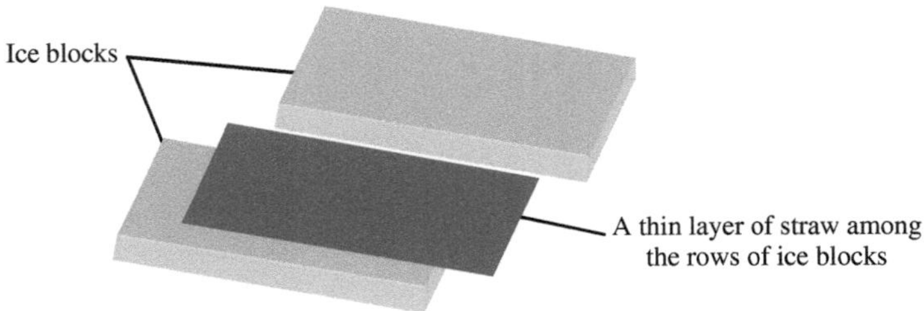

Fig. 5.19 Overview showing a thin layer of straw or other fibers between layers of ice blocks.

It is worth noting that the importance and glory of the ice-houses' architecture can be observed in their cone-shaped domes. Their shape was such that it was completely suitable for the ice-houses' purposes; warm air, because of its higher density, rose up and collected below the dome, keeping it far from the ice reservoir. The ice blocks stored in the reservoir experienced only minimal exposure to heat from winter to summer, consequently delaying their melting.

– *The influence of the angle of sunlight on location of the ice-house's dome relative to main shadowing wall*

  The location of the dome relative to the main shadowing wall differed among the ice-houses. In some cities, the dome was located north of the main shadowing wall, and in others was placed to the south of the wall. The location of the dome made a difference in the positioning of the entrance to the ice-house reservoir. The dome of ice-houses of the southern parts of Iran at lower latitude was placed north of the main wall because the angle of sunlight was closer to the vertical. In some ice-houses, to reduce exposure to sunlight and subsequently heat, the architect constructed a dome in the northern part of the main shadowing wall to block sunlight from hitting the dome directly. In such cases, the reservoir entrances remained in shadow.

5.1.4 Subcomponents

In some ice-houses, in addition to the main components, there might be a number of subcomponents, such as a guard room (Fig. 5.4), a room for a caretaker or ice-house operator, a storeroom, a waiting room, entrance doors, and a water inlet duct. It should be noted that their location on the ice-house plan was influenced by how the shadowing wall was connected to the dome. Some of these options are briefly discussed in the following subsection.

- *Caretaker's/operator's room*
  This small space served as a shelter for the ice-house operator and sometimes as a storehouse for ice-making tools and equipment. For example, Fig. 5.20 shows the position of the room of the ice-house operator in the plan of an ice-house in the city of Ravar, Kerman [42].
- *Ice-house entrance doors*
  As mentioned earlier, the position of the dome in the ice-house structure affected the number of ice reservoir entrances. In some cases, there were as many as four, such as in the Barzan ice-house in the city of Abarkouh, Yazd. In some ice-houses, like one in the city of Ravar, Kerman (Fig. 5.20), the ice reservoir had a subsidiary entrance from the caretaker's room. In cities like Yazd and Kashan, due to the construction of ice-houses south of the main shadowing wall, an entrance was created on the north side between the ice-making pond and the ice reservoir to take advantage of wall's daytime shadowing effect. Another entrance, built on the south side, was exposed to sunlight (Fig. 5.21). In

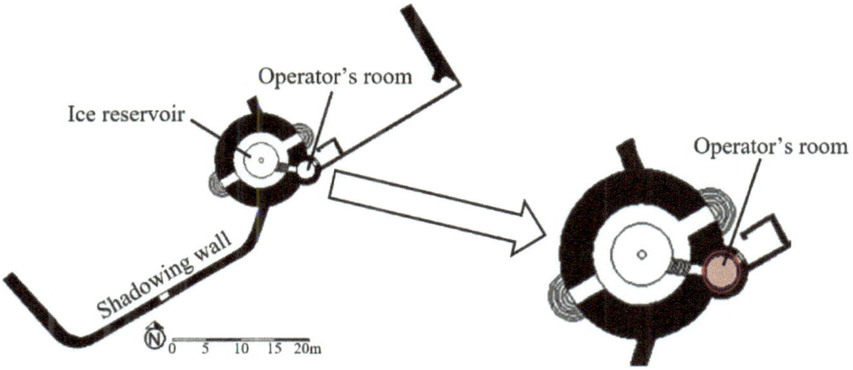

Fig. 5.20 Operator's room in the plan of Ravar ice-house in Ravar City, Kerman, Iran [23, 24].

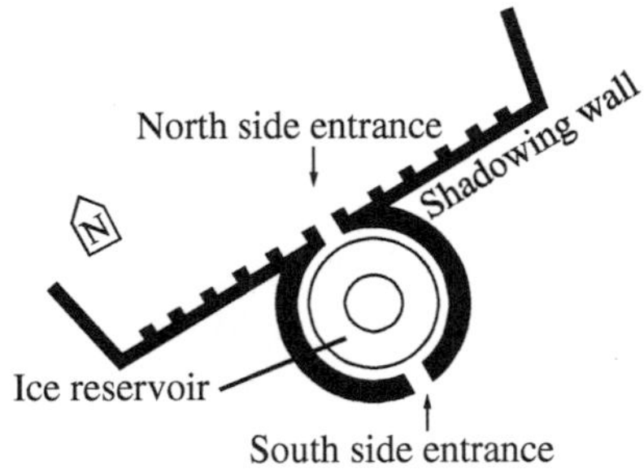

Fig. 5.21 Plan of Meybod ice-house located in Meybod City, Yazd, Iran [23].

winter, the north side entrance was used to transfer ice blocks from the ice-making pond to the reservoir. Workers slid the ice into the reservoir down a steep ramp between the entrance door and the pit. In summer, the ice-house operator mainly delivered ice to customers by opening the south side entrance door and using the staircase [43].

In cities like Kerman and Sirjan, where the ice-house domes were placed north of the main shadowing wall, the entrance doors were built on the eastern and western sides of the dome, so that the entrances always stood in shadow. Thus, less warm air could penetrate the reservoir [23]. Moayedi ice-house located in Kerman City, Kerman, has three entrance doors, one each in the eastern, western, and southern sides of its dome (Fig. 5.22). The eastern entrance has a steep ramp that was only used to transfer the ice blocks from the ice-making ponds into the reservoir. The western entrance in addition to its steep ramp has a staircase, indicating that it was a place for the ice to exit from the reservoir. The south entrance door gave access to the operator's room, and through it, the ice reservoir. This entry structure and its location ensured that warm air and the sun's rays would never penetrate to the ice reservoir [23]. The height of the entrance doors was low, offering less opportunity for warm air to affect the stored ice (Figs. 5.23 and 5.24).

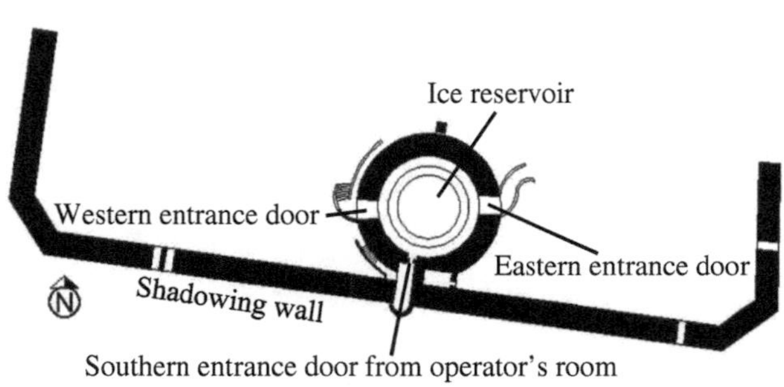

Fig. 5.22 Plan of Moayedi ice-house in Kerman City, Kerman, Iran [23].

Fig. 5.23 View of one entrance of Ravar ice-house in Ravar City, Kerman, Iran.

Fig. 5.24 View of an entrance of Mahmoud Abad ice-house located in Mahmoud Abad village, Sirjan City, Kerman, Iran.

– *Water inlet duct*:

  To transfer water to the ice-making pond(s), the architect(s) created a waterway at the bottom of the shadowing wall (Figs. 5.24 and 5.25). After each step directing water into the ice-making pond(s), this waterway was blocked, so that animals (for instance, dog, cat, fox, and jackal) could not enter the pond(s) through it (Fig. 5.26).

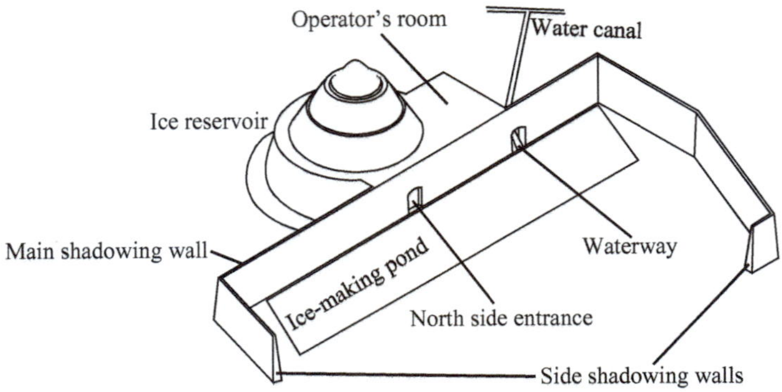

Fig. 5.25 Water transfer pathway to the ice-making pond of Bafran ice-house located in Bafran village, Naein City, Isfahan, Iran [21, 22].

Fig. 5.26 View of a waterway in Kabutarkhan ice-house, in Kabutarkhan village, Rafsanjan City, Kerman, Iran.

5.2 Typology of ice-houses

Traditional Iranian ice-houses can be categorized into several types in terms of their architectural structure [17, 20]: (1) ice-houses with no vault, (2) ice-houses with a domed roof, (3) ice-houses in the shape of a tunnel, and (4) ice-houses with no shadowing walls. The four types are briefly described in the following subsection.

5.2.1 Ice-houses with no vault

Ice-houses without vault were constructed only in the city of Isfahan and its suburbs, and were used until a few decades ago. This type of ice-house generally had a shadowing wall of 4 to 5 m in height and 12 m in length.

Next to the north side of the wall, a pit was dug, roughly 5 to 6 m deep, around 12 m long, and about 5 m wide (Figs. 5.27 and 5.28). In Chapter 4 of Chardin's diary, he describes how small chunks of ice were thrown into the pit and arranged compactly beside one another. Laborers then poured water over the chunks, so that freezing would create an integrated ice mass. Thereafter, they covered this mass with a mat and heaped a thin layer of straw on top of the mat. The process was then repeated to make the next layer of ice mass, building up layers to near the top of the pit. Therefore, among the layers of integrated ice mass were the mats and the layers of straw. This layering technique facilitated breaking up and removing the ice from the pit to sell during summer. Second, the mat and straw between layers of ice insulated the remaining ice mass. After ice production season was over, workers covered the top surface of the ice with straw to the height of 1 to 2 m, and left the mass until ice was needed during the warmer months. Fig. 5.29 shows an old photo of ice-making in the city of Isfahan.

The role of the shadowing wall in these types of ice-houses was mostly to shade the ice pit during the day. It should be noted that, in addition to the large number of ice-houses in the city of Isfahan (around twenty and

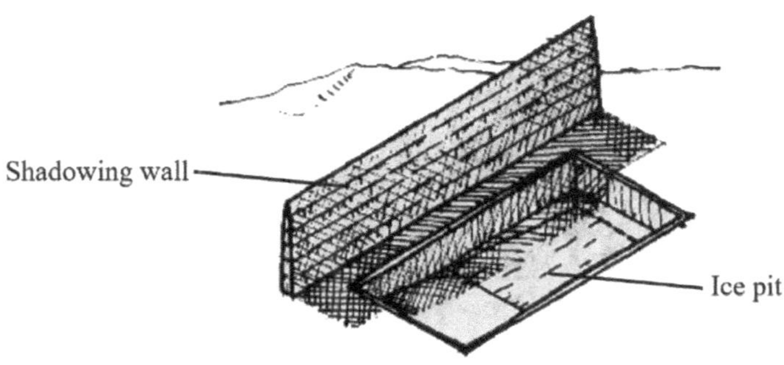

Fig. 5.27 Schematic illustration of an ice-house with no vault, including a shadowing wall and a large ice pit located in Isfahan City, Isfahan, Iran [20].

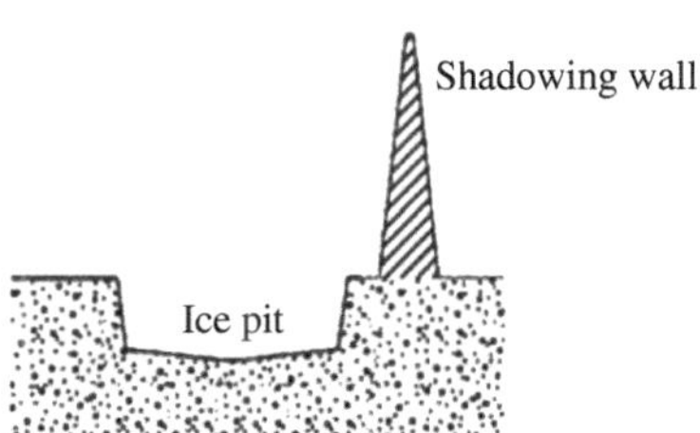

Fig. 5.28 Cross-section of an ice-house without vault in Isfahan City, Isfahan, Iran [44].

Fig. 5.29 An old photo of ice production in Isfahan City, Isfahan, Iran, taken by Hoeltzer [45].

even more in some periods – see Fig. 4.11), the capacity of some ice-houses was high, so much so that some ice might remain in the pit for the next year [20, 45, 46].

5.2.2 Ice-houses with a domed roof

A specific type of ice-house design with a large dome made of brick, baked brick, or mud-brick over the ice pit/reservoir was constructed in response to hot and dry desert conditions. Most were located on the outskirts of the central desert and northeastern parts of Iran. They have the following components [20]:

- Shadowing wall(s),
- Ice-making pond(s),
- Large ice pit/reservoir with a domed roof,
- Waiting room, mostly, as shown in Fig. 5.30.

Common features of the ice-houses with a domed roof are as follows [17]:

1. The pit/reservoir was usually greater than 5 m deep and was generally entirely filled with ice. Among them, as an example, the depth of the ice reservoir of Bafran ice-house in Bafran village, Naein City, Isfahan, was 5.5 m.
2. A staircase, usually situated at the south entrance door and the waiting room, connected the surface structures to the ice reservoir.
3. Coverage of these types of reservoirs in most parts of Iran was in the form of either a "Bastoo" or "Rok," was generally made of stratum or mud-brick.

Fig. 5.30 View of a waiting room in an ice-house with domed roof.

4. Usually, there was about 1.5 m distance between the inner wall of the dome and the upper edge of the ice pit/reservoir, called the "ambulatory."
5. The walls of the ice pit/reservoir were built of baked brick, or brick, or stone, with a layer of Sarooj overall as a coating where there would be contact with the ice stored.
6. Ice reservoirs had one to four entrance doors, most designed to transfer ice to the reservoir, located next to the ice-making pond(s), so that they had a steep ramp to facilitate ice transfer into the pit/reservoir.
7. The shadowing walls were high, but never taller than the domed roof of an ice-house.
8. These ice-houses had a waiting room, usually clinging to the entrance door, which acted as a buffer between the inner side of the ice reservoir and the surroundings air.
9. The exterior walls of domed roofs rose from 1.5 to 2.5 m above the ground, probably to lessen the roof thrust. The arch of the dome generally started at a distance of about 0.5 to 1 m above the ground.
10. Ice-making ponds generally had a depth of between 30 and 60 cm.
11. The conical dome was usually constructed with mud-brick, the entrances were mostly built with baked brick, and the walls were generally made of stratum.
12. For the stability of shadowing walls and their resistance to wind, a number of buttresses were generally added; where they were not utilized, architects were forced to increase wall thicknesses, although they consistently made them thinner toward the top.

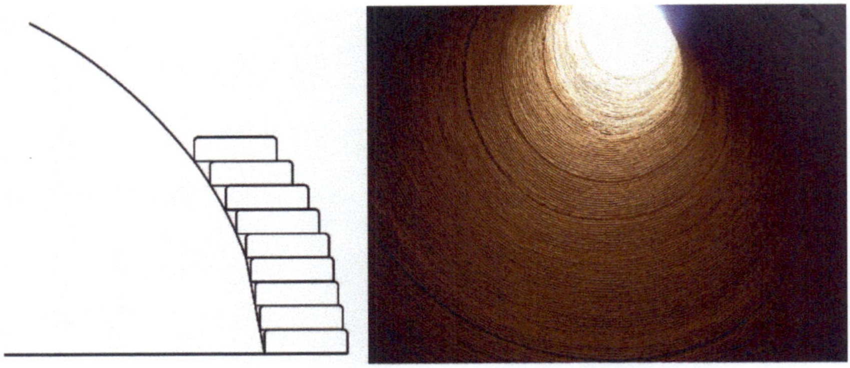

Fig. 5.31 View of the layout of bricks (or bricklaying) in the Ragchin pattern [47].

13. At the bottom of the ice pit/reservoir was a waterway that drained the water from melting ice into a well.
14. The dome of these types of ice-houses was always constructed in the Ragchin[l] configuration, meaning layers of brick or mud-brick laid parallel to the horizon line (Fig. 5.31) [48].
15. To increase the dome's resistance and stability, as well as to diminish its construction cost, the thickness of the dome shell decreased from the bottom to the top. Doing so reduced the weight of the dome, lowered its center of gravity, and consequently reduced construction material use.
16. The exterior surface of the dome was mainly in the form of a staircase (stepped), which facilitated maintaining and repairing these large structures.

5.2.3 Ice-houses in the shape of a tunnel

Another form of ice-house, developed in the north-central parts of Iran, such as Tehran and Saveh, and northwestern areas, like Zanjan, Hamedan, Tabriz, and Urmia, operated similar to the dome-shaped ice-houses, although their physical configuration varied. They were constructed in the shape of tunnels. Most parts of these ice-houses were under ground, and their thick walls were made of rubble or brick and water mortars such as Sarooj. Their brick ceilings were generally built in the form of either ribbed vaults or barrel vaults[m] (also known as "tunnel vaults") [20].

As shown in Fig. 5.32, after water was frozen in the shallow ice-making ponds, the ice produced was thrown into the reservoir through a vent lo-

[l]This is a Persian word and a term in Iranian architecture.

[m]This vault is in the shape of a hollow semi-cylinder and is created by the movement of an arch along a line; in other words, a series of arches are placed side by side (i.e., one after another).

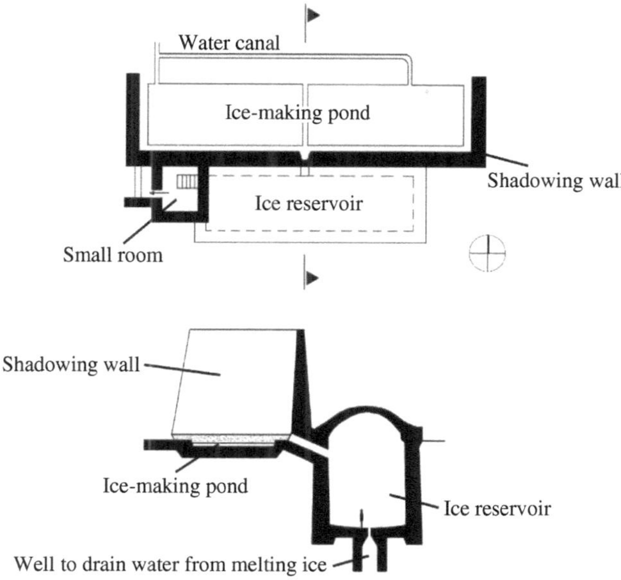

Fig. 5.32 Plan and cross-section of an ice-house in the shape of a tunnel [20].

cated under the vault. Afterward, each time that ice was thrown into the reservoir, water was poured over it to create an integrated ice mass. These ice-houses did not require layers and covers of insulating straw because they were in areas where winter weather was relatively very cold; hence, the ice stored in the reservoir remained frozen well into the warmer months [20].

Given that in the western and northwestern parts of Iran, the summer weather was cooler than that of the desert outskirts, so the consumption of ice was less; consequently, there were fewer of this type of ice-house than of the domed structures in the areas edging the desert. It should be noted that in colder regions, the use of ice produced naturally in the mountains was common, so the need for man-made ice-houses was lower. One example of a tunnel-shaped ice-house is located in the south of the city of Saveh, Markazi. The reservoir of this ice-house has a width of about 3.5 m and a length of 12 m, and part of it lies 5 m below the ground level. The walls are around 80 cm thick and made of a particular brick, called "Kazakh brick" in Persian, each with the dimensions of 25 × 25 cm, and a mortar mixture of sand and lime. The distance between one vault rib and another is about 3 m (Fig. 5.33) [20].

Most tunnel-shaped ice-houses or those with domed roofs, because of their large dimensions, required a relatively large area for ice production, and due to the low price of land in the suburbs, they were built outside cities/villages. Once purchased, the ice was transported to people's houses and

Fig. 5.33 Schematic front view of an ice-house in Saveh City, Markazi, Iran [20].

kept in straw-lined boxes. The insulating effect thus provided helped to preserve the ice for four to five hours [20]. In the majority of houses in Tehran that had a domestic cistern, people kept their ice in a "Pasheer," the place where water could be drained from the domestic water reservoir, whose temperature was about 10°C colder than the ambient air. Additionally, in Tehran, ice was put in sacks full of straw. Most of the ice-houses in Tehran have been almost completely demolished, but some areas such as Kan, Khani Abad, and Islamshahr still have remains of these types of structures (Figs. 5.34 and 5.35).

Fig. 5.34 View of Kan ice-house that is being destroyed, Tehran City, Tehran, Iran.

Fig. 5.35 View of Jafar Abad-e-Jangal ice-house located near Tehran City, Tehran, Iran [44].

5.2.4 Ice-houses with no shadowing walls

In some desert areas of Iran such as Sabzevar, Semnan, and Shahroud, the length of ice-making ponds sometimes reached a few hundred meters. These ponds were not sheltered from the sun by shadowing walls, but their adjacent ice reservoirs were covered with large brick/mud-brick and stratum domed roofs, and several ice-houses were constructed next to one another (Fig. 5.36). The nature of these ice-houses made it necessary to produce and store ice as quickly as possible; otherwise, the daytime's high temperatures would quickly undo a night's ice making. To facilitate the process, the ice-making ponds were quite shallow.

The ice-houses of the city of Sabzevar had dome-shaped covers, made of bricks laid in the Ragchin pattern, and constructed over a large pit, with a main entrance that was also the main door to the reservoir. In front of this door, large terrace-like broad trenches were established slightly below the ground level. To produce ice, sufficient water was directed through small canals to fill these shallow trenches. The trenches were blind; that is, there was no outflow for the water, so it was retained, to freeze on cold and clear winter nights, as shown in Fig. 5.36. Laborers then entered the trenches, cut the ice into large but portable pieces, and brought them inside the reservoirs. The entrance of the ice-house was then closed, sealed, and insulated from the outside. In the summer, the caretakers/operators of these ice-houses demolished the main entrance in order to enter the ice reservoir and break up the large pieces of ice into small chunks for sale (Fig. 5.37).

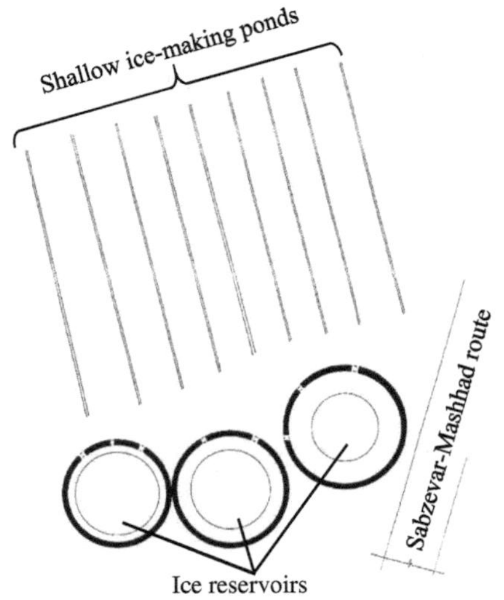

Fig. 5.36 Plan of three adjacent ice-houses located alongside Sabzevar-Mashhad route in Sabzevar City, Razavi Khorasan, Iran, known as the Triplets.

Fig. 5.37 Carrying small pieces of ice from pit/reservoir to outside space for sale [20].

5.3 Construction materials used in ice-houses

Mud-brick and clay (or mud) constituted the most important construction materials in Iranian desert architecture, especially for building ice-houses. Not only were they readily available, but they had insulating qualities appropriate for reducing the exchange of thermal energy (heat and cold) between inside and outside spaces. In addition, with a Kahgel coating, they were the most-resistant construction materials then available in desert areas. Stone and brick were other important local construction materials because of their strength, availability, and cheapness, and so they were also often used in the construction of ice-houses. Builders usually employed stones at the base of the structures and bricks for the vaults. The exterior facade of the ice-house was then coated with Kahgel. The walls of the ice pit/reservoir were also made of stone (rubble), solid packed earth, or brick/mud-brick and lined with Sarooj [15–17]. Kahgel's insulating properties helped prevent the penetration of moisture from snow and rain. Most importantly for ice-houses, the light khaki color of Kahgel coating and mud-brick also helped deflect the areas' intense sunlight, thus lessening heat absorption. The uniform color of the buildings' facades plus their graceful uncluttered shapes made them attractive and often dramatic to look at. Their builders, however, often went even further in adding visual appeal to these highly utilitarian structures.

5.4 Decoration of ice-houses

Ice-houses, despite their practical function, towering size, and apparent simplicity, were generally graceful in form, in keeping with the elegant taste of Iranian architectural tradition. The harmony of their shapes and clean lines was enhanced by their mainly being coated in one material, of one color. The impression of austerity was avoided, however, through low-relief decoration in their otherwise smooth walls, but here again, the builders chose the simplest forms of decoration, such as laying the mud-bricks in zig-zag or other patterns, creating small depressions or bumps to add texture, and establishing a number of blind arcades along the walls, thereby avoiding boring uniformity [15]. The shadowing walls of some ice-houses were decorated with repeating geometric motifs and drawings, and those of others with stylized patterns of trees and plants [49]. Most of these decorations were located on the upper parts of the shadowing walls. Symmetrical patterns of holes (variously shaped openings) created to border the upper parts of some walls both added beauty and reduced the effects of wind pressure, as well as lightening the overall weight of the structure. In the ice-houses of Kerman Province, these decorations are still easily observed (Table 5.5).

Table 5.5 Decorations of shadowing walls of ice-houses located in Kerman Province, Iran [23].

Ice-house name	Type of decorations: Geometric motifs	Type of decorations: Lightening	Decorations design	Photo
Anar	–	–	–	–
Arg-e Bam	–	–	–	–
Haj Rashid (or twin Yakhdans)	√	√	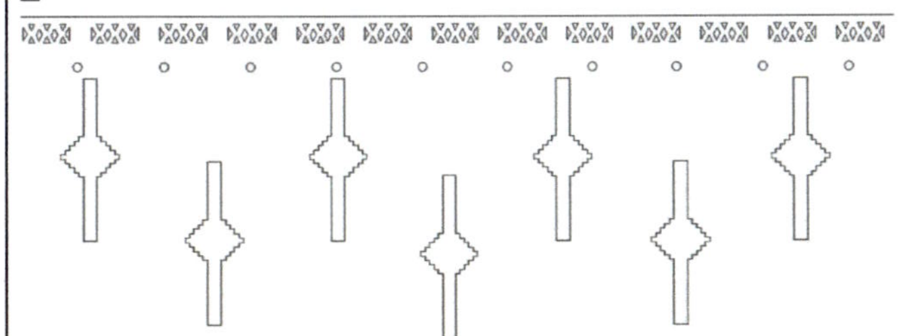 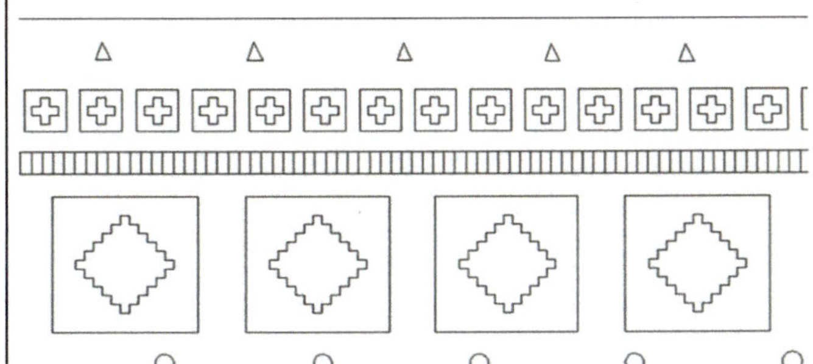	 

Ravar		√	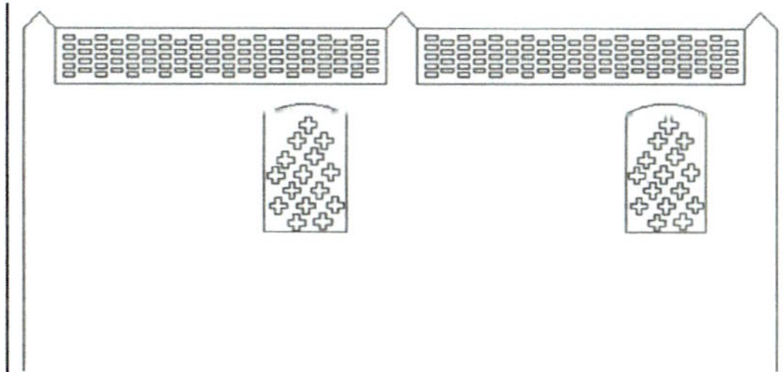	
Rig Abad	√	√	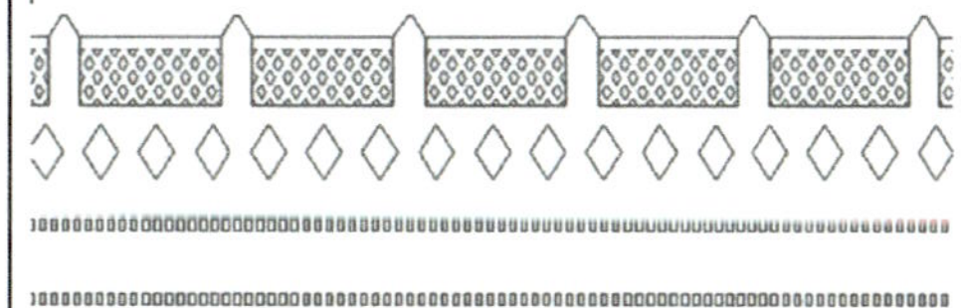	
Haji Agha Ali	–	–	–	–
Kabutarkhan	–	–	–	–
Zarisef	√	√	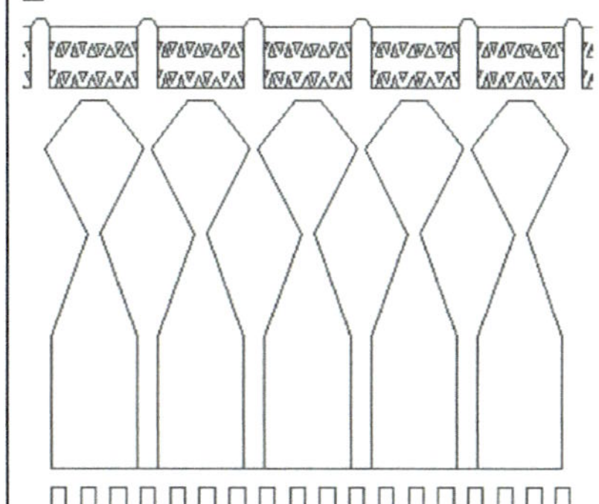	

Continued

Table 5.5 Decorations of shadowing walls of ice-houses located in Kerman Province, Iran—cont'd

Ice-house name	Type of decorations: Geometric motifs	Type of decorations: Lightening	Decorations design	Photo
Langar	√		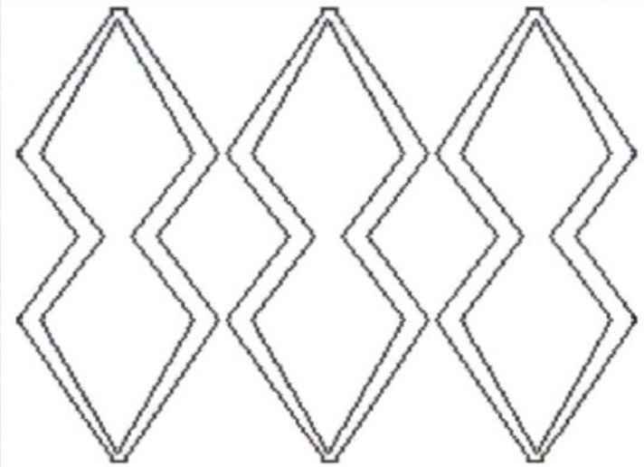	
Mahmoud Abad	–	–	–	–
Moayedi		√	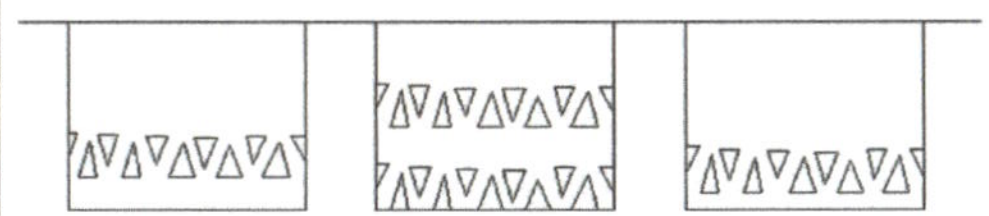	

5.5 Traditional methods of ice production

In the past, ice was naturally produced in various ways for summertime use. In the first method, local workers collected compressed snow or ice from different locations (e.g., from mountains) and added it to chilled water in order to increase the speed of freezing and produce ice faster [17]. For example, in the city of Isfahan, the snow and ice around the ice-house were collected and compacted, then transferred to the ice pit to accelerate the ice making [17], as shown in Figs. 5.38 and 5.39.

In another method of producing ice, briefly described in Chapter 2, shallow ice-making ponds, located next to the shadowing wall(s), were filled with fresh water up to a certain level (mostly equal or lower than 1 cm, depending on the ambient temperature) on cold and clear winter nights. Severe cold during the night froze the thin layer of water in these ponds [16]. It should be noted that the amount of water directed onto the frozen mass on subsequent nights was carefully judged to be only as much as could freeze each night. Over time, a thick layer of ice gradually filled the ponds. Thereafter, in the early morning, laborers broke up the ice mass into smaller pieces using sledgehammers or other tools, and then dragged the ice, piece by piece, into the pit/reservoir (Fig. 5.40).

Fig. 5.38 Collecting and compacting snow in an ice-house in Isfahan City, Isfahan, Iran, taken by Hoeltzer [45].

Fig. 5.39 Collecting snow and ice and placing them in an ice-house's pit in Isfahan City, Isfahan, Iran, taken by Hoeltzer [45].

Fig. 5.40 An old photo of workers and their tools producing ice in Jolfa district, Isfahan City, Isfahan, Iran, taken by Hoeltzer [45].

When the weather warmed up and the need for ice was felt, laborers wearing boots and rubber vests pulled out the ice blocks from the pit/reservoir using iron hooks or other tools, and then brought these blocks to a large scale that was hung from the ceiling of the waiting room beside the ice pit/reservoir by strong chains. To retain cold, the door of this room, which usually was on the southern-facing side of the structure, was opened only in the early morning (mainly) or late afternoon. The majority of ice-houses had enough ice for most of the year.

5.6 Vulnerability of ice-houses

Despite the use of resistant and climate-friendly materials in traditional ice-houses, a number of hazards threatened these structures, some of which are listed as follows [15, 16]:

1. Due to the singularity of the structure in relatively open spaces, severe desert winds and storms, usually accompanied by sand, contributed to the erosion of ice-houses.
2. Humidity had a devastating effect on ice-houses because of (1) the gradual penetration of water from ice melting in the reservoir and soaking the surrounding ground and (2) water entering from the ice-making ponds due to the short distance between them and the shadowing wall(s), although this seepage was partially prevented by the architecture of traditional ice-houses.
3. The growing popularity of the modern ice-producing industry diminished the cultural, financial, and economic standing of ice-houses over time, so they were eventually abandoned and left desolate. In addition, lack of attention to them as sites of cultural heritage and beauty meant that their value was ignored. Therefore, conditions became appropriate for their neglect and even demolition.

Nevertheless, these structures made of mud-brick/brick, stratum, and Kahgel, and have a natural proportionality with the country's desert conditions, and their historical and technical significance should motivate efforts to maintain, protect, and renovate them. As an example of what can be done, Fig. 5.41 shows the large interior space of the reservoir of Arg-e Bam ice-house in the city of Bam (old part), Kerman, which was converted into a conference hall, after repair in 1993 and before an earthquake in 2003.

5.7 Effect of climate on the distribution of ice-houses

Among the features of the Iranian plateau's climate are its cold winters and hot summers [14, 50–52]. They make it possible to produce ice in winter and necessitate its use in summer. Ice-houses have been found in the outskirts of the desert, where there are at least 10 to 20 days of frost during the winter [42]. However, at some altitudes and in the foothills, these structures

Fig. 5.41 Interior view of the Arg-e Bam ice-house's reservoir, after transformation into a conference hall in 1993 and before an earthquake in 2003, Bam City, Kerman, Iran.

(although simpler than those in desert and semi-desert areas) were built for the following reasons [14]:

- There was no snow in the surrounding mountains.
- Mountains were impassable, and transporting ice from glaciers was difficult and costly.
- Farmers owned certain glaciers and prevented people from taking their ice.
- Ice traders also owned some glaciers and wanted to offer their ice to other regions.
- Ice and snow collected from glaciers was expensive.

For example, in the cities of Kangavar and Hamadan, despite the existence of glaciers, man-made ice-houses were constructed [53], consisting only of a main pit and a roof. They had no ice-making pond(s) and shadowing wall(s) because workers directly filled the main pit with snow and then vigorously beat the snow to compact it into a dense mass. The fierce cold of these areas facilitated transition of the mass into ice [54]. It should be noted that the number of frost days in the city of Hamadan was high, so with the extreme cold, there was no need for shadowing wall(s).

At higher altitudes in southern parts of Khorasan, a number of cisterns were filled with snow during winter [55] and served a storage role somewhat similar to that of ice-houses. Approaching the central desert areas of Iran from the north, despite these areas experiencing fewer frost days, a large number of ice-houses were found that could produce ice by means of ice-making pond(s) and with no shadowing wall(s). Such structures, observed in cities such as Semnan and Damghan as well as Shirvan and Bojnurd [36, 56], represent an intermediate stage between incomplete (usually providing storage only) and complete (full feature) ice-houses. Finally, in areas such as Yazd, Kerman, Gonabad, and Sirjan, with their warmer winters and locations on the outskirts of the desert, complete ice-houses started to appear, with all the typical features: the main pit, shadowing wall(s), ice-making pond(s), and subcomponents [36, 57, 58]. In these areas, the shadowing walls were essential to ice making, because the temperature difference between sunny and shady places was high.

In hot and humid areas of Iran, such as regions next to the Caspian Sea, Persian Gulf, and Oman Sea, either there is no frost or the number of frost days is low, so the production of ice naturally in winter is impossible; consequently, there are no ice-houses. Therefore, in these areas, a large number of cisterns of different shapes and capacities were traditionally used to meet communities' cold water demands during the hot/warm months. It should be noted that, in certain regions, in addition to the above-mentioned reason, other factors, including water salinity, few frost days, scant population, and the financial status of residents, contributed to a lack of ice-houses. However, rich people in these regions usually migrated to the mountain areas in summer to enjoy the weather and natural ice. In some cases, compressed snow or ice was also transported to these areas for their use.

The construction of ice-houses, in addition to being directly affected by climate, might also be influenced by the architectural style of an area. Thus, the ice-houses located on the outskirts of the desert followed the architectural style of the area, which differed from that of mountainous regions. For example, the architectural style of the Jaban ice-house situated in Jaban village in the central district of Damavand county, Tehran, resembled that of its residential houses, and the construction also followed local traditions that combined beaten mud and frames made of black poplar wood [42]. Note that the architectural style itself was influenced by and harmonized with local climate conditions.

5.8 Locations of ice-houses in cities

Ice-houses were usually built on the outskirts of cities and villages, first, because these structures covered a large area, and second, because the relatively low price of ice made it economically sensible to build them in the suburbs where land was cheaper. In some areas, cities' defensive walls doubled the shadowing walls of ice-houses. Many ice-houses were constructed alongside traditional economic routes such as the "Silk Road," "Spice Road," and the "Shahi Road," to take advantage of this proximity by selling ice to travelers. Sabzevar ice-houses and Jaban ice-house were examples of this type. For instance, the Jaban ice-house was built to meet the ice needs of people from Mazandaran Province (especially the cities of Qaemshahr, Amol, Babol, Sari) who migrated to cooler and more pleasant areas in summertime [42].

The following factors also affected the location of ice-houses [18]:

- Ice-houses had to be far enough away from cemeteries to keep the water used to make ice free from contamination.
- The majority of ice-houses were built near water resources (such as Qanats) to ensure easy access to water.
- Typically, ice-houses were built outside city walls to keep their product far from heavy traffic and health problems.
- To take advantage of chilly winds blowing from the north, trap their cold air above ice-making ponds, and thus accelerate ice production, ice-houses were built in open spaces away from other structures.
- The sites for construction had to have a solid and firmly compressed soil that could withstand ice-houses' heavy mass.
- The depth of groundwater under a site had to be greater than the depth of the ice pit/reservoir to prevent groundwater seeping into the pit/reservoir and affecting stored ice.

5.9 Construction of an ice-house

Once the site and building orientation had been carefully chosen, ice-house construction on the outskirts of the desert involved a series of distinct stages. First, the architect directed building of the ice reservoir and its dome, then the shadowing wall(s), and finally, the ice-making pond(s). These stages are briefly described in the following subsection.

5.9.1 Construction of the ice-house reservoir and its dome

Constructing the ice reservoir and its dome involved the following stages:

1. *Implementation of the dome plan:* An architect first drew a plan of the ice reservoir on the ground, including the dimensions of the main pit and the ambulatory (walkway) around it, and the thickness of the supporting pier and the dome walls. Typically, the dimensions of the ambulatory features and the wall pier were, respectively, two-and-a-half and seven-to-nine mud-bricks. Note that a mud-brick was generally square in shape and had a length of about 20 cm [59] (Fig. 5.42).

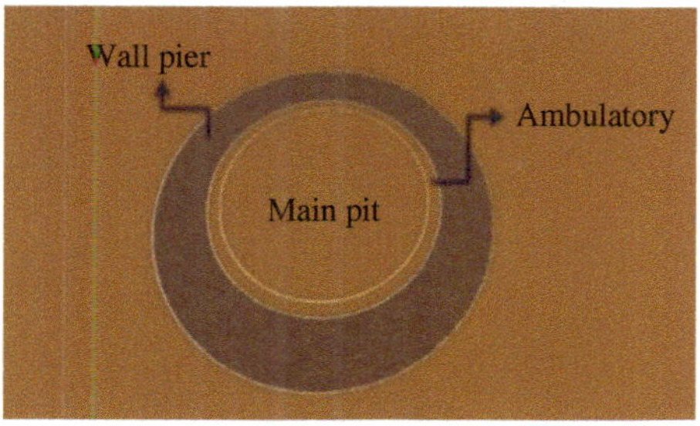

Fig. 5.42 Schematic drawing of a dome plan [18].

2. *Construction of the dome's foundation:* At the architect's discretion, if the structure needed a foundation, a hole was dug by adding 30 cm to the width of the dome's wall, around the outside perimeter at a height of 0.6 to 1 m. It was then filled with lime concrete (Fig. 5.43). The ratio of materials used in the lime concrete was 40% lime, 40% clay, to 20% sand and rubble [60].
3. *Digging the main pit:* In the next stage, the main pit was dug to a depth of about 1 m. Thereafter, laborers made mud-bricks using the excavated soil. They then used these mud-bricks to build the domed roof over the ice pit/reservoir [59] (Fig. 5.44).
4. *Construction of the dome body:* In accordance with the importance of the ice-house (e.g., the political and social position of its owner and its location in a city or village), the opinion of its owner, and the available facilities, in these ice-houses, bricks were used instead of other materials to build the dome's wall up to a height of 1 to 1.2 m. Then, mud-bricks were laid to complete the rest of dome. After finishing the construction of the ice-house's pier, the builders left two to three weeks for the structure settle, an important step because of its weight. Thereafter, the architect directed building of the ice-house's dome. Its outer surface was often in the form of a staircase, or "stepped" [59] (Fig. 5.45).

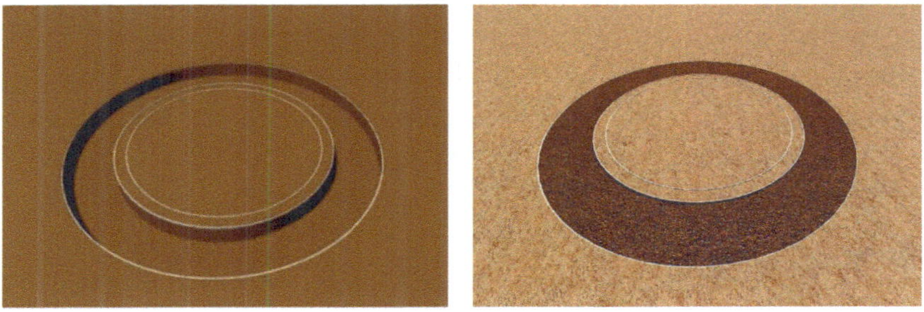

(A) (B)

Fig. 5.43 Schematic drawing of (A) digging the foundation of an ice-house's dome and (B) filling the foundation trench with lime concrete [18].

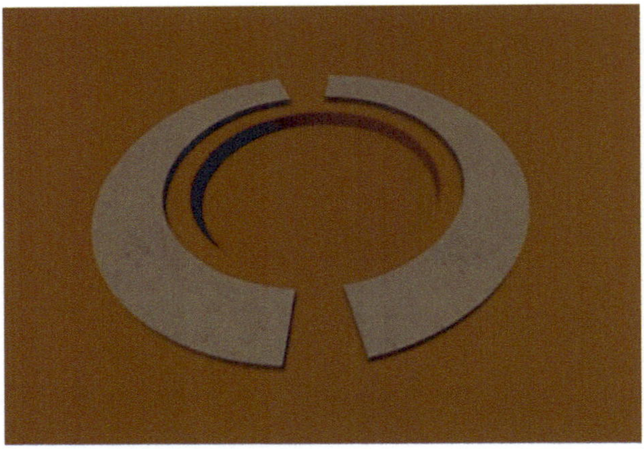

Fig. 5.44 Schematic drawing of digging the main pit [18].

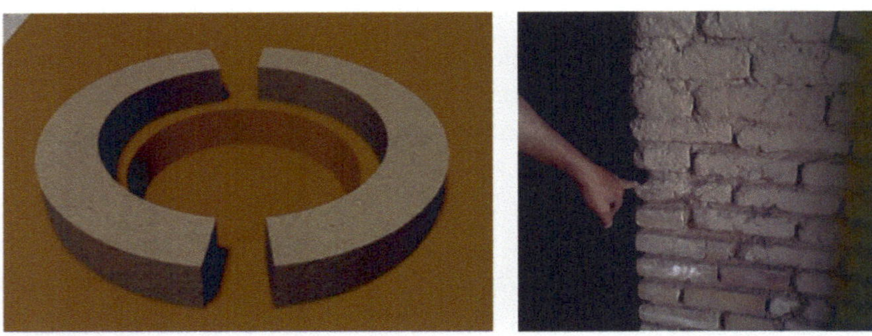

(A) (B)

Fig. 5.45 (A) Schematic drawing of the construction of dome wall and (B) picture of the brick wall of Bafran ice-house [18].

5. *Construction of dome with an outer surface in the form of a staircase:* To build a stepped dome, the builders/architects stood facing the outer wall of the circular base and carefully arranged bricks or mud-bricks, several bricks deep according to the desired wall thickness, in a particular engineered style around the whole circle. After setting out one or two courses[n] of brick or mud-brick directly atop one another, they positioned the next ones a few centimeters in from the previous perimeter (about 4 to 7 cm) and continuing this process for 5 to 7 courses of bricks/mud-bricks (about 35 cm). Thereafter, one brick was taken out to

[n]Course is an architectural term meaning a layer of the same unit running horizontally in a wall. In addition, it can be defined as a continuous row of any masonry unit such as bricks, mud-bricks, stones, and so on [61].

reduce the thickness of the dome's wall. This technique did not change the layout of the mud-bricks as seen from the dome's interior, but created a stepped exterior. The height of the steps and the number of brick/mud-brick courses depended on the architect's style and skill. If the architect wanted more courses of brick/mud-brick, he had to reduce the number of steps, and vice versa. Bricklaying continued until the thickness of the dome's wall reached one-and-a-half mud-bricks. The upper part of the dome was finished with 2 to 1.5 mud-bricks [62] (Fig. 5.46). It should be noted that all stages of the dome construction were based on the architect's experience, skill, and chosen style, and did not follow any particular technique. For example, the depth and height of the steps depended on (1) the number of courses of bricks/mud-bricks laid and (2) the number of bricks/mud-bricks removed (Fig. 5.47).

6. *Construction of the pit walls:* Different methods and materials were utilized to build the pit walls, which were constructed of solid packed earth, brick/mud-brick, or rubble, and usually covered with Sarooj or less often with Kahgel [63, 64]. For example, the ice pit walls of some ice-houses in Kerman were made of brick, and the pit walls of an ice-house located in the city of Abarkouh were built of rubble. Fig. 5.48 indicates two examples of the materials used in constructing ice pit walls: (A) solid packed earth and (B) rubble.
7. *Digging a well to drain water:* After finishing the construction of the ice pit, a well with dimensions of 15 to 20 cm in diameter and 1 to 1.5 m deep was dug at the bottom of the pit to drain away water from melting ice (Fig. 5.49), so the ice would remain frozen longer.

(A) (B)

Fig. 5.46 Schematic (A) of digging the pit and constructing a stepped dome and (B) view of the outer surface of the dome under construction [18].

Fig. 5.47 (A) Schematic drawing of the layout of dome mud-bricks, (B) Zarisef ice-house's stepped dome at a depth of half mud-brick (before repair), and (C) Kabutarkhan ice-house with the dome in the stepped form at a depth of one mud-brick [18].

Fig. 5.48 View of the ice pit walls of (A) Firuzabad ice-house in Damghan City, Semnan, made of packed earth, and (B) an ice-house in Abarkouh City, Yazd, Iran, made of rubble [44].

Fig. 5.49 Schematic drawing of a well placed at the bottom of an ice pit to drain off meltwater [18].

8. *Entrances of ice-house:* Most of ice-house entrances were constructed in the form of barrel vaults, usually using pitched (or barrel) bricklaying techniques (Fig. 5.50A) by which masonry units were laid in a series of rings side by side so that the arch grew longitudinally from one end to the other [65–67]. Occasionally, Roman bricklaying (Fig. 5.50B) methods were adopted, with bricks being laid in horizontal courses with their beds radiating from the center to make the arch [65–67]. Both pointed arches and elliptical arches can be found. Fig. 5.51 illustrates two examples of this type of vault to show the two different methods of bricklaying used for the entrances of ice-houses.

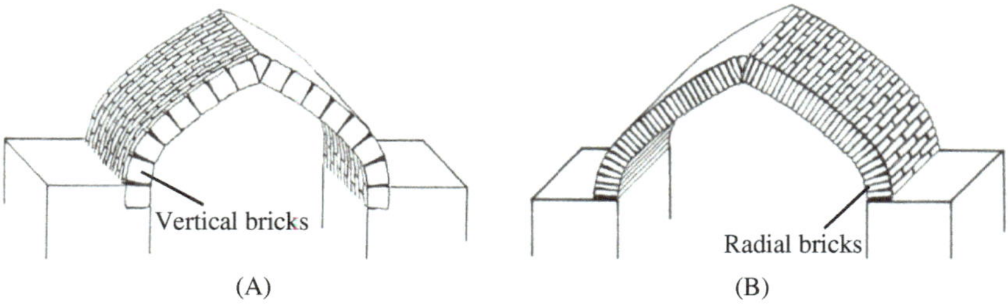

Fig. 5.50 Schematic illustrations of vaults created with (A) barrel bricklaying and (B) Roman bricklaying [65, 66].

(A) (B)

Fig. 5.51 Picture of the entrance of (A) Langar ice-house with barrel bricklaying and (B) Haj Rashid ice-house with Roman bricklaying [18].

9. *Covering the dome:* When the structure of a dome was completed, a layer of Kahgel made of half-chopped straw was used to protectively coat and insulate its outer surface.

– *Differences in construction methods for the main pit and the dome of ice-houses in central, eastern, and southeastern parts of Iran.*
In cities such as Naein, ice pit excavation and dome construction were carried out at the same time, meaning that the soil obtained from digging out the pit was employed to make mud-brick for simultaneously constructing the dome. In many cases, workers even prepared the mortar needed to build the dome while they were working on the pit. This method was mostly implemented in southeastern areas of Iran such as the city of Kerman [59]. However, according to some modern architects, the more common practice was to leave a "settling" period after the construction of the dome, before laborers started digging the pit. This method was mostly used in central Iran, for example, the city of Kashan [62]. The differences in construction practices can probably be attributed to differences in soil makeup and stability. Table 5.6 schematically shows how the construction of the main pit and its dome differed between the cities of Kerman and Kashan.

5.9.2 Construction of shadowing walls

The construction of shadowing walls, like that of other ice-house components, was performed in several steps, described in the following subsection.

1. *Implementation of the shadowing wall's plan and its foundation:* After construction of the ice reservoir and its dome, the builders started to create the shadowing wall, the second most important component of the ice-house. Construction materials were chosen based on the importance of

Table 5.6 Differences in ice pit and dome building in the cities of Kerman and Kashan [18].

Kashan City	Kerman City
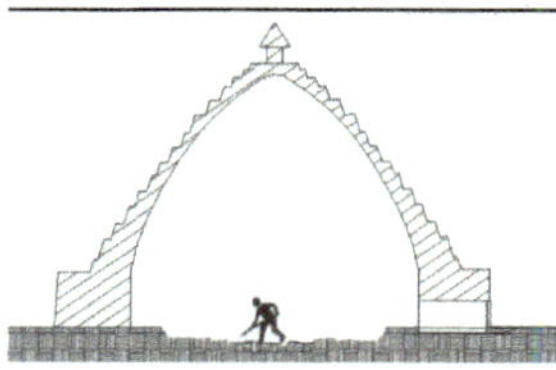	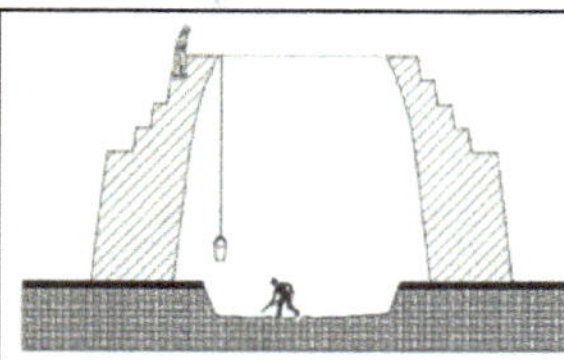
Schematic cross-section of dome construction in both cities	
 Schematic cross-section of stepped dome construction	 Schematic drawing of simultaneously construction of pit and dome
 Schematic cross-section of digging the pit after dome construction	 Schematic cross-section of dome construction at the same time as digging the pit

the ice-house and the architect's preferences and style. In most cities, walls were usually made mainly of stratum, although in some cases, mud-brick was used. The first step was to determine the optimum wall orientation so that it would face the cold winter winds. The width of these walls at the bottom varied within the range of 1.5–3.5 m (Table 5.1), and like garden walls, they generally lacked a foundation. However, in some rare cases, they had a foundation made of lime concrete. After the architect had marked out the footprint for the wall on the ground, if a foundation was deemed necessary, workers dug a trench as wide as the intended wall plus an extra 30 cm on either side, to a depth of 0.6 to 1.2 m. The trench was then filled with lime concrete (Fig. 5.52) [18].

2. *Construction of the shadowing wall's body:* In some ice-houses, rubble or several courses of mud-brick were used to create the wall base.

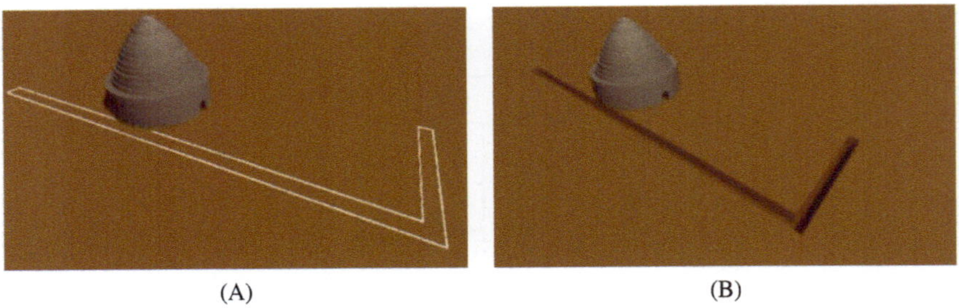

Fig. 5.52 Schematic drawing of (A) a shadowing wall's footprint marked on the ground, and (B) the wall's foundation trench [18].

In addition to these materials, some builders employed saltwater instead of fresh water when making the mud-brick or stratum, to diminish water penetration of the structure. This approach led to stronger walls and better moisture resistance. For example, the Bafran ice-house's shadowing walls in Bafran village, the city of Naein, were made using this method, as shown in Fig. 5.53 [64]. The height of the shadowing walls was generally between 7 and 15 m, and their thicknesses decreased from bottom to top, to add stability, reduce weight, and use less material. On the wall's surface, a number of holes about 10 to 15 cm in diameter were created approximately every 2 m to support scaffolding that would be installed for wall repair. Fig. 5.54 schematically shows the construction of a shadowing wall's body.

Fig. 5.53 Picture of Bafran ice-house's shadowing wall's base, with mud-bricks made with saltwater [18].

(A) (B)

Fig. 5.54 Schematic drawing of (A) the construction of the shadowing wall's base and (B) the wall completed with a conical cross-section [18].

3. *Constructing openings in the shadowing wall:* For ease of movement through the shadowing wall, builders created several openings in it. The arched roofs of these openings were in the form of barrel vaults made using either pitched or Roman bricklaying techniques (Fig. 5.55).
4. *Construction of the stratum shadowing wall:* To build the shadowing wall of stratum, first, a layer called a "Lad" in Iranian architecture was made to a height of about 40 to 80 cm, and then, one or two courses of mud-brick were laid on the surface of the Lad to flatten it; thereafter, the next Lad was constructed. This work was continued until the desired height was reached (Fig. 5.56).
5. *Construction of shadowing wall buttresses:* In most ice-houses, once construction of the main and side shadowing walls was complete, a number of

(A) (B)

Fig. 5.55 Picture of a shadowing wall's opening in (A) Moayedi ice-house with a vault in the Roman bricklaying form and (B) Langar ice-house with the pitched bricklaying form [18].

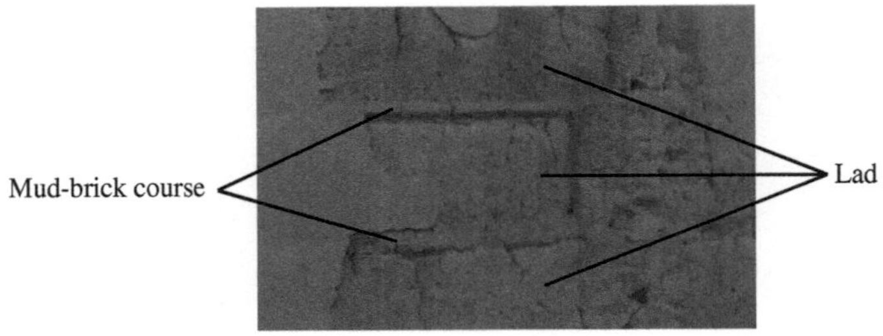

Fig. 5.56 Picture of Lad layers with one course of mud-brick between them in the stratum shadowing wall of Kabutarkhan ice-house, in Kabutarkhan village, Rafsanjan City, Kerman, Iran [18].

supportive buttresses were added along the walls. However, in ice-houses for which particularly high shadowing wall(s) were planned, wall building was periodically interrupted so that buttresses could be added as soon as the walls had reached a certain height. Once these stabilizing features were in place, the walls could safely continue to grow taller [68] (Fig. 5.57).

6. *Covering shadowing wall surfaces:* After the shadowing wall(s) and buttresses were built, their surfaces were covered with Kahgel made of chopped straw.

5.9.3 Construction of shallow ice-making ponds

Finally, after completion of the two main components of the structure, the builders proceeded to create the ice-making pond(s). To do so, a shallow rectangular pit 30 to 60 cm deep was dug beside and parallel to the length

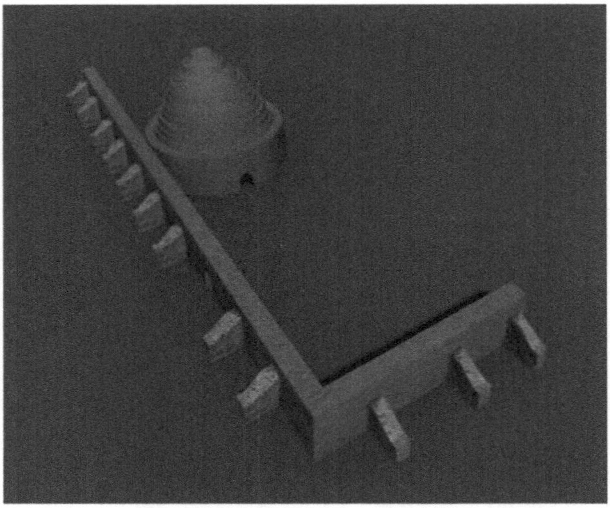

Fig. 5.57 Schematic drawing of shadowing wall buttresses [18].

of the main shadowing wall. Generally, no materials were used to build these pond(s) [62]. However, in some cases, for reasons already mentioned, the bottoms and walls of ponds were lined with Sarooj to reduce water seepage and keep the water relatively free of soil particles. In rare cases, the pond walls were reinforced with bricks, such as in Meybod ice-house, to prevent loose soil from friable edges pouring into the pond and contaminating the ice. It also prevented the entrance of saltwater when it was present.

5.10 Ownership of ice-houses and socioeconomic relationships

One of the most famous quotes from Karim Khan Zand (1705–1779), the founder of Iran's Zand Dynasty, is as follows [14, 69]: "*Two things bring joy to me: one is ice water in winter and another is the cooked rice for the night of Nowruz [the Iranian New Year], because only in winter is ice available to everyone.*" This description and many other examples allude to the fact that the consumption of ice, even in the hot/warm months, was generally restricted to the aristocracy. Ice was not considered to be an essential commodity and so ordinary people could live without it. Ice buyers, like ice-house owners, were mostly rich people, generally living in cities. Hence, the majority of ice-houses were located in cities and their suburbs, and the minority in rural areas because their owners and masters themselves lived in rural villages.

Whenever a community's economic worth rose, its number of ice-houses increased subsequently. For instance, when social and economic conditions became favorable for the construction of ice-houses, rich people, local governors, and rulers proceeded to build them, and a wider range of society started to enjoy their product. In contrast, during economic downturns, the number of ice-houses in service dropped, and ice consumption was again limited to the aristocracy. Al-Isfahani, in his book entitled "Nesf-e Jahan Fi Tarif-e Esfahan," stated that [70] "*There are about forty large ice-houses in the city [Isfahan] and around it, and in the boom time of the city, the number of ice-houses was more than this, but they are now left workless and ruined due to the lack of need for them.*" It should be noted that, in addition to social and economic conditions, the majority of Iran's aristocratic and prosperous families migrated from the cities to cooler, more pleasant areas in the summer months, which also affected the number of working ice-houses. Despite the privilege of the wealthy and a source of profit of their owners, ice-house ownership nevertheless sometimes had a role in public charity.

Iran has a long history of charitable giving of various sorts [71]. Endowments (called "Waqf" in Persian) refer to donations (e.g., a structure, a piece of land or property, other assets [72]) made by those more fortunate for the good of those less well off. There were long-standing religious, cultural, and civic traditions for doing so. For instance, the construction and supply of cisterns and public drinking water sources (e.g., Qanats, Sangabs) in cities and villages were undertaken by some rich people as direct acts of charity for all to enjoy and benefit from. That is, the building and supply themselves were the charitable endowment. The case for many ice-houses was somewhat different in that the monies obtained from selling their product constituted the charitable donation, not the ice itself. In practice, owners allocated to charity all or a portion of the proceeds they received from the sale of ice.

Sound economic reasons supported this generosity: once an ice-house existed, because the water needed for ice-making in winter was generally free and labor was cheap, ice sales represented almost pure profit for the owners [73]. Since both ice customers and ice-house owners, in general, had substantial financial resources, they were free to indulge their charitable impulses and were encouraged to do so by their culture and religion. Donations improved the donor's status both as a member of society and as a member of the faith, much as medieval European aristocrats were encouraged to help the poor as an act of faith and to confirm their lofty status and piety. Overall, only rarely were the products of ice-houses used directly for public welfare. One exception was when ice was donated for specific religious occasions.

In years when the official mourning months (Muharram[o] and Safar[p]) for Shia Muslims occurred in the warm season, ice-houses devoted to charity had to provide the ice needed for the mourning ceremonies. In other years, proceeds from the sales of ice after any expenses associated with the ice-houses such as repair and maintenance, laborers' salary, and so on, were designated for mourning and commemoration of the martyrs of Karbala [31]. For example, in the text of the deed of endowment of an old ice-house in the city of Khansar, Isfahan, it is written that [74] "*Proceeds obtained from the sales of ice after expenses of the ice-house and drinking fountain*

[o]The first month of the Islamic calendar.

[p]The second month of the Islamic calendar.

Fig. 5.58 View of an old Saqqa-khaneh in Naqsh-e Jahan Square, Isfahan City, Isfahan, Iran [45].

*[called a 'Saqqa-khaneh[q]' in Persian (*Fig. 5.58*)] are spent for Imam Hussein's[r] mourning ceremonies any way the trustee knows is proper, provided that ice is not taken free of charge, which is forbidden.*" In addition, in the deed of endowment of an ice-house in Zafaraniyeh village, the city of Sabzevar, Razavi Khorasan, it is written that "*There is an ice-house in the place that manufactures ice, located in the village of Zafaraniyeh, where the produced ice must be donated to the travelers and pilgrims who go through the village.*" (Fig. 5.59).

The salaries of laborers who worked in the private or endowed ice-houses were paid from ice sales. Private ice-houses were more common, and usually had an employer and a number of permanent and seasonal/temporary workers. Many owners engaged in producing ice by traditional means did so after inheriting their ice-houses and the associated practices through a long line of ancestors. For example, most of the ice-houses in the city of Qom were private [77]. Their owners constructed, repaired, and maintained them at their own expense. The water required for ice production was taken either from Qanats or from streams whose water was not needed for crop

[q]A Saqqa-khaneh was situated in all important locations of cities and villages (alongside mosques, shrines, bazaars, even at the intersections of alleys) to provide cold water for thirsty people, and mainly made possible by endowments with a religious and functional role [75, 76].

[r]Hussein (or Husayn) ibn Ali (626–680 CE) was the third *Imam* of Shia Islam and the grandson of the Islamic prophet Muhammad.

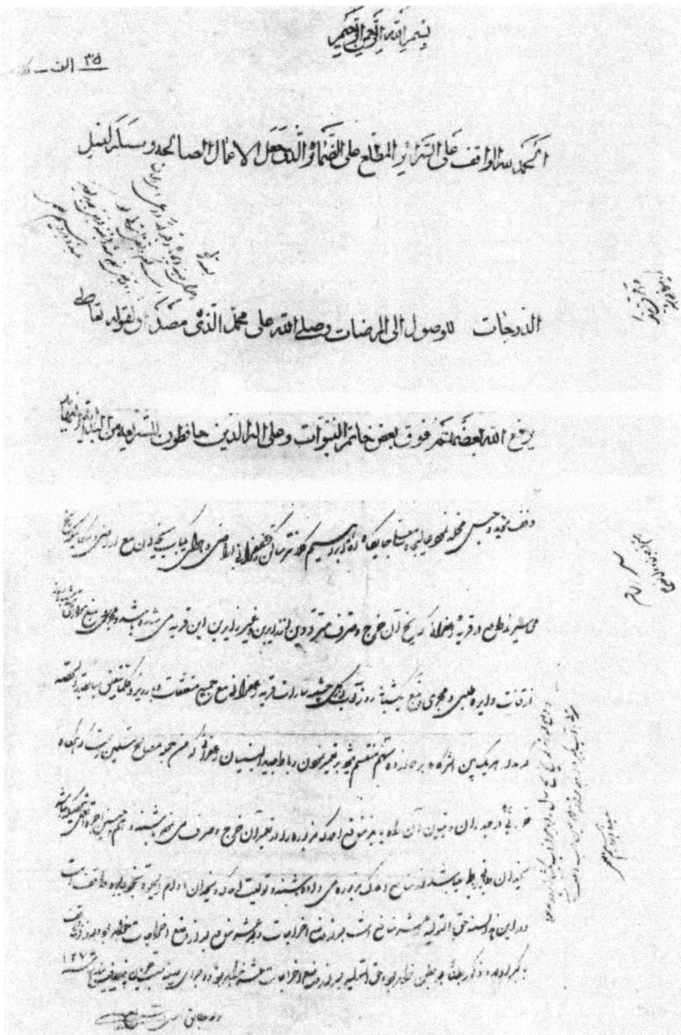

Fig. 5.59 Deed of endowment of an old ice-house in Zafaraniyeh village, Sabzevar City, Razavi Khorasan, Iran.

irrigation in winter, called "Khordeh Aub[s]" in Persian [14]. This matter will be explained further in Section 5.12.

In some cases, such as that of the Haj Mohammad Khan ice-house in Shemiran, Tehran City, Tehran, ice production was carried out by workers of the ice-house owner, but the structure itself might be rented out for

[s]"Khordeh" means "a small part," and "Aub" means "water" in Persian. In the past, it was also the unit of water distribution.

Fig. 5.60 Deed of endowment of Zarisef ice-house, engraved in stone, located in Kerman City, Kerman, Iran.

distribution and sales of ice; the renters were again usually the owner's workers [78]. The majority of traditional Iranian ice-houses belonged to traders, who utilized them as commercial enterprises up to the late 1970s and delivered ice to the people and market directly. However, some were built by benefactors to provide free ice to people for humanitarian purposes, especially in villages and small towns [42]. Fig. 5.60 shows the deed of endowment for Zarisef ice-house, engraved in stone, located in Kerman City, Kerman. This stone deed of endowment, unlike some others, has been installed in the ice-house.

5.11 Ice sales and ice-house customers

Once the warm season began, the entrance doors of ice-house reservoirs, sealed at the completion of ice collection, were reopened. For example, Qeytarieh ice-house in Tehran City, Tehran, was opened on approximately

June 1, but a thick curtain always hung in the opening to the reservoir to protect its interior and ice from the warmer ambient air. Early every morning, usually at around 5 am, the ice-house operator would remove the layer of straw insulation from the ice mass inside the reservoir, then use a hatchet-like tool to break off enough ice for one day's sales. The ice chunks were kept near the entrance door or waiting room, and protected under straw or sacking. The ice-house operator and/or ice seller was usually a worker employed by the ice-house owner, but sometimes the job was leased out to someone else [31]. Ice sales were generally conducted in a waiting room located next to the south door and usually involved weighing with large scales [30]. Similar to market practices worldwide before standardization was enforced, ice-houses had an inaccurate scale tool to weigh the ice when selling—usually to the seller's advantage. According to historical evidence, the unit of measurement in many ice-houses was the capacity of a particular sack (called the "djowall" or "juval"), which was woven from goat hair or wool [42]. Naturally, the ice-house owner or ice sellers supplied this sack and specified its dimensions.

The first morning customers of ice-houses were usually itinerant ice-cream sellers, who worked around ice-houses [31]. Itinerant ice sellers were the second group of customers; they supplied the daily ice needs of markets and businesses such as ice-cream shops, grocery stores, dairies, and butcher shops. These ice sellers carried their ice with the help of either working animals (for instance, donkeys, mules, horses) or carts. There was also another group of ice sellers who sold the ice for domestic use. They stationed themselves in their customers' neighborhoods or at intersections to catch passing trade [31].

- *Ice transportation*: Ice was transported from ice-houses to different cities and villages in djowalls. For this purpose, ice traders or sellers filled the djowalls with ice, packed the sacks together, and thereafter, loaded them onto working animals (mostly donkeys). Moisture from ice melting inside the goat hair or wool sacks first soaked them, and then evaporated from their outer surfaces, thereby keeping the ice cool for at least a few hours during transport [42].
- *Ice use*: Ice was mainly used to cool water and other beverages throughout the hot/warm months of the year, especially for national and/or religious ceremonies, and even weddings. It also cooled the water in stone troughs (or "Sangabs" in Persian) and Saqqa-khanehes provided throughout cities for public consumption [42], and was utilized to make ice cream and Faloodeh (or Paloodeh), which is a traditional Iranian

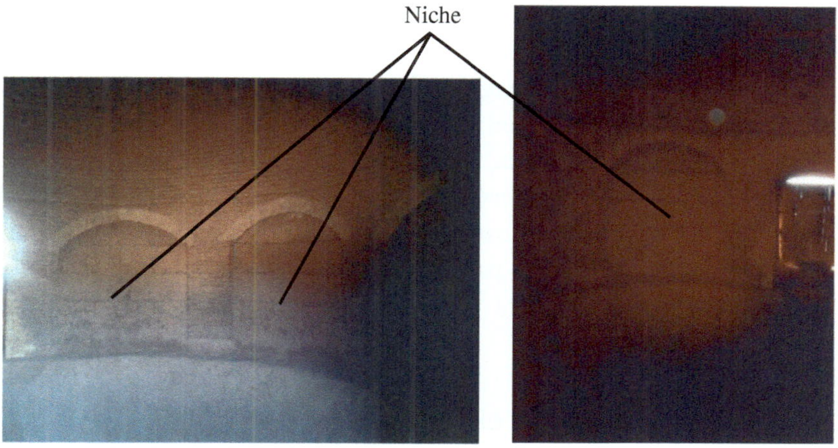

Fig. 5.61 View of niches located in the interior space of the ice reservoir of Bafran ice-house, in Bafran village, Naein City, Isfahan, Iran, for the preservation of perishable foods [21, 22]. In these two figures, the niches are currently filled with mud.

cold dessert similar to a sorbet [79]. Pomegranate seeds were occasionally poured into the freezing layers to make the ice more beautiful and pleasant when used [80]. People sometimes also preserved their food, fruits, and vegetables using ice within the home. In rare cases, inhabitants around an ice-house (mainly in rural areas) gave their perishables to the ice-house operator to store inside the ice reservoir and thus prevent corruption by summer heat. Features found in some ice-houses attest to this practice. For example, niches built into the inner dome wall of Bafran ice-house, in Bafran village, the city of Naein, Isfahan, were fitted out with shelves for storing food (Fig. 5.61).

5.12 Water sources of ice-houses

The availability of clean and fresh water for producing ice was an important issue in choosing locations for ice-house construction in each city/village. For this reason, not only did the ice-house have to be built in open space around a city/village, but the construction site had to be near a reliable water source, and also upstream of farms and gardens, so that the water was still fresh and clean. The water needed to produce ice was mainly supplied from local Qanats, springs, or dams. Additionally, it was sometimes diverted from streams whose water was not needed for agriculture in winter [42]. Qanats, however, were the major and preferred water source because their water had lower levels of salinity and contamination than other sources did. Moreover, a Qanat's water had traversed a long distance underground before reaching the ice-making pond(s), so its temperature was closer to freezing point, which facilitated ice production [30]. As an example, one of Cheshmeh Ali's waterways was used to feed an ice-house near the city of Damghan, Semnan. This waterway, called "Khordeh Aub," supplied clean water for mosques and public bathrooms throughout the year. In winter, the ice-house owners could use a part of this Khordeh Aub for the production of ice, although they were not the legal owners of the water and had no legal right to sell it [14].

5.13 Study of ice-house examples

In the following subsections, four traditional ice-houses, exemplifying the typical types, are described in detail.

5.13.1 Meybod ice-house in Meybod City, Yazd, Iran

The city of Meybod is one of the very old and rare Iranian cities lying on and thus part of the ancient caravan routes. Over the ages, these routes were traversed by riders of the Achaemenid (550–330 BCE), Parthian (247 BCE–224 CE), Sasanian (224–651 CE) eras, and in later periods, by caravans crossing the central deserts of Iran. At cities like Meybod, the travelers could rest, safe from the dangers and stresses of the desert. The ancient route passing through Meybod City was the Ray-Kerman main way. Alongside this route in the city, there was a complex that included a caravanserai, a "Chapar Khaneh,[t]" a cistern, and an ice-house. Meybod ice-house was a huge structure made of mud-brick and mud, and located in front of the caravanserai but separated from the complex by a street. This ice-house has a long history of quenching the thirst of passengers coming in from the desert [17]. Fig. 5.62 schematically shows the plan and cross-section of Meybod ice-house.

The ice-making area of Meybod ice-house was a shallow pond about 50 cm deep and with an approximate area of 800 m^2. The ice-house's shadowing walls were 8 m tall and 2 m thick at the base. The length of the south wall was 42 m, and the east and west walls were both 20 m long (the east wall has been completely destroyed). A primary entrance door built into

[t] This Persian term means the "house of a courier," referring to the postal service used during the Achaemenid era [81–83].

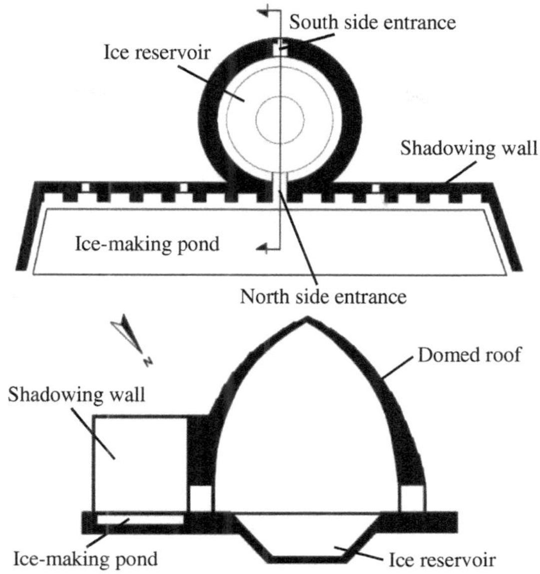

Fig. 5.62 Schematic illustration of plan and cross-section of Meybod ice-house located in Meybod City, Yazd, Iran [84].

the middle of the main-south shadowing wall had a width and height of 2.2 m and 2 m, respectively. This door was used to connect the ice-making area to the reservoir, and allowed for storing ice in winter and removing it during summer (Fig. 5.62). Another entrance door constructed opposite the primary entrance door seems to have played only a minor role.

The reservoir of Meybod ice-house was dug in the solid clay ground in the form of an inverted and truncated cone (Fig. 5.63). Its diameter and depth were 13 m and 6 m, respectively. There is a 1.5-m-wide walkway around the reservoir. The large conical roof of the ice-house's reservoir was built right on the solid soil, with the first mud-bricks of the dome put directly on that soil, thus avoiding the need to construct a foundation. The surface area of the circular base of this dome was 300 m^2, and the structure rose to a height of 15 m above the ground. The width of the dome's circular mud-brick wall was skillfully reduced from the bottom to top, so that it varies from 2.40 m at the base to 20 cm at the vertex (the size of one mud-brick). The number of mud-brick courses laid to build the dome is 250 [84]. Meybod ice-house remains one of the tallest buildings in the desert city of Meybod, and dates back to the Safavid period.

Fig. 5.63 View of the ice reservoir of Meybod ice-house in Meybod City, Yazd, Iran.

5.13.2 Arg-e Bam ice-house in Bam City, Kerman, Iran

Arg-e Bam ice-house is located in the northeast of Arg (the largest adobe building in the world), outside the fortified wall of this UNESCO World Heritage Site city but attached to it, and is one of the largest ice-houses in Iran. It had fallen into ruin over the centuries since its ancient beginnings, but was restored in the late twentieth century. Despite the collapse of a part of its domed roof during an earthquake in 2003, its beauty and grandeur are still impressive [85]. Fig. 5.64 shows the cross-sectional views of the ice

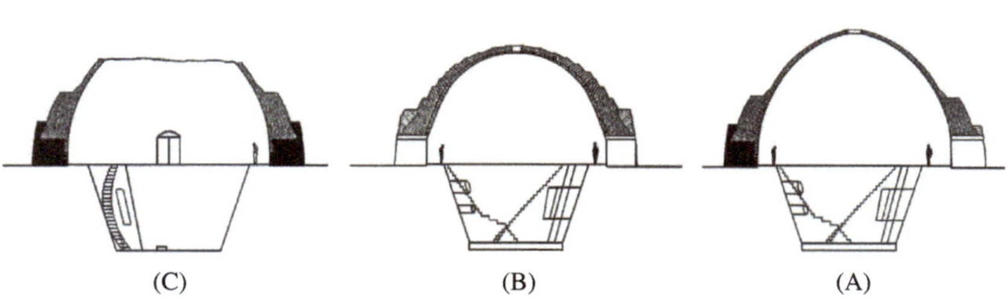

Fig. 5.64 Cross-section of ice reservoir and dome of Arg-e Bam ice-house located in Bam City, Kerman, Iran, showing (A) before repair, (B) after repair in 1993, and (C) after earthquake in 2003 [85].

reservoir and the dome of Arg-e Bam ice-house in the city of Bam (old part), Kerman, for three cases: before repair, after repair in 1993, and after the 2003 earthquake. According to registered documents, this ice-house was also built in the Safavid period. In keeping with the era's traditions, the construction materials were a combination of stratum, mud, and mud-brick, except for its reservoir, and it was coated over all with Kahgel, as are most of the buildings and walls of ancient Arg. The uniformity of style and material gives a particular harmony to this site, and helps the whole citadel blend in well with its desert environment.

The dome is 15.8 m in diameter, and it rises about 11 m above the ground. The diameter of the ice-house reservoir is 12.7 m, and its depth is 6 m. The wall on which the dome is built is 2.8 m wide, the thickness of the dome at the impost (or "Pakar" in Persian) is 2 m, and the structure reaches 30 cm at the vertex of the dome. The ice reservoir is made of brick, and a staircase built into its wall provided access to its contents. The operator's room, built between the shadowing wall and the ice reservoir, has the dimensions 2×2.6 m, and was made of mud-brick. It is covered by a cloistered arch,[u] as depicted in Fig. 5.65.

The length and height of the shadowing wall of this ice-house are, respectively, 65 m and 9 m, and it has a number of blind arcades running along its length to enhance its solidity, stability, and beauty. No traces of the ice-making pond remain, but according to the location of the water canal

Fig. 5.65 View of bricklaying of operator's room vault, Arg-e Bam ice-house, Bam City, Kerman, Iran.

[u]A cloistered arch (or vault) is composed of four triangular coves rising from a square plan in corbelled courses to an apex, and meeting in vertical diagonal planes. The axial sections resemble arcs, but are actually pseudo-vaults [86].

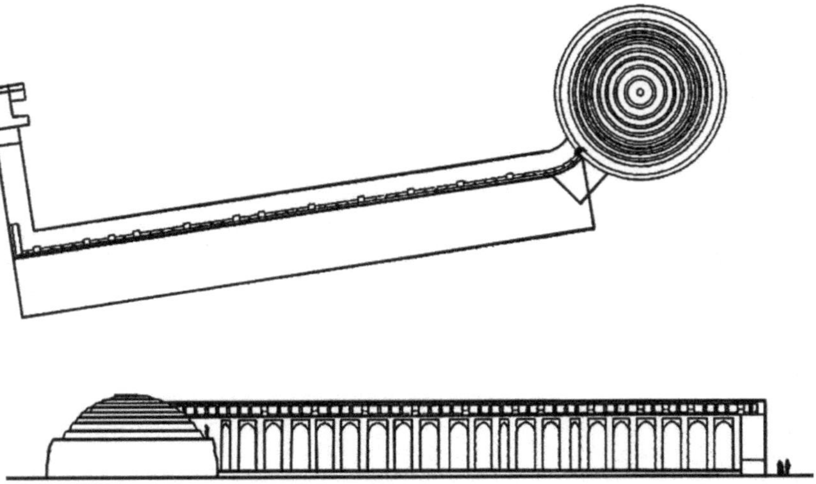

Fig. 5.66 Schematic illustration of plan and north-side view of the repaired Arg-e Bam ice-house in 1993 [85].

and the configuration of the shadowing wall, its approximate dimensions were 15 × 50 m, and its distance from the wall was about 1 m. Fig. 5.66 schematically indicates the repaired Arg-e Bam ice-house in 1993.

According to the available evidence, the repaired dome differs from the original one (Fig. 5.64). During the repair, the modern architect reduced the dome's height, subsequently increasing the number of stairs on the outer surface, and incidentally the weight of the dome. Although this ice-house was a functioning structure and its users were mostly local laborers, its initial architect paid little attention to exterior decoration. He did, however, decorate the interior space of the reservoir by creating simple and beautiful designs (such as rhombi) with different courses of mud-brick.

The important point about this ice-house is its operation: Was it possible to produce ice in Bam's climate in winter? Were there enough frost days to manufacture ice due to the capacity of the ice-making pond to fill the reservoir? Note that the elevation of the cities located on the outskirts of the Lut Desert (like Bam City) was about 800 m lower than that of the cities around the central desert of Iran; accordingly, these cities have both fewer frost days, and in summer, more rapidly reach the season's maximum ambient air temperature. A study of the climate of the region and the climate adaptation features of the building shows that it was possible to produce ice in winter, although the number of frost days totaled less than one month. In addition, the situation of sunlight in summer relative to the shadowing wall's position and also the thickness of the ice reservoir's walls indicates that, when a reservoir was filled with ice, it could be kept frozen until the end of June.

It should be noted that a seasonal life pattern prevailed for many in the city of Bam. That is, most residents migrated to warm and cold regions during the cold and hot months, respectively. In the hot months, most of those who could afford to do so migrated to villages around the Jebal Barez mountain chain, which was known to have pleasant weather. Only those whose job it was to irrigate the palms had to stay in the city and tolerate the hot weather. Therefore, the ice-house was generally efficient enough to produce ice for at least as long as most residents needed it. When winter production of ice was not possible some years, compressed snow and natural ice was transported from the Jebal Barez mountains to the city.

5.13.3 Barzan ice-house in Abarkouh City, Yazd, Iran

Abarkouh is one of the cities of Yazd Province that lies beside the Marvast Desert. This city has a desert climate: hot summers and cold winters. Frequent winter frosts meant that people could produce as much ice as they wanted. Barzan ice-house, built during the Qajar period [6], has a 50 cm wide walkway around its ice reservoir to allow workers to move around and access its stored ice, as shown in Fig. 5.67. The reservoir of this ice-house is 4.5 m deep. Barzan ice-house has four entrance doors, three of which were to transfer ice into the reservoir. These three entrances have a steep ramp that facilitated the movement of ice blocks into the reservoir. The architect added the fourth door for bringing out the ice, replacing the steep ramp with a staircase so that laborers would not slip and fall into the reservoir. A well at the bottom of the reservoir drained off and collected any meltwater from the ice reservoir. Fig. 5.67 schematically illustrates Barzan ice-house in the city of Abarkouh, Yazd.

With the coming of the warm/hot months, the door designated for ice removal was opened, and workers started chipping away at the ice mass stored in the reservoir with tools such as sledgehammers and hatchets. They then brought out the pieces of ice required through the staircase. In some ice-houses like Barzan, where the daily ice sales were high, a waiting room was built near the exit door (Barzan's waiting room, measuring $2.6 \times 2.8\,m^2$, was 4 m from the door, as indicated in Fig. 5.67), and the ice for a day was stored in it.

The extreme cold of winter nights in Abarkouh made an ideal situation for ice-making. The only requirements for a successful ice-house were a supply of fresh, clean water and a number of laborers. Today, three ice-making ponds can still be observed beside the ice reservoir and its domed roof (Fig. 5.68). Two of these ponds were fed from one Qanat and the third from another Qanat. On the south side of each pond, running east–west, a mud-brick shadowing wall was constructed, rising to a height of 4 m and extending from 30 m to 50 m long. The prevailing climate ensured that the water inside

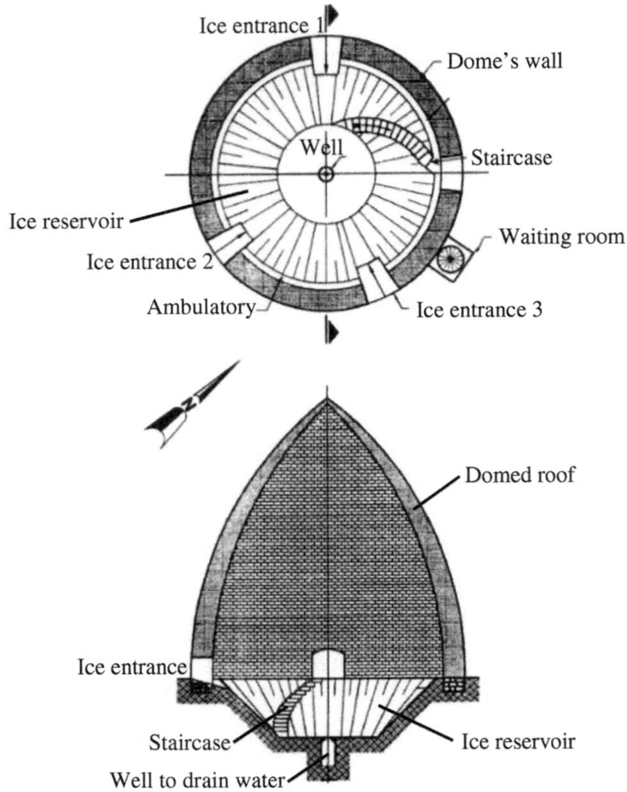

Fig. 5.67 Schematic illustration of plan and cross-section of Barzan ice-house, Abarkouh City, Yazd, Iran [27].

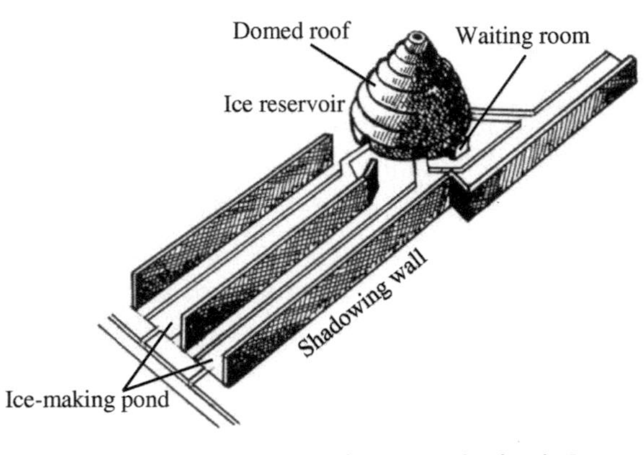

Fig. 5.68 Schematic illustration of Barzan ice-house in Abarkouh City, Yazd, Iran [20, 27].

the ponds started to freeze early in the night, and the depth of ice increased readily over time. When the reservoir was filled with ice, all doors were closed and not opened again until the next winter, apart from the one used for ice removal. This practice prevented warmer ambient air entering the reservoir.

5.13.4 Haji Agha Ali ice-house in Abbas Abad Haji village, Rafsanjan City, Kerman, Iran

Most of Kerman region, with its suffocatingly hot summers, high elevation, and near mountainous areas, is an appropriate place for the construction of ice-houses [42]. Haji Agha Ali ice-house located in Abbas Abad Haji village, the city of Rafsanjan, Kerman, is one of the most important and famous ice-houses in Kerman Province. It was built by Haji Agha Ali, who owned considerable land in the city of Rafsanjan and also in the villages near Qasem Abad and Esmaeil Abad [42]. Free ice was given to all farmers who worked on Haji Agha Ali's lands. This aid, which was included in the contract between the owner and the farmers, was also extended to daily workers. The additional ice produced was sold to other villagers.

As illustrated in Fig. 5.69, the ice reservoir is cylindrical in shape, with a depth of 4.5 m and a diameter of 9.5 m. The reservoir's floor and walls are made of brick, and it has a capacity of ~290 m^3. The height from the bottom of the reservoir to the vertex of its conical dome is about 16 m. A staircase clinging to the side of the reservoir wall allowed workers or the ice-house operator to reach the bottom of the pit as stocks diminished. As usual, the architect has constructed a small well at the bottom of the reservoir to drain

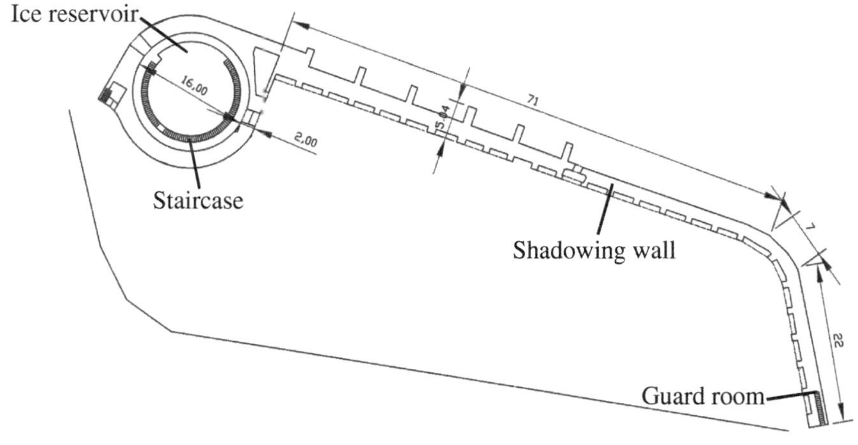

Fig. 5.69 Schematic illustration of plan of Haji Agha Ali ice-house, located in Abbas Abad Haji village, Rafsanjan City, Kerman, Iran [42]. All numbers in this figure are in meters.

away the water from melting ice. Less commonly, he has also created a small hole at the tip of the dome to allow warm air collected under the dome to escape into the surrounding environment during the warmer months when ice-house was used for ice sales. At its base, the dome is more than 2 m thick, creating stability and improving the cold retention of the chamber. The various features of this structure ensured that ice stored in the reservoir was well protected from warm air, and so remained intact for longer.

The length and height of the shadowing wall of Haji Agha Ali ice-house are 100 m and about 13.6 m, respectively. The wall is 2.4 m thick at its base and 22 cm at the top. The ice-making pond walls were made of brick to a depth of about 30 cm, but their floors were natural beaten earth. To facilitate ice-making, the ponds needed to be supplied with water of the lowest salinity, which was difficult to find in desert areas. Thus, the water needed for this ice-house was supplied from the best Qanats.

5.14 Conclusion

Designing efficient and enduring buildings that optimize energy consumption and provide thermal comfort for inhabitants [87], as well as harmonizing with their surroundings, needs insights into various factors such as local climate and geographical conditions (principal ones), plus material availability and local cultural and religious beliefs. Most traditional Iranian structures (for example, buildings, ice-houses, and cisterns) were designed with these considerations in mind, using indigenous knowledge and well-tested technologies and methods. Therefore, these structures, in addition to immediately benefiting the people they served, were commercially viable, environmentally sustainable, culturally appropriate, and all together valuable resources for their communities over hundreds of years. Hence, in view of the long history of proven worth for these structures and their technology, it makes sense to investigate whether they have anything to teach us about addressing today's energy crises and environmental challenges. Iran's long history of sustainable architecture certainly has lessons that can benefit modern buildings. A combination of traditional and modern technology is also an efficient, practical, and even cost-effective approach to meeting people's requirements and achieving a degree of sustainability.

References

[1] T. Fardpour, Energy-saving patterns of Iranian traditional architecture in various climates, Adv. Mat. Res. 689 (2013) 66–70.
[2] I. Yüksek, T. Esin, Analysis of traditional rural houses in Turkey in terms of energy efficiency, Int. J. Sustain. Energy 32 (6) (2013) 643–658.

[3] A. Ebrahimi, A. Shayegani, M. Mahmoudi Zarandi, Thermal performance of sustainable element in moayedi ice-house in Iran, Int. J. Archit. Heritage 13 (2019) 1–17.
[4] S. Asadi, M. Fakhari, M. Sendi, A study on the thermal behavior of traditional residential buildings: rasoulian house case study, J. Build. Eng. 7 (2016) 334–342.
[5] H. Shahamat, Formal sustainability in traditional architecture of Iran according to five principles of traditional architecture of Iran, J. Appl. Environ. Biol. Sci. 4 (1) (2014) 100–110.
[6] J. Eiraji, S.A. Namdar, Sustainable systems in Iranian traditional architecture, Procedia Eng. 21 (2011) 553–559.
[7] M. Nikfetrat, M.A. Tayyeb, Sustainability importance of vernacular (native) architecture with cultural approach, J. Soc. Issues & Humanit. 1 (5) (2013) 159–163.
[8] N. Valibeig, S. Nasekhian, N. Nazarieh, Analysis of effect of climatic conditions on the general form and components of traditional ice-houses in Iran, J. Geogr. Space, Islamic Azad Univ., Ahar Branch 61 (2018) 141–159 (in Farsi).
[9] B. Hosseini, A. Namazian, An overview of Iranian ice repositories, an example of traditional indigenous architecture, METU J. Fac. Archit. 29 (2) (2012) 223–234.
[10] G.B. Gholikandi, M. Sadrzadeh, S. Jamshidi, M. Ebrahimi, Water resource management in ancient Iran with emphasis on technological approaches: a cultural heritage, Water Sci. Technol. Water Supply 13 (3) (2013) 582–589.
[11] M.K. Moosazadeh, Y. Zandi, Comparative evaluation of Sarooj mortar mix design according to various sources and best mix design determination, J. Maremat-e Asar Baft-haye Tarikhi-Farhangi 9 (17) (2019) 63–74 (in Farsi).
[12] M.M. Masoumi, H. Banakar, B. Boroomand, Review of an ancient Persian lime mortar "Sarooj,", Malays. J. Civil Eng. 27 (1) (2015) 94–109.
[13] T. Tavousi, Urban Construction and Architecture in Hot and Arid Climate of Iran, Payam Publications, Tehran, Iran, 2002 (in Farsi).
[14] M.H. Papoli Yazdi, M. Labaf Khaniki, Ice-house and Artificial Ice Production, Second Regional Climate Change Conference, Iran National Meteorological Organization, Zanjan, Iran, Nov. 13–14, 1999 (in Farsi).
[15] M.A. Mokhlesi, The traditional ice-houses, the forgotten architectural masterpieces, in: Proceedings of the Second Congress on Architecture and Urban Planning History, vol. II, Cultural Heritage Organization of Iran, 1995, pp. 690–695 (in Farsi).
[16] N. Alizadeh Gohari, M. Latifi, Ice-House Architecture, a Masterpiece in the Heart of the Desert, the Regional Scientific Conference on Desert Architecture, Islamic Azad University, Ardestan Branch, Ardestan, Isfahan, Iran, 2006 (in Farsi).
[17] J. Abazari, B. Tashakori, Qanat, Cistern and Ice-House, Project of Theoretical Foundations of Architecture, Islamic Azad University, Taft Branch, Taft, Yazd, Iran, 2008 (in Farsi).
[18] N. Nazarieh, N. Valibeig, S. Nasekhian, Analysis of local architects' capability in process of building transitional ice-houses (comparing executive methods of center to eastern and southeastern Iran), Maremat Me'mari-e Iran 7 (14) (2018) 45–60 (in Farsi).
[19] A.R. Dehghani-Sanij, in: A. Sayigh (Ed.), Cisterns: Sustainable Development, Architecture and Energy, River Publishers, Denmark, 2016.
[20] V. Ghobadian, Climatic Analysis of the Traditional Iranian Buildings, University of Tehran Press, Tehran, Iran, 1998 (in Farsi).
[21] M. Soltani Mohammadi, M. Boluri Bonab, Study of physical structure of ice-houses (Yakhdans) of Naein City, Asar J. 73 (2016) 43–58 (in Farsi).
[22] M. Azad, M. Soltani Mohammadi, Study of Yakhchals of Naein with analytical viewpoint based on elements of vernacular architecture, J. Hous. Rural Environ. 38 (166) (2019) 95–108 (in Farsi).
[23] N. Valibeig, S. Nasekhian, N. Nazarieh, Analysis of the impact of shading wall geometry on formation of ice-houses in Southeast Iran (case study: Kerman), J. Archit. Urban Plan. 18 (2017) 147–166 (in Farsi).
[24] N. Valibeig, N. Nazarieh, Analysis of the role of physical geometry in formation of ice-houses dome in Kerman, Asar J. 75 (2017) 67–84 (in Farsi).

[25] S. Shams, Beauties of Iranian Architectural Art, Traditional Dictionary of Iranian Architecture, Elm-o Danesh Publications, Tehran, Iran, 2009 (in Farsi).
[26] A. Zamani, Blind arcade and arch-facade decoration in Islamic historical monuments, Art & People Mag. 127 (1973) 25–38 (in Farsi).
[27] P. Javaheri, M. Javaheri, Challenge for Water in Ancient Fars, vol. 2, Iranian National Committee on Irrigation & Drainage (IRNCID) and Water National Museum of Iran (WNMI), Tehran, Iran, 2000 (in Farsi).
[28] J.J. Morier, A Journey through Persia, Armenia, and Asia Minor, to Constantinople, in the Years 1808 and 1809, Adamant Media Corporation, Boston, MA, 2004.
[29] M. Siroux, Caravansérails d'Iran et petites constructions routières, de l'Institut Français d'Archéologie Orientale, Série t. 81, Le Caire, Egypt, 1949 (in French).
[30] T. Tavousi, Effective climatic conditions on the evolution of the ice-house architecture, J. Iran. Stud., Shahid Bahonar Univ. Kerman 20 (Fall 2011) 169–192 (in Farsi).
[31] Q. Mousavi, Traditional ice-houses of Abhar City and way of ice-formation in them, Farhang Mardom Mag. 10 (2005) 33–36 (in Farsi).
[32] F.J. Pettijohn, Sedimentary Rocks, Harper & Brothers, New York, NY, 1957.
[33] S. Shafaghi, Traditional ice-houses of Central Iran, an example of architectural genius, J. Isfahan Cult. 35 (2008) 42–49 (in Farsi).
[34] M.R. Maheri, A. Maheri, S. Pourfallah, R. Azarm, A. Hadjipour, Improving the durability of straw-reinforced clay plaster cladding for earthen buildings, Int. J. Archit. Heritage 5 (3) (2011) 349–366.
[35] P. Marzban, H. Ma'rouf, An Illustrated Dictionary of Visual Arts (Architecture, Sculpture, Painting): English-Persian & Persian-English, second ed., Soroush Press, Tehran, Iran, 1992 (in Farsi).
[36] J. Safinejad, The Fundamentals of Human Geography with Reference to Human Geography of Iran, University of Tehran Press, Tehran, Iran, 1984 (in Farsi).
[37] N. Valibeig, A. Rahimi Ariaei, S. Rahravi Poodeh, Survey of the outer-shell structure of the double-shell discrete Nar domes, Iran 56 (2) (2018) 161–172.
[38] M.K. Pirnia, Stylistics of Iranian Architecture, Soroush-e Danesh Publications, Tehran, Iran, 2001 (in Farsi).
[39] E. Filehkesh, N. Afchangi, Study of traditional methods of coping with water shortage in arid and semi-arid regions (case study: Sabzevar), in: The 1st National Conference on Drought Mitigation and Water Shortage, Kerman City, Kerman, Iran, Feb. 2001 (in Farsi).
[40] J. Ghazbanpour, New Life—Old Structure: A Selection of Valuable Historical Structures, second ed., Ministry of Roads and Urban Development in collaboration with Cultural Heritage, Handicrafts and Tourism Organization of Iran, Tehran, Iran, Aug. 1995 (in Farsi).
[41] M.R. Moghtader, Ab-anbar: Conservation de l'eau Sur le Plateau Iranien, UNESCO, Paris, France, Dec. 1982 (in French).
[42] B. Hourcade, M.P. Berthou, Les Glaciéres du Plateau Iranien, Luqman, Numéro 20 (1994) 87–98 (in French).
[43] A. Alaei, Sh. Ranjbar, Z. Sabzi, Pleasant water in the desert, the study of two examples of the masterpieces of desert architecture: Cistern and traditional ice-house, in: The First National Conference on Architecture, Civil, and Urban Environment, Hamadan, Iran, 2014 (in Farsi).
[44] H. Jorgensen, Ice Houses of Iran: Where, How, Why, Mazda Publishers, Costa Mesa, CA, 2012.
[45] E. Hoeltzer, Iran by One Hundred and Thirteen Years Ago (Part I: Isfahan), Collected and Translated by M. Asami, Ministry of Culture and Art and Anthropological Center of Iran, Tehran, Iran, 1976 (in Farsi).
[46] J.E. Polak, Iran and Iranians, Translated by K. Jahandari, Kharazmi Publications, Tehran, Iran, 1982 (in Farsi).

[47] N. Valibeig, S. Rahravi Poodeh, A. Rahimi Ariaei, Structural and geometric analysis of discontinuous double-shell Persian domes in Isfahan and Nain dome-building schools, Int. J. Archit. Heritage 11 (8) (2017) 1101–1120.
[48] J. Ahmadi, Design of domes in Islamic-Iranian architecture, Environ. Conserv. J. 16 (2015) 591–600, special ed.
[49] A.R. Dehghani, Water in the Plateau of Iran: Qanat, Cistern and Ice-House, Yazda Publications, Tehran, Iran, Oct. 2009 (in Farsi).
[50] G.A. Olivier, Olivier's Travelogue (Socio-Economic History of Iran in the Early Era of the Qajar), Translated by M.T. Mirza, Ettela'at, Tehran, Iran, 1992 (in Farsi).
[51] E. Beazley, M. Harverson, Living with the Desert: Working Buildings on the Iranian Plateau, Orchid Press Publishing Ltd., Bangkok, Thailand, 2006.
[52] A. Zargar, An Introduction to the Iranian Rural Architecture, Shahid Beheshti University Press, Tehran, Iran, 1999 (in Farsi).
[53] M. Golzari, Kermanshahan—Kurdistan, vol. 1, Society for the National Heritage (Bita), Tehran, Iran, 1978 (in Farsi).
[54] A. Nehchiri, Historical Geography of Cities, Madreseh Publications, Tehran, Iran, 1997 (in Farsi).
[55] A. Gabriel, Durch Persiens Wüsten, Strecker und Schröder Verlag, Stuttgart, Germany, 1935 (in German).
[56] E. Seyedizadeh, A.A. Abbasian, Bojnoord: Khorasan Northern Passageway (Historical and Religious Monuments and Structures), Ardeshir Publications, Mashhad, Iran, 1993 (in Farsi).
[57] H. Mojtabavi, Historical Geography of Gonabad, Marandiz Publications, Gonabad, Iran, 1995 (in Farsi).
[58] E. Beazley, Some vernacular buildings of the Iranian plateau, Iran: J. Br. Inst. Persian Stud. 15 (1) (1977) 89–102.
[59] N. Nazarieh, Interview with Gh. Farzanehie Mohammadi (Master of Restoration), Construction Technology of Traditional Ice-Houses of Naein City, April 2015.
[60] N. Nazarieh, Interview with H. Mohseni (Master of Restoration), Construction Technology of Traditional Ice-houses of Naein City, June 2015.
[61] C. Harris, Dictionary of Architecture and Construction, fourth ed., McGraw-Hill Education, New York, NY, 2006.
[62] N. Nazarieh, Interview with GH. Salehi (Master of Restoration), Construction Technology of Traditional Ice-houses of Kashan City, April 2015.
[63] N. Nazarieh, Interview with J. Nazarieh (Expert of Restoration), Construction Technology of Traditional Ice-houses in Kerman, Aug. 2015.
[64] N. Nazarieh, Interview with J. Pourbafran (Master of Restoration), Construction Technology of Traditional Ice-houses of Naein City and the Repaired Bafran Ice-house, Feb. 2016.
[65] M.K. Pirnia, Arches and vaults in Iranian architecture, Compiled by Z. Bozorgmehri, Asar J. 24 (1994) 5–45 (in Farsi).
[66] H. Zamrashidi, Arch and Vault in Iran's Architecture, Kayhan Publications, Tehran, Iran, 1993 (in Farsi).
[67] M. Hejazi, F. Jafari, Structural effects of brick arrangement and span length on mid-pointed arches, Adv. Mat. Res. 133–134 (2010) 411–416.
[68] N. Nazarieh, Interview with N. Rezayat (Expert of Restoration), Construction technology of traditional ice-houses of Isfahan, July 2014.
[69] M.E. Bastani-Parizi, Epic of the Desert, Amir Kabir Publications, Tehran, Iran, 1978 (in Farsi).
[70] M.M. Al-Isfahani, in: M. Sotudeh (Ed.), Nesf-e Jahan Fi Tarif-e Esfahan, Amir Kabir Publications, Tehran, Iran, 1989 (in Farsi).
[71] E. Mostafavi, M. Taheri, The introduction of benefactors of Pasteur Institute of Iran and the role of the Waqf tradition in the establishment and promotion of this institute, Iran. J. Med. Ethics Hist. Med. 12 (1) (2019) 287–296.

[72] J.J.S. Marie, S. Naghshpour (Eds.), Revolutionary Iran and the United States: Low-Intensity Conflict in the Persian Gulf, Ashgate Publishing Ltd., Farnham, UK, 2011.
[73] A. Zarei, H. Hashemi Zarjabad, Study on water structures of Birjand City—case study: ice-houses, Soc. Cult. Stud. Khorasan 2 (2006) 5–16 (in Farsi).
[74] H.R. Mirmohammadi, Geography of Khansar, vol. 1, Secretary of Congress of Agha Hossein Khansari Publications, Qom, Iran, 1999 (in Farsi).
[75] M. Rezaei Badafshani, Small and sustainable city spaces, in: A. Atkinson (Ed.), Techniques and Technologies for Sustainability, Proceedings of International Conference and Summer School 2007, Topic 2—Improving the Management of Urban Infrastructure Systems, Univerlagtuberlin, Berlin, Germany, 2008, pp. 136–155.
[76] H. Zahedi, R. Khorshidian, Saqqa-Khaneh School as Iranian contemporary art movement: post colonialism or orientalism? in: Proceedings of International Conference on Arts, Social Sciences, History and Interdisciplinary Studies, 2017, pp. 226–231, Kyoto, Japan, April 20–21.
[77] H. Modarressi Tabatabaei, Geographical Guide of Qom's History, vol. 1, Hekmat-Qom Publications, Qom, Iran, 1956 (in Farsi).
[78] M. Asghari-Moghadam, Ice-houses as forgotten near-city phenomena, Develop. Geogr. Educ. 59 (2001) 48–52 (in Farsi).
[79] M. Krondl, Sweet Invention: A History of Dessert, Chicago Review Press, Chicago, IL, 2011.
[80] M. Barshan, History of Water and Irrigation in Kerman Province, Kerman Studies Center Publications, Kerman, Iran, 2008 (in Farsi).
[81] M. Barghandan, Iran's new social media-friendly approach, Turk. Policy Q. 14 (1) (2015) 137–145.
[82] L. Williams, Kind Regards: The Lost Art of Letter Writing, Michael O'Mara Books, London, UK, Oct. 2012.
[83] J. Patil, A. Chaudhari, The study of an impact of individuals' income on the perception about financial, retail and premium services by Indian post, Int. J. 5 (5) (2017) 67–73.
[84] A.A. Pooya, The Archaic Image of Meybod: A Historical Survey of the Organism and Structure, Islamic Azad University Press, Meybod Branch, Yazd, Iran, 1992 (in Farsi).
[85] F. Parsi, M. Nezari, Ice-house, vol. 42, Memar Magazine, 2007, April & May (in Farsi).
[86] J.S. Curl, S. Wilson, The Oxford Dictionary of Architecture, third ed., Oxford University Press, Oxford, UK, Jan. 2016.
[87] M. Soltani, A.R. Dehghani-Sanij, A. Sayadnia, F.M. Kashkooli, K. Gharali, S.B. Mahbaz, M.B. Dusseault, Investigation of airflow patterns in a new design of Wind tower with a wetted surface, Energies 11 (2018) 1–23.

CHAPTER 6

Saving energy through semi-natural and low-energy production of ice

6.1 Introduction

The traditional and natural methods of producing ice in winter, described in Chapter 2, and of storing it for use in the hot/warm months (primarily to cool drinking water) have now been completely abandoned. Among the reasons for abandoning these long-standing practices is their lack of adequate sanitation according to modern standards. As mentioned, in the heyday of the ice trade, attempts were made to avoid obvious contamination in the water supplied for ice-making, for instance, by relying on Qanats as a water source and by locating the ice-making ponds away from cemeteries and heavy traffic pathways. However, the unlined nature of most ponds, the practice of insulating layers of ice under mat, straw and other dry vegetation, and the transportation of ice to consumers mostly in sacks woven from goat hair or wool would all have added contaminants to the ice. When nothing better was possible, naturally produced ice was of course a great and welcome luxury, but it became less attractive once less-natural methods were available to produce a cleaner product. Other drawbacks to the continued use of natural methods were the high costs of constructing and maintaining traditional ice-houses and the large expanses of land they occupied. With the advent and common usage of mechanical and artificial refrigeration systems, traditional ice-making was soon abandoned, and the buildings were left to deteriorate over time. In their place, modern refrigeration systems of different types, sizes, and capacities are widely used for various applications around the world, including to cool drinking water and other types of beverages; preserve, transport, and store (short- or long-term) perishables and medicines; provide air-conditioning (AC); use in special industrial processes (e.g., chemicals, plastics, and natural gas liquefaction), and so on, as detailed in Chapter 3 [1–8]. In addition, the demand for these systems is constantly and rapidly increasing worldwide for a range of reasons, such as the growth of the

Ice-Houses
https://doi.org/10.1016/B978-0-12-822275-1.00011-4
Copyright © 2021 Elsevier Inc.
All rights reserved.

world's population; the improvement of people's living standards; the rising demands for thermal comfort; the development of industries; the increase in global ambient air temperatures; and the trend toward greater personal disposable income in more members of populations worldwide [9–12].

Despite having many advantages and applications, refrigeration and AC systems consume large amounts of electrical energy [13–15], accounting for about 20% of the total global electric power generation [16], mostly supplied by fossil-fuel power plants [17–19]. It is also anticipated that the worldwide demand for electrical energy to supply the refrigeration and AC sectors will more than double by 2050 [16]. Therefore, these systems are both directly and indirectly involved in creating global environmental challenges and concerns. For instance, they are responsible for nearly 10% of the world's emissions of greenhouse gases (GHGs) [20, 21], especially CO_2, a major contributor to global warming and associated climate change [22–26]. Table 6.1 illustrates the percentage of electrical energy consumption and GHG emission from the refrigeration and AC sectors in several countries for different years. Countries with large populations (e.g., China and the United States) and those located in hot regions of the world, for example, the Kingdom of Saudi Arabia (KSA) and the United Arab Emirates (UAE), are increasingly demanding and normalizing the use of AC and

Table 6.1 Electric power consumed and GHGs emitted by refrigeration and AC systems in some countries.[a]

Country	Power consumed by refrigeration sector (%)	GHG emissions by refrigeration sector (%)	Reference
United States	~20% in 2015	–	[27]
United Kingdom	~16% in 2011	Over 10% in 2011	[28]
Japan	~25% in 2015	–	[27]
China	Over 15% in 2019	–	[29]
Germany	14% in 2006	–	[30]
Iran	–	~10% in 2017	[31]
KSA	Over 30% in 2018	–	[32]
Indonesia	–	~15.4% in 2015	[33]
UAE	Over 29% in 2013	–	[34]

[a] The percentages provided in this table will change over time.

refrigeration systems. In fact, these systems (particularly AC) have a vital role in the social development and economic growth of countries situated in hot climates [35]. As a result of this widespread reliance on thermal comfort, the International Institute of Refrigeration (IIR) [16] reported that greater than 15 million people currently work in the refrigeration industry worldwide.

In addition to trying to improve the energy efficiency of conventional refrigeration and AC systems, researchers are interested in finding and developing sustainable and eco-friendly energy technologies that can support efficient, practical, and even cost-effective solutions for both meeting people's thermal comfort requirements and reducing environmental problems. Although most studies are looking forward to new discoveries, there is no reason to ignore traditional approaches, as they may still have something to offer us if we revisit and adapt them. For example, the knowledge and technology used for the natural production of ice and its storage through traditional ice-houses can be adapted to supply all or part of the space cooling demands of buildings. This knowledge, coupled as before with the cold temperatures of winter, can be utilized to manufacture ice via a semi-natural method (also called a low-energy technology) for a variety of purposes, such as to cool confined spaces and to precool fruits and vegetables between being harvested and entering cold storage in the summer [36–43]. This adaptation of ancient methods will consume only the small amount of energy that is required to produce ice and circulate coolness,[a] in comparison with conventional refrigeration systems (for instance, vapor compression and absorption refrigeration systems). This chapter formulates and describes how to produce ice through such a semi-natural/low-energy approach.

6.2 Semi-natural and low-energy production of ice in underground ponds

Due to the high price of the land that would be needed for the semi-natural production of ice, as well as to minimize melting of the ice produced (or losing coolness), the ice-making and storage pond should be built at a sufficient depth underground. This subterranean space can then be left open or entirely covered. Moving operations totally underground will reduce the expenses of production, storage, and use of ice. The most important advantage of using covered ponds instead of noncovered or open ones is

[a]Here, the word "coolness" means internal energy at low temperatures. It is similar to the popular use of the expression "the source of heat" utilized for internal energy at higher temperatures and usually employed for energy analysis [43, 44].

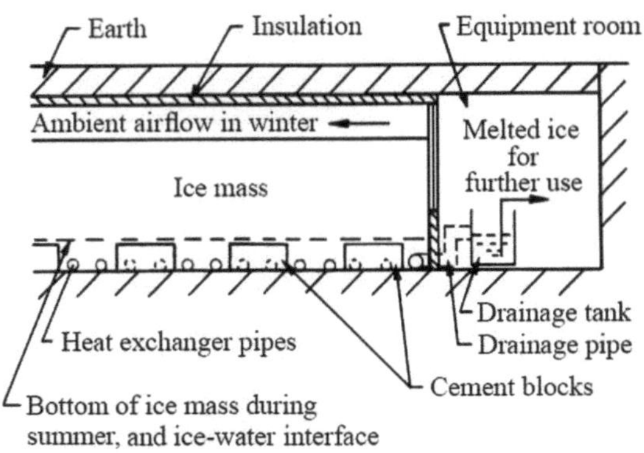

Fig. 6.1 Cross-section of a covered underground pond for producing ice [36, 43, 45]. The ice manufactured in summer is placed on cement blocks.

that covered ponds would be completely protected from any sources of heat (like sunlight), contamination (e.g., blowing dust, pollen, or leaves), and precipitation (e.g., rain and snow). In addition, the land covering underground ice ponds can be used for a variety of purposes such as the parking area and landscaping [36, 43]. They will also take advantage of the insulating qualities and lower temperature associated with moving underground. Fig. 6.1 indicates a schematic design of semi-natural production of ice in an underground pond with the cover on. The method of semi-naturally producing ice is described below in detail.

6.2.1 Description of a semi-natural ice-producing scheme

Ice can be produced in underground spaces such as the basement of a residential house or the underground parking lots of residential, commercial, industrial, and institutional complexes, or shopping centers if they are situated in an area with cold winters. The process is as follows: First, pipes (e.g., made of copper) of appropriate diameter and long enough to meet the cooling demands of the building are installed on the floor of the basement (or parking lot) for later exchange of heat, and then, parallel rows of cement blocks, about 5 cm high and around 10 cm wide, are laid out on the floor. These rows, one block high only, extend roughly the length of the room and are evenly spaced apart, about the same width as the width of the individual blocks. Some of the heat exchanger pipes actually run through the openings in the blocks, as shown in Fig. 6.1. To one side, there will be a small equipment room that houses the pump and other equipment, separated from the space used for the production and storage of ice by a robust wooden or brick wall. Producing and storing ice in the basements or parking lots of buildings has the additional advantage of the very solid foundations used to support such structures; clearly, there is no fear of basement floors collapsing under the weight of water/ice.

In the cold months of the year or when the temperature of the ambient air is sufficiently low to produce ice (e.g., lower than $-2°C$), a set amount of freshwater (for instance, to a standing depth of around 1 mm)[b] is poured on the floor of the pond, and then, cold ambient air is directed over it by a blower or fan, mounted above the ice-making surface. Thus, the layer of the water can exchange heat with the cold stream of air. Consequently, the water will first reach freezing point and then freeze. Subsequently, another set amount of water is poured on that ice layer to be frozen. This action is repeated whenever the temperature of the ambient air is low enough, until a huge mass of ice is gradually produced to a desired height. Efficiency depends on the coolness of the ambient air. It should be noted that to prevent damage from the volume expansion of water during the process of freezing, only a small amount of water is added to the ice-making pond at a time [43]. Eventually, the ice will be stored in the same space in which it has been produced, to wait until it is needed in the hot/warm months. In this way, with only the small amount of energy needed to work the blower or fan, and using the coolness of the ambient air, the amount of ice needed for any purpose/application can be manufactured by means of a semi-natural method. The process can even be automated for efficiency [36, 43].

During the summer or when the building occupants need cooling for thermal comfort, the cold water required for cooling coils or AC units (e.g., fan coils) is supplied via the circulation of water flowing inside the heat exchanger pipes located on the floor of the underground pond and under the mass of stored ice. This exchange of heat between the pipes and the ice mass will lead part of the ice to melt, mainly from the bottom, resulting in water lying on the bottom of the pond. When the height of this water reaches the height of the cement blocks (about 5 cm above the floor), the water will be drawn off into the drainage tank located in the equipment room, where it can be further utilized for AC. Continuously draining the water ensures that the huge mass of ice remains resting on the cement blocks, and so no water pressure will be exerted on the walls of the underground pond. The height at which the drainage pipe is linked to the drainage tank, situated in the equipment room, is such as to ensure the existence of an ice/water interface at the bottom of the ice mass [42].

[b]The amount will vary from one region to another region depending on weather and geographical conditions.

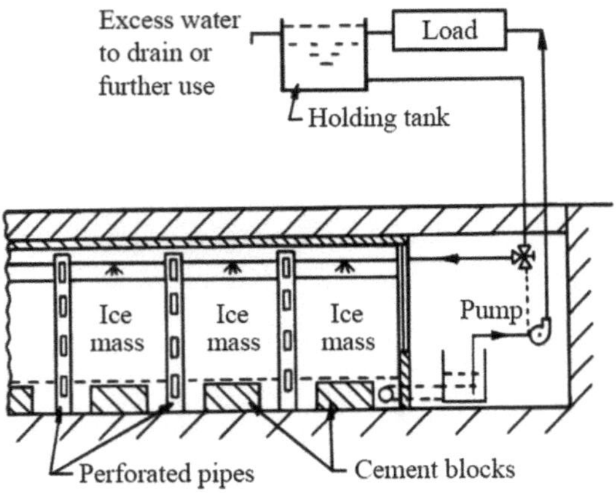

Fig. 6.2 Cross-section of a covered underground ice pond; water is sprayed on the surface of the stored ice during the summer [39].

Another way to cool the water returning from the AC units (e.g., fan coils) is to pour it directly on the surface of the mass of ice produced and stored in the underground pond, as shown in Fig. 6.2. The water moves down through a number of narrow vertical pathways (or perforated pipes), which have been installed inside the ice mass during the ice production, and then evacuates via the drainage pipe, coming back to the AC units [38]. With this method, the mass of ice stored in the underground pond melts more rapidly than with the previous technique.

6.3 Governing equations of semi-natural method of ice production

Using the first law of thermodynamics, which is based on the law of conservation of energy, the governing equations for the semi-natural method of ice production can be written as follows [36, 43]:

(a) To cool a thin layer of water that has been sprayed on the pond and bring it to freezing point (or 0°C), in fact, the process of sensible cooling[c] of that water layer to its freezing point:

$$\left(\dot{Q}_{c1}+\dot{Q}_{e1}+\dot{Q}_{r1}\right)\Delta t_1+Q_{k1}=\rho_w A_i S C_w\left(T_{wi}-0\right) \tag{6.1}$$

[c]Sensible heat, or sensible energy, is defined as the amount of thermal energy (cooling or heating) that is needed to alter the temperature of a substance or thermodynamic system [46, 47].

(b) To freeze the thin layer of water at 0°C, in other words, the process of latent cooling[d] or freezing of that water layer:

$$\left(\dot{Q}_{c2} + \dot{Q}_{e2} + \dot{Q}_{r2}\right)\Delta t_2 + Q_{k2} = \rho_w A_i S h_{if} \quad (6.2)$$

(c) To cool the layer of ice from 0°C to a temperature less than it (or the process of sensible cooling of the ice layer):

$$\left(\dot{Q}_{c3} + \dot{Q}_{e3} + \dot{Q}_{r3}\right)\Delta t_3 + Q_{k3} = \rho_i A_i S_i C_i \left(0 - T_{if}\right) \quad (6.3)$$

In the above equations, $\dot{Q}$ and Q are, respectively, the rate of heat transfer (in Watt or kW) and the total heat transferred (in J or kJ); the subscripts c, e, r, and k represent the heat transfer by convection, evaporation (or sublimation), radiation, and conduction, respectively; Δt_1, Δt_2, and Δt_3 are, respectively, the time intervals required for the cooling of water, freezing of water, and subcooling of the ice to its final temperature (T_{if}); ρ and C are the density and specific heat capacity, respectively; S and S_i are the depth (or thickness) of the water layer before and after freezing; A is the area of the water/ice layer; T_{wi} is the initial temperature of water; h_{if} is the enthalpy of the solidification of water; the subscripts w and i refer to water and ice, respectively; the subscripts 1, 2, and 3 represent cooling water from its initial temperature (T_{wi}) to 0°C, freezing the water, and cooling the ice to its final temperature (T_{if}), respectively.

6.4 Determining the rate of ice production by semi-natural method

If the area of the ice-making underground pond, the S thickness (considering a constant amount for it), the ambient air conditions (such as temperature, pressure, and relative humidity), and the velocity of airflow that is blown on the surface of the pond by the blower or fan are known when the temperature of ambient air is low enough, the heat transfer terms in Eqs. (6.1)–(6.3) can be solved, and accordingly, the time intervals Δt_1, Δt_2, and Δt_3 can be calculated. Thus, the rate of production of ice can be determined by the following equation [36, 43]:

$$\dot{S}_i = S_i / \left(\Delta t_1 + \Delta t_2 + \Delta t_3\right) = \rho_w S / \rho_i \left(\Delta t_1 + \Delta t_2 + \Delta t_3\right) \quad (6.4)$$

[d]Latent heat, or latent energy, is defined as thermal energy released (cooling) or absorbed (heating) by a substance or thermodynamic system, during a process of constant temperature. In other words, the latent heat means the thermal energy in a hidden form that is supplied or extracted to alter the state of a substance or system with no alteration of its temperature [48, 49].

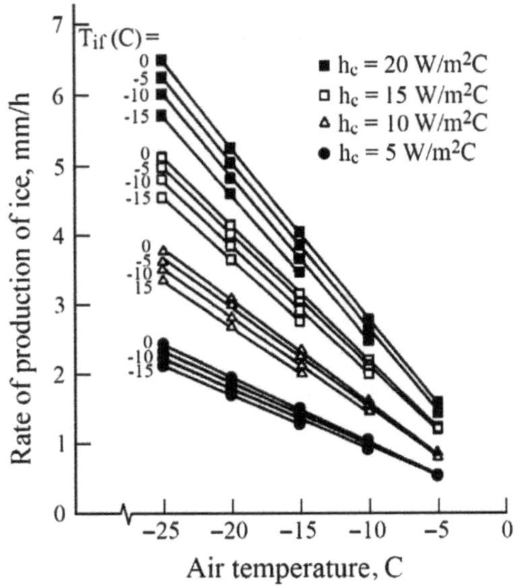

Fig. 6.3 Rate of production of ice in a covered underground pond as a function of the temperature of the ambient air, final ice temperature (*Tif*), the convective heat transfer coefficient (*hc*), and *Twi* = 10 °C [36, 43].

Fig. 6.3 shows the rate of production of ice (in mm/h) for different convective heat transfer coefficients (h_c), the final ice temperatures (T_{if}) 0°C, −5°C, −10°C, and −15°C, and the initial temperature of water (T_{wi}), considered to be 10°C, when the ambient air temperature (T_a) that must be blown over the pond's surface is equal to or lower than −5°C. As illustrated in this figure, when h_c = 20W/m^2°C, T_a = −5°C, and T_{if} = 0°C, ice can be produced at the rate of about 1.75 mm/h. In addition, at the same convective heat transfer coefficient (h_c = 20W/m^2°C), when T_a = −2°C and −10°C, ice can be manufactured at the rate of around 0.9 mm/h and 2.8 mm/h, respectively, by means of the semi-natural method. For the conditions T_{if} = 0°C, T_{wi} = 10°C, and h_c = 10W/m^2°C, the rates of production of ice ($\dot{S}_i$) for the ambient air temperatures −2°C, −5°C, and −10°C would be about 0.4 mm/h, 0.9 mm/h, and 1.7 mm/h, respectively.

6.5 Estimating the maximum height of ice that can be produced by semi-natural method

If the ambient air temperature (T_a) is known (a main factor in making ice), as well as choosing the appropriate velocity for propelling air over the surface of water/ice in the covered underground ice pond with a blower or fan, determining the convective heat transfer coefficient (h_c), selecting the final ice temperature, and solving Eqs. (6.1)–(6.4), the rate of production of ice ($\dot{S}_i$) can first be computed for any time of the day or night in cold months of the year. Then, the maximum height of ice production can be estimated. If the temperature of the ambient air is not known, the following relationship can be used to approximate it for different times of the day or night [36, 43, 50]:

$$T(t) = \frac{1}{2}(T_x + T_n) + \frac{1}{2}(T_x - T_n)\cos\frac{2\pi}{24}(t - 15) \tag{6.5}$$

where T_x and T_n are, respectively, the maximum and minimum temperatures of the ambient air for the day of interest. For many of the world's cities, the maximum and minimum ambient air temperatures for all days of the year are measured and recorded by the meteorological organizations of those cities. Choosing these values and substituting in Eq. (6.5), the temperature of the ambient air can be estimated for any time of the day or night for the desired day. Afterward, the maximum height to which ice can be produced is computed.

6.6 Choosing specifications for a semi-natural ice-house to meet the building's annual cooling demand

The dimensions required for the ice-making covered underground pond of a semi-natural ice-house can be determined from the following equations [36, 43]:

$$Q_i F_r = Q_L \tag{6.6}$$

$$\rho_i A_i h_i \left(h_{if} + C_w \Delta T_w\right) F_r = Q_L \tag{6.7}$$

In the above equations, Q_i is the total coolness that can be used if there has been no heat absorbed by the mass of ice (or the total coolness stored in the ice mass); Q_L is the annual cooling demand; F_r is the coolness recovery factor; h_i is the height of the ice mass; and ΔT_w refers to the temperature increase of melted ice before it leaves the pond (or is disposed of). In Eq. (6.7), it is assumed that the ice is at the initial temperature of 0°C.

In Eqs. (6.6) and (6.7), F_r is a function of the amounts of all the heat absorbed (or coolness lost) by the ice mass from its surroundings during the periods of storage and usage. In other words, the higher the amounts of heat absorbed, the lower the F_r will be. Another point is that if the ice had been produced at a temperature less than freezing point (or 0°C), due to the heat absorbed from its surrounding environment, it first moves from its final temperature (T_{if}) to 0°C and then begins to melt at this temperature.

To determine the coolness recovery factor (F_r) and the area of the ice-making covered underground pond (A_i), the following procedure is employed:

1. The maximum height of ice that can be manufactured in the desired location by means of the semi-natural approach is determined.
2. The height of the ice mass (h_i) is chosen, although it cannot exceed the maximum height obtained. The amount of this height depends on the ground conditions and properties, plus the total initial (such as design, construction, and purchasing the required equipment) and operating (like maintenance and labor) costs of the pond. It is from this point of view that as the height of ice increases, the area of the underground pond becomes smaller.
3. A quantity for F_r is assumed, and then, the area A_i is calculated from Eq. (6.7). Subsequently, the pond's dimensions and specifications are determined.
4. Using numerical methods, the heat transfer from floor, ceiling, and sidewalls of the ice pond can be calculated, and F_r is specified by the following equation [36, 43]:

$$F_r = \left(Q_i - Q_g\right) = \frac{Q_L}{Q_i} \tag{6.8}$$

In the above equation, Q_g is the total heat transferred from the sidewalls, floor, and ceiling of the pond to the ice mass at the times of production, storage, and usage. By means of proper insulating materials, heat transfer through the surroundings of the covered underground pond to the ice mass throughout the stages of producing, storing, and utilizing ice can be minimized. Nevertheless, the total heat absorbed (or heat gained) by the mass of ice from its surrounding environment must be calculated, if necessary, which is complicated. This heat depends on various factors such as the earth's thermal regime; the soil type, density, and moisture content; and the thermal properties of soil (thermal conductivity, specific heat capacity, and thermal diffusivity) around the underground pond, as well as freezing of the soil moisture in winter [43] and the thermal characteristics of insulation materials used. The heat absorbed melts part of the ice produced and stored in the pond, thereby reducing the amount of space cooling possible in the summer.

5. If the F_r obtained is different from the quantity initially assumed (from stage 3), the trial and error procedure will be continued until the two quantities of F_r are finally equal to each other. The expected quantity for F_r is between 0.8 and 0.9 [36, 38]. Eventually, the quantities of F_r and A_i are determined.

6.7 The coefficient of performance (COP) of a semi-natural ice-house

After the design of the semi-natural ice-making pond and the selection of the operating conditions of the system, the amount of electrical energy required to blow air into the covered underground pond can be calculated. If this amount of energy is indicated by W, the coefficient of performance (COP) of the semi-natural refrigeration system (which can also be called the semi-natural ice-house) can be determined by means of Eq. (6.9). The COP of this system can be between 20 and 25 [36] and is much higher than that of the conventional vapor compression refrigeration systems, which is around 3 [43][e]—based on using a similar procedure in calculating both COPs.

$$\beta = \frac{Q_L}{W} \tag{6.9}$$

In the above equation, β represents the COP. Due to the high values of COP, the semi-natural ice-house can easily compete economically and environmentally with the conventional vapor compression and absorption refrigeration systems used in areas that have relatively cold winters. It should be noted that the electric power needed for the semi-natural ice-house can also be supplied by a small solar energy system or other available renewable energy technology.

Example 6.1 In a semi-natural ice production factory located in the city of Hamedan, Iran, the temperature of the ambient air (T_a) that would blow over an ice-making pond on a clear winter night is equal to $-10°C$, and due to the wind speed, the convective heat transfer coefficient (h_c) is considered to be $15 W/m^2 °C$. Determine:

(a) The rate of production of ice ($\dot{S}_i$) in mm/h, and

(b) If the ice production took place in an open pond (naturally), what the rate of production of ice ($\dot{S}_i$) would be.

[e]Under normal circumstances, the cooling COP for AC ranges from approximately 3 to 5 [51].

Solution:

(a) Referring to Fig. 6.3 and considering the final ice temperature (T_{if}) to be equal to 0°C, the rate of production of ice ($\dot{S}_i$) would be 2.25 mm/h.

(b) If the ice production is carried out naturally in an open ice-making pond, referring to Table 2.4 (in Chapter 2) as well as considering a relative humidity (ϕ) equal to 35% and a convective heat transfer coefficient (h_c) equal to 15 W/m^2 °C (by twice averaging, once between the ambient air temperatures of −8°C and −12°C, and another time between the relative humidities of 20% and 50%), the rate of production of ice ($\dot{S}_i$) acquired would be 4.22 mm/h. The increase observed in the production of ice in the open pond, case (b), can be attributed to the radiation heat exchange between the pond's water/ice surface and the sky on a clear winter night.

6.8 Conclusion

Refrigeration and AC systems have become a necessary part of today's world, covering a wide range of domestic, agricultural, and industrial applications all over the globe. According to a report provided in 2019 by the IIR [16], approximately 5 billion refrigeration, AC, and heat pump systems operate throughout the world to meet the cooling needs of various sectors. Correspondingly, these systems contribute substantially to the global consumption of electrical energy (generated largely by fossil-fuel power plants), and the subsequent environmental problems and challenges, especially through emitting GHGs and polluting the air. The use of ecologically friendly, sustainable, and low-energy approaches/technologies (e.g., producing ice semi-naturally for use in space cooling) offers an effective, economical, and practical alternative for meeting humans' thermal comfort demands, managing energy consumption and costs, reducing local and global environmental footprints, and achieving sustainable development.

References

[1] A. Ameen, Refrigeration and Air Conditioning, Prentice-Hall of India Pvt. Ltd., New Delhi, Delhi, India, Jan. 2006.

[2] S.N. Sapali, Refrigeration and Air Conditioning, second ed., PHI Learning Pvt. Ltd., New Delhi, Delhi, India, Nov. 2013.

[3] C. Xu, J. Zhang, H. Dinh, Q. Xu, Process synthesis of mixed refrigerant system for ethylene plants, Ind. Eng. Chem. Res. 56 (28) (2017) 7984–7999.

[4] J. Wagner, M. Schäfer, A. Schlüter, L. Harsch, J. Hesselbach, M. Rosano, C.-X. Lin, Reducing energy demand in production environment requiring refrigeration—a localized climatization approach, HVAC&R Res. 20 (6) (2014) 628–642.

[5] A. Dehghani, H. Bridjanian, A. Farshi, Selection separation method, for hydrogen and light hydrocarbons gases from waste-gas streams of a petroleum refinery, Pet. Sci. Technol. 30 (21) (2012) 2196–2207.
[6] C.W. Gellings, New electrification technology, Strat. Plann. Energy Environ. 31 (2) (2011) 69–78.
[7] N.M. Shah, A.F.A. Hoadley, G.P. Rangaiah, Inherent safety analysis of a propane precooled gas-phase liquified natural gas process, Ind. Eng. Chem. Res. 48 (10) (2009) 4917–4927.
[8] I. Dincer, Refrigeration Systems and Applications, John Wiley & Sons, Inc., New York, NY, 2017.
[9] N. Kalkan, E.A. Young, A. Celiktas, Solar thermal air conditioning technology reducing the footprint of solar thermal air conditioning, Renew. Sustain. Energy Rev. 16 (2012) 6352–6383.
[10] J. Aman, D.K. Ting, P. Henshaw, Residential solar air conditioning: energy and exergy analyses of an ammonia–water absorption cooling system, Appl. Therm. Eng. 62 (2) (2014) 424–432.
[11] A.R. Dehghani-Sanij, E. Tharumalingam, M.B. Dusseault, R. Fraser, Study of energy storage systems and environmental challenges of batteries, Renew. Sustain. Energy Rev. 104 (2019) 192–208.
[12] J. Villarroel-Schneider, A. Malmquist, J.A. Araoz, J. Martí-Herrero, A. Martin, Performance analysis of a small-scale biogas-based trigeneration plant: an absorption refrigeration system integrated to an externally fired microturbine, Energies 12 (2019) 3830.
[13] D. Jung, R. Radermacher, Performance simulation of single evaporator refrigerator with pure and mixed refrigerants, Int. J. Refrig. 14 (1991) 223–232.
[14] M. Santamouris, A. Argiriou, Renewable energies and energy conservation technologies for buildings in southern Europe, Int. J. Sol. Energy 15 (1994) 69–79.
[15] D. Hong, L. Tang, Y. He, G. Chen, A novel absorption refrigeration cycle, Appl. Therm. Eng. 30 (14–15) (2010) 2045–2050.
[16] J.L. Dupont, P. Domanski, P. Lebrun, F. Ziegler, 38th Note on Refrigeration Technologies: The Role of Refrigeration in the Global Economy: Introduction, The International Institute of Refrigeration (IIR), Paris, France, June 2019.
[17] L.O.S. Buzelin, S.C. Amicoa, J.V.C. Vargasa, J.A.R. Parise, Experimental development of an intelligent refrigeration system, Int. J. Refrig. 28 (2005) 165–175.
[18] S.M. Jadhav, M. Arulprakasajothi, U. Chandrasekhar, D. Yuvarajan, Experimental investigation of vapour absorption refrigeration cycle for automobile cabin cooling, in: U. Chandrasekhar, L.J. Yang, S. Gowthaman (Eds.), Innovative Design, Analysis and Development Practices in Aerospace and Automotive Engineering (I-DAD 2018), Springer, Singapore, 2019, pp. 41–51.
[19] K.F. Fong, T.T. Chow, C.K. Lee, Z. Lin, L.S. Chan, Comparative study of different solar cooling systems for buildings in subtropical city, Sol. Energy 84 (2) (2010) 227–244.
[20] D. Cowan, G.G. Maidment, I. Chaer, Estimation of cooling energy demand and carbon emissions from urban buildings using a quasi-dynamic model, NY-14-C056, ASHRAE Transactions, 120 (1) (2014) 1–8.
[21] V. Morlet, D. Coulomb, J.L. Dupont, 35th Informatory Note on Refrigeration Technologies: The Impact of the Refrigeration Sector on Climate Change, the International Institute of Refrigeration (IIR), Paris, France, Nov. 2017.
[22] A. Khajuria, N.H. Ravindranath, Climate change vulnerability assessment: approaches DPSIR framework and vulnerability index, J. Earth Sci. Clim. Chang. 3 (2012) 1–6, 1000109.
[23] A.R. Kazemi, S.B. Mahbaz, A.R. Dehghani-Sanij, M.B. Dusseault, R. Fraser, Performance evaluation of an enhanced geothermal system (EGS) in the Western Canada Sedimentary Basin, Renew. Sustain. Energy Rev. 113 (2019) 109278.

[24] A.Y. Htut, S. Shrestha, V. Nitivattananon, A. Kawasaki, Forecasting climate change scenarios in the Bago river basin, Myanmar, J. Earth Sci. Clim. Chang. 5 (2014) 1–11, 1000228.
[25] C. Kinney, A.R. Dehghani-Sanij, S.B. Mahbaz, M.B. Dusseault, J.S. Nathwani, R.A. Fraser, Geothermal energy for sustainable food production in Canada's remote northern communities, Energies 12 (2019) 1–25, 4058.
[26] J.M. Calm, Emissions and environmental impacts from air-conditioning and refrigeration systems, Int. J. Refrig. 25 (3) (2002) 293–305.
[27] M. Almanza, A. Kedous-Lebouc, J.P. Yonnet, U. Legait, J. Roudaut, Magnetic refrigeration: recent developments and alternative configurations, Eur. Phys. J. Appl. Phys. 71 (2015) 10903.
[28] I.P. Koronaki, D. Cowan, G. Maidment, K. Beerman, M. Schreurs, K. Kaar, I. Chaer, G. Gontarz, R.I. Christodoulaki, X. Cazauran, Refrigerant emissions and leakage prevention across Europe–Results from the Real Skills Europe project, Energy 45 (1) (2012) 71–80.
[29] Z. Xin, China Boosts Energy Efficiency Efforts, China Daily, 2019. https://www.chinadaily.com.cn/a/201907/02/WS5d1abf6aa3103dbf1432b6b0.html. (Accessed 21 July 2020).
[30] E. Way, A. Change, S. Likely, When is enough, enough?: IKK or Chillventa—either way a change is likely, Industry News, ASHRAE J. 48 (Dec. 2006) 6.
[31] I. Papst, I. Croiset, S. Mischel, Refrigeration and Air Conditioning Greenhouse Gas Inventory for Iran, p. 15, in: Deutsche Gesellschaft für Internationale Zusammenarbeit (GIZ) GmbH, Bonn and Eschborn, Germany, 2018.
[32] M. Krarti, N. Howarth, Transitioning to high efficiency air conditioning in Saudi Arabia: a benefit cost analysis for residential buildings, J. Build. Eng. 31 (2020) 101457.
[33] D. Oppelt, H. Herlianika, I. Papst, Refrigeration and Air Conditioning Greenhouse Gas Inventory for Indonesia, p. 8, Deutsche Gesellschaft für Internationale Zusammenarbeit (GIZ) GmbH, Jakarta, Indonesia, Aug. 2017.
[34] A. Mokri, M.A. Ali, M. Emziane, Solar energy in the United Arab Emirates: a review, Renew. Sustain. Energy Rev. 28 (2013) 340–375.
[35] D. Coulomb, J.L. Dupont, A. Pichard, 29th Informatory Note on Refrigeration Technologies: The Role of Refrigeration in the Global Economy, The International Institute of Refrigeration (IIR), Paris, France, 2015.
[36] M.N. Bahadori, M. Yaghoubi, Ventilation and Natural Cooling System in Traditional Buildings of Iran, Markaz-e Nashr-e Daneshgahi, Tehran, Iran, 2006 (in Farsi).
[37] A.J. Gorski, Compiler, in: Proceedings of the First Annual Workshop on Ice Storage for Cooling Applications, ANL-81-45, Argonne National Laboratory, Lemont, IL, June 1981.
[38] M.N. Bahadori, K.G.T. Hollands, P.F. Galpin, A feasibility study of seasonal storage of coolness in Canada, in: Solar Energy Program, vols. I & II (Appendices), National Research Council Canada, Ottawa, ON, Canada, Jan. 1983. Report No. STOR-11.
[39] M.N. Bahadori, Production and storage of ice for cooling, in: Proceedings of the Second International Passive and Low Energy Architecture Conference, Published by Pergamon Press, Crete, Greece, June 1983.
[40] C.E. Francis, A.J. Gorski, Passive and hybrid ice systems for comfort conditioning and chilling applications, in: Proceedings of the 7th National Passive Solar Conference, American Section of the International Solar Energy Society, Knoxville, TN, Aug. 1982, pp. 923–926.
[41] E.L. Morofsky, Long-term latent heat storage, the Canadian perspective, in: Proceedings of the First U.S.—China Conference on Energy, Resources and Environment, Pergamon Press, Beijing, China, Nov. 1982, pp. 405–412.

[42] M.N. Bahadori, D.K. Pugsley, Long-term storage of coolness through a gradual production of ice in winter, in: Proceedings of the First U.S.—China Conference on Energy, Resources and Environment, Published by Pergamon Press, Beijing, China, Nov. 1982.
[43] M.N. Bahadori, Natural production, storage, and utilization of ice in deep ponds for summer air conditioning, Sol. Energy 34 (2) (1985) 143–149.
[44] M.N. Bahadori, Natural air-conditioning systems, in: K.W. Böer (Ed.), Advances in Solar Energy, Springer, Boston, MA, 1986, pp. 283–356.
[45] M.N. Bahadori, Natural cooling systems and an economic feasibility study of long-term storage of coolness, Energy 9 (7) (1984) 589–604.
[46] C.J. Adkins, Equilibrium Thermodynamics, second ed., McGraw-Hill, London, UK, 1975.
[47] P.T. Landsberg, Thermodynamics and Statistical Mechanics, Oxford University Press, Oxford, UK, 1978.
[48] P. Perrot, A to Z of Thermodynamics, Oxford University Press, Oxford, UK, 1998.
[49] J.O.E. Clark, The Essential Dictionary of Science, Barnes & Noble Books, New York, NY, 2004.
[50] M.N. Bahadori, M.J. Chamberlin, A simplification of weather data to evaluate daily and monthly energy needs of residential buildings, Sol. Energy 36 (6) (1986) 499–507.
[51] R.Z. Wang, Z.Y. Xu, T.S. Ge, Introduction to solar heating and cooling systems, in: R.Z. Wang, T.S. Ge (Eds.), Advances in Solar Heating and Cooling, Woodhead Publishing Ltd., Cambridge, UK, 2016, pp. 3–12.

CHAPTER 7

Ice production to reduce peak power use for space cooling

7.1 Introduction

The ability to produce cold artificially and use it for purposes such as cooling buildings is one of the most important developments in the history of engineering. With this invention, humans have been able to keep their living and working environments at a suitable temperature and relative humidity, thus providing thermal comfort for themselves during the hot/warm months. It also allows them to live comfortably in areas that would otherwise be uninhabitable because of high ambient air temperature and relative humidity. Overall, the artificial production of cold through various approaches has brought about an enormous change in the welfare, well-being, and development of human societies. Nowadays, the use of gas coolers and refrigeration and air-conditioning (AC) machines (e.g., chillers) is significantly increasing due to the worldwide expansion of construction, the global rise in temperature (by more than 2°C above the preindustrial era's level [1]), the rising living standards of greater numbers of people, a general failure to supply thermal comfort in summer by evaporative water coolers (particularly in hot and humid climates), and other reasons. According to a report published in 2019 by the International Institute of Refrigeration (IIR) [2], over 2.6 billion AC units (stationary and mobile) now operate worldwide. These air conditioners consume more than 8% of the total global electric power, although this percentage differs broadly from one country/region to another as a result of local weather conditions and the level of development [2, 3]. Fig. 7.1 shows the global stock of AC systems at the end of 2016. China, the United States, and Japan hold about 67% of the world's air conditioners. China is currently the biggest consumer of room AC units worldwide and responsible for around 25% of global space cooling CO_2 emissions [4].

The rapid, growing, and continuous trend in using refrigeration and AC machines of various types, sizes, and capacities is increasing the consumption of electrical energy, especially during the warmer months of the year.

https://doi.org/10.1016/B978-0-12-822275-1.00009-6
Copyright © 2021 Elsevier Inc.
All rights reserved.

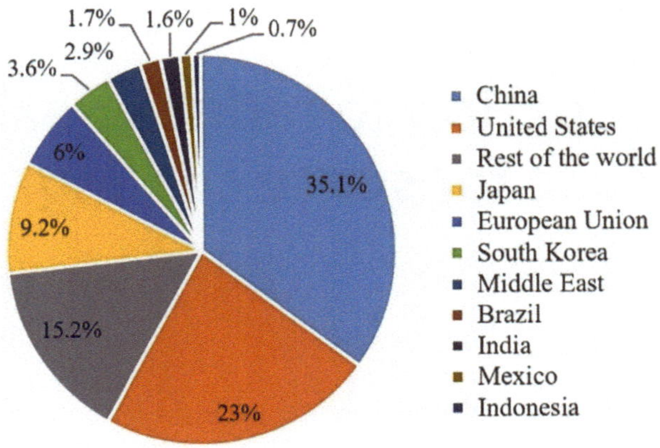

Fig. 7.1 Stock of AC systems based on country/region, late 2016 [2, 3].

In fact, it is projected that electric power requirements for space cooling and AC will have tripled by 2050 [2, 3]. Hence, new power plants (both renewable and conventional) must be established, requiring high initial investment of money and resources, and adding to environmental degradation. One way to prevent the construction of new power plants and enhance the load factor[a] of existing ones is peak shaving, meaning reduction of the maximum load, by cold thermal energy storage (CTES)[b] [10, 11]. The stored coolness can be used to meet part of the cooling needs of the building, thus shaving the peak electrical demands. CTES, also called cold storage, exists in various forms: sensible storage (for instance, by water, rocks, and bricks), latent storage (e.g., by ice, salt hydrates, and paraffin), and thermochemical/sorption storage (e.g., by inorganic materials) [8, 12–14]. The aim of using cold storage systems is to reduce the maximum demand for electric power, plus support uniform electricity consumption throughout the 24-h of the day by the refrigeration and AC systems of buildings [8, 10, 11, 15–18], with no drop in the quality of energy services [11, 19]. The end result is diminishing electricity bills. In general, CTES can primarily be utilized for

[a]In the terminology of electrical engineering, the "load factor" is defined as the ratio between the average load (or demand) and the peak load (or maximum demand) in a specified time period (normally for the duration of a whole 24-h day) [5–7] and is a dimensionless number. This factor is a measure of the rate of utilization.

[b]CTES means a process that includes adding cold energy to a storage medium (such as water, ice, and other phase change materials), and extracting the energy from that medium whenever it is required for usage [8, 9].

AC purposes in different types of buildings (for instance, residential, office,[c] institutional, commercial, and industrial buildings, and sport stadiums); processes that need a large amount of cold energy for a short time (e.g., food processing); and certain chemical plants. Other applications of CTES are district cooling, combustion turbine inlet air cooling, and mission-critical operations [20, 21].

Since the 1970s, when electricity-generating organizations in the United States recognized the need to reduce the consumption, generation, transmission, and distribution of electric power during peak hours,[d] they have paid great attention to cold storage for commercial applications. These organizations found that if they could produce and store electrical energy during off-peak hours when there is less demand and use it during on-peak hours, they could reduce the consumption of electricity at peak periods and utilize that capacity for other applications. Some of these organizations have provided various motivations to encourage end users to modify their electricity utilization patterns, including [28]:

- Fiscal tariffs[e] in the form of special rates. In electricity markets, tariffs can mainly be divided into two categories [29, 30]: Time of Use (TOU) tariffs and Real Time Pricing (RTP). The former are determined based on how much and when electric power is used by consumers, and the latter alters according to the electricity market price.
- Charges and restrictions imposed on consumption during peak times.
- Discounts and financial assistance to encourage consumers to shift electric power usage from on-peak periods to off-peak periods.

Peak-hour electricity demand can also be reduced by physical means at the sites of consumption. Of interest here is studying the different methods that have been suggested for producing and storing cold for space cooling uses. In general, the refrigeration and AC system of a building is designed or chosen to meet the maximum cooling demand of the building. This

[c]Office buildings are one of the best options for installing CTES systems for cooling and ventilating space, since these types of buildings generally have short occupancy hours.

[d]Peak hours (or times) of electric power consumption refer to the hours when consumer demand for electricity is at its highest within a specified period of time. They generally occur during the day (between 2 pm and 6 pm in the summer and between 6 am and 8 am and 6 pm and 8 pm in the winter months [22, 23]) and are determined based on daily, monthly, seasonal, and annual time scales. Peak periods vary from one area to another, depending on factors such as climate, demographic, economic, social, and industrial conditions, as well as the time of day, day of the week, and season of the year [24–27].

[e]A tariff is defined as a tax or duty to be paid on a particular thing (for instance, goods and services).

maximum demand may occur for just a few hours of the year, and at other times, the refrigeration and AC system works at lower capacity. Therefore, a smaller system can be chosen that operates continuously at a constant capacity. During the hours when the cooling needs of a building are lower than the capacity of the refrigeration and AC system, ice (as a storage medium) can be produced by the system for later use when the building's cooling needs are higher than the system's capacity. This cold storage in the form of ice can be employed to reduce the consumption of electricity during on-peak hours and shift it to non-peak hours [13, 14].

It should be noted that no refrigeration and AC system works at its full capacity for its entire period of operation. In addition, the ventilation and cooling load varies at different times of the day and night. The full capacity of cold-producing machines like chillers can only be used for a limited number of hours a day (on-peak hours) in summer. If electrical energy and ventilation and cooling loads can be transferred to off-peak times or reduced to a smaller value, a chiller with a lower capacity can be employed and would have a much lower initial cost. This chapter focuses on the production and storage of cold in the form of ice, because ice CTES systems have a much denser storage capacity than other storage mediums do, despite the fact that they need more sophisticated refrigeration and AC systems. In addition, they have the potential to be used in low-temperature air systems. In ice-based CTES systems, the latent heat of fusion of water is used to store cooling capacity [12, 13].

7.2 Cold production and storage in the form of ice

In the past, ice blocks were cut from frozen lakes or rivers in winter and stored in suitable insulated places such as ice-houses. Later, the stored ice was utilized as the source of cold in refrigeration systems during the warmer months. With the advent of mechanical refrigeration (in approximately the last quarter of the nineteenth century [31–33]), new systems replaced the old cold storage systems, supplying "instantaneous cooling load," meaning the rate at which cold thermal energy is convected to the zone air at a specified point in time [34, 35]. Cold storage is actually utilized in refrigeration and AC systems to shift electricity load (or demand) over time, thus flattening the load curve, which significantly diminishes the costs of electricity production in summer.

There are four methods by which cold storage can be used in reducing the electric power consumption of the refrigeration and AC system in an office, institutional, commercial, or industrial building, and shifting it to off-peak times [10]:

1. Working the refrigeration and AC system continuously at low capacity during business days by storing cold energy at night and using it during the day.
2. As with method (1), with the difference that the refrigeration and AC system does not operate during on-peak hours, but meets the cooling needs of the building through this nocturnal storage.
3. As with method (2), with the difference that the refrigeration and AC system works during weekends or other times when the building is closed or unoccupied in order to produce and store ice.
4. As with method (3), with the difference that the refrigeration and AC system operates all year round, producing and storing ice throughout the winter and utilizing it for space cooling in the summer.

Ice can be produced and stored in a variety of ways and used for cooling and AC purposes as required. The ice-based CTES technologies can be divided into two main groups, according to the process of heat transfer in the storage tank, as in [20, 21, 28]:

1. Static process based, including ice-on-coil (with both internal and external melting) and encapsulated ice storage, and
2. Dynamic process based, such as ice slurry and ice harvesting storage.

The semi-natural or low-energy approach for production of ice, described in Chapter 6, combines both processes. Two types of ice CTES techniques commonly utilized in commercial and industrial applications, namely internal and external melt ice-on-coil, are briefly explained next.

In the ice-on-coil with external melt method, the heat exchanger, submerged inside a rectangular or cylindrical (dominant shapes) tank filled with water, is in the form of parallel coils (or tubes). A refrigerant or secondary coolant fluid such as an ethylene glycol solution, at a temperature below the freezing point of water (0°C), passes through inside the coils, causing ice to form around their exterior surfaces. During the daytime or whenever a building needs to be cooled, water flows over and around the ice produced and melts it from the outside, cooling the water so that it can be used to cool the building when pumped to the fan coil units.

In the ice-on-coil with internal melt method, the heat exchanger is in the form of parallel and spiral coils (or tubes), located inside the tank in compressed form, and the fluid stream in adjacent coils flows in opposite directions (Fig. 7.2). The water around the coils in the tank is stationary. Through the tubes flows a refrigerant or secondary coolant fluid such as an

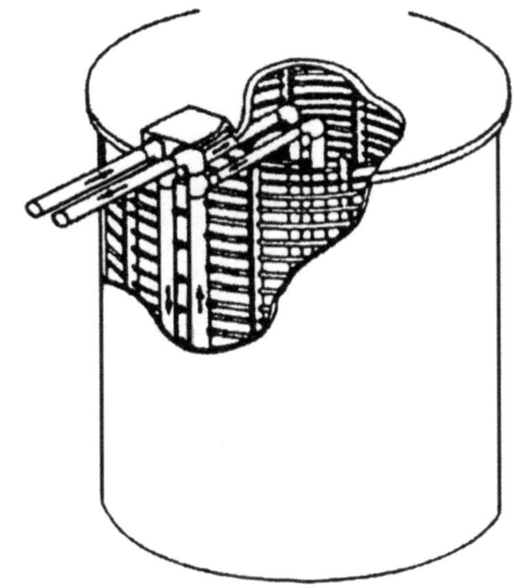

Fig. 7.2 Tank and coils for producing and storing ice using the internal melt ice-on-coil method.

ethylene glycol solution, which is cooled by passing through an evaporator. At the time of charging (ice producing), the fluid circulates in a closed loop between the evaporator of the refrigeration and AC machine and the storage tank, freezing the water around the outer surface of the coils in the tank. During the discharge (ice melting) phase, the fluid circulates in a closed loop between the tank tubes and the fan coil units used for cooling the building, causing the ice to melt from the inside.

Since the storage tank is separate from the chiller, depending on the storage capacity of the ice and the cooling demand of the building, several tanks can be used in parallel. To optimize system space, storage tanks of small dimensions can be employed, or regular-sized tanks can be placed underground. Respectively, Figs. 7.2 and 7.3 schematically represent the tank and the coils located inside it, and the circuits of the system of ice production and storage through the internal melt ice-on-coil method. As shown in Fig. 7.3, in this method, the chiller can be placed either upstream or downstream in the system.

Many studies have investigated storing cold within tanks of different shapes, sizes, and materials, using different storage techniques and various antifreezes and storage mediums, as well as the performance of different kinds of CTES systems. In addition, there have been economic assessments on topics such as diminishing the initial costs of such a system, and related

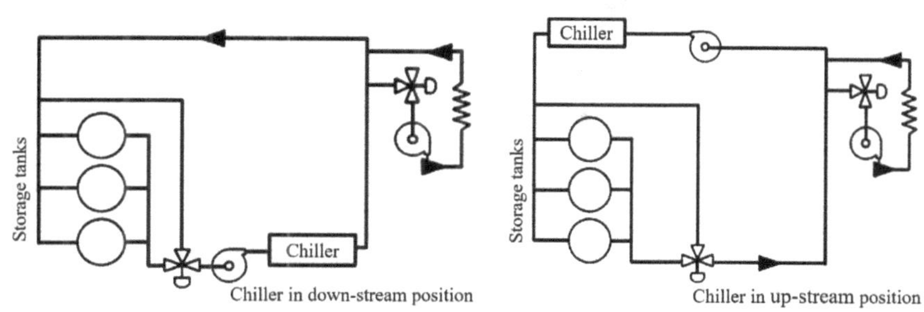

Fig. 7.3 Circuits of ice production and storage system for internal melt ice-on-coil method.

expenses, including for operation, repair and maintenance, and electric power consumption [10, 28, 36–52]. These studies and activities have become more widespread in recent years because of the significantly increasing demand for electrical energy and space ventilation and cooling in buildings [8, 12, 53–68], particularly in combination with solar AC systems and cold storage because of the intermittent nature of solar irradiation [69–79]. A few of the investigations are briefly described below.

MacCracken [36] has done considerable research on reducing the cost of producing and consuming energy, and storing ice in fully plastic tanks. Pearson [37] assessed and compared an ice storage system with five conventional AC systems for three office buildings: (1) a four-floor building with 8000 m^2 infrastructure, (2) a ten-floor building with 20,000 m^2 infrastructure, and (3) a twenty-floor building with 40,000 m^2 infrastructure. The storage system used by Pearson consisted of the means for storing ice and a combination of air and chilled water at low temperatures. The results obtained indicate that this system reduced the construction cost of the office buildings, as well as the capacity of electric power generation. Bahadori [10] proposed four approaches for storing cold energy for an office, commercial, or industrial building in order to reduce electrical energy consumption at on-peak times and shift it to off-peak ones. Sohn and Tomlinson [38] investigated cooling systems with daily energy storage at military installations in the United States, plus the efficiency of the storage tank used and that of the system's performance overall.

Tran et al. [39] studied and evaluated the performance of one of the existing systems for storing chilled water in commercial buildings. This study consisted of eight storage systems from four projects situated in Canada and the United States. The parameters of water volumetric flow rates, and the inlet and outlet temperatures of the storage tank were measured throughout the tank charging and discharging periods in order to determine the cooling capacity of the charging and discharging processes. In addition, the storage efficiency was assessed for the charging and discharging periods. Sohn [40] examined the performance of a cooling system with ice storage for a dental clinic building at Fort Bliss, TX, the United States. This system included a 26-ton ice-maker and a 300-ton-h aboveground steel storage tank. The peak cooling load of the building was 70 ton. This cold storage system was considered to supply the building's cooling need between 12 pm and 4 pm. Outside of these hours, the building was cooled via an existing chiller. In this study, the performance of the ice CTES system was compared with that of the existing chiller. Nasrollahi [50] designed a new cold storage system that relies on producing ice-on-coil with external melt.

Sehar et al. [55] studied the effect of ice storage cooling systems on the electricity consumption of a chiller for large- and medium-sized office buildings in various climate areas. The results show that weather conditions play an important role in the electrical energy consumption of chillers, both with and without storage systems. Areas with hot summers and minor changes in temperature between day and night are associated with greater chiller electricity consumption. Arcuri et al. [62] carried out technical and economic assessments of the use of ice storage for commercial buildings in five Brazilian cities. They utilized two operational methods and two kinds of chillers for new installations and retrofits. Their results indicate that ice storage can effectively shape peak cooling demand in such buildings. MacPhee and Dincer [67] evaluated the operation of four primary kinds of ice production and storage systems that are used for space cooling: ice slurry, ice-on-coil with external melt, ice-on-coil with internal melt, and encapsulated ice thermal storage. They also compared these ice CTES systems based on energy and exergy efficiencies for the charging, storage, and discharging processes.

Sun et al. [69] investigated and analyzed an attempt to reduce conventional summer energy use through the integration of a solar AC system and cold storage in an office building in Beijing, China. This one-floor building consisted of 40 offices and 5 conference rooms, with a 4025 m^2 infrastructure. Chilled water was employed as a storage medium. The researchers used the dynamic transient software TRNSYS to simulate the performance of the solar AC system. A comparison of the simulation results with empirical data showed good agreement. Using the data obtained from their simulation model, the researchers also examined the capacity of cold storage in solar AC systems for China's various climate areas and found that using properly designed solar AC systems integrated with cold storage would benefit China. Ibrahim et al. [75] experimentally studied the operation of a solar absorption cooling system integrated with ice CTES for an office space appropriate for seven people, located in Dhahran, Saudi Arabia, for two subsequent-day periods, one in March and the other in October. The cooling load for this office space was 10 kW, and the ice-on-coil method was used to produce ice inside a storage tank. The ice CTES, charged on the first day and discharged the next, was able to supply a back-up time of around 5–6 h for discharging, which was adequate for cooling the office space in the early hours of chiller warm-up (1–2 h) and during the low solar irradiation hours of the evening.

7.3 Working methods used in cold-producing systems to create and store ice

One of the key issues in designing cold storage systems for buildings is how these systems are to be operated to achieve the desired objectives [80]. They can be worked on a daily (24-h), weekly, or seasonal cycle, depending on the building's size and use, local climate, and the level of service required. For example, some large buildings like hotels need a CTES system based on a weekly cycle. This type of storage system is charged on the weekend, when the electricity rates are low, and is progressively discharged during the other days, resulting in diminishing initial expenses [80]. There are approaches for charging and discharging cold storage on a daily basis to supply cooling requirements during on-peak times. These approaches are usually classified into two main categories according to their mode of operation [12, 13, 67]: full storage or partial storage. In other words, the difference between the two categories is in the amount of on-peak cooling load transferred to off-peak hours. Fig. 7.4 illustrates three operating approaches for CTES using a chiller. Other strategies are possible by means of various sequences of multiple chillers.

In the full storage approach, the whole on-peak cooling load shifts to non-peak times (Fig. 7.4A). The refrigeration unit of the full storage system operates at its full capacity during off-peak hours and is switched off at on-peak hours, so the entire building's cooling need is met during these hours by the ice in the storage tank. This system requires a higher refrigeration capacity compared to the partial storage systems and is typically designed to work on the hottest predicted days of the year. Therefore, its storage tank and refrigeration equipment must be larger. The full storage system is the most attractive if on-peak time periods are relatively short or peak demand

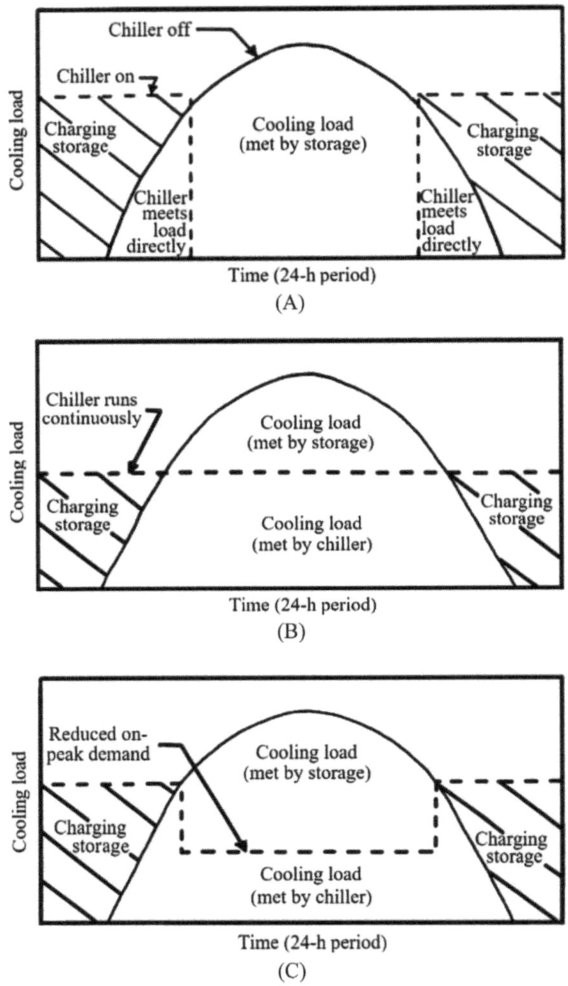

Fig. 7.4 Three operating approaches for cold thermal energy storage (CTES): (A) full storage, (B) partial storage load-leveling, and (C) partial storage demand-limiting [10, 12, 13, 28].

rates are high [12, 13]. With this kind of storage system, the peak cooling load may be reduced by up to 80%–90% less than needed with a conventional cooling system [81, 82]. The control of this system is also relatively simple [80, 83].

In the partial storage approach, part of the cooling load during peak times is met by the refrigeration and AC machine (or chiller), and the remainder is supplied by the CTES system. Partial storage systems can work with a chiller that is smaller and of lower capacity relative to the design load and can be divided into two subcategories in terms of operating conditions [12, 13]: load-leveling and demand-limiting. In partial storage load-leveling, the chiller runs continuously at its full capacity (designed capacity) 24-h a day (Fig. 7.4B). When a building does not require cooling energy or this need is less than the capacity of the chiller, the production and storage of cold is done in the form of ice within the storage tank. When the load (or demand) is higher, the surplus cooling demand is met by melting the ice stored in the tank. In this system, the required capacity for refrigeration reaches its lowest amount. A partial storage load-leveling operation can save around 40%–60% of peak cooling power load [81, 82]. This type of operation is very appropriate for applications where the on-peak cooling load is much higher than the average load [82, 83].

In partial storage demand-limiting, the refrigeration unit runs at a lower capacity during on-peak hours to keep energy consumption costs below the maximum specified level of consumption (Fig. 7.4C). This approach is a middle ground between the full storage and the partial storage load-leveling methods and is also the most economical option in terms of energy consumption cost [28, 83]. However, this method of storage requires more-sophisticated control systems than other storage approaches, to directly monitor the peak cooling demand of the building, as this peak demand must be supplied by the CTES [80, 82]. The partial storage demand-limiting operation is most appropriate for buildings with considerable demand rates and short occupancy hours [82]. Overall, partial storage is a more favorable alternative in terms of economics, due to the lower initial expenses, and so, it is the most commonly installed type of ice-based CTES system [12, 13].

7.4 The advantages of partial storage systems

Compared to other systems, partial storage systems have a number of advantages, the most important of which are summarized below.

1. Reduced capacity of chillers, and consequently, decreased associated costs (e.g., capital and operating expenses).
2. Transfer of electrical energy consumption associated with ventilation and cooling load from on-peak (or high demand) periods to off-peak (or low demand) periods, or significant reduction during on-peak hours and improved load factor.
3. No need to build new power plants (renewable and/or conventional) to meet electricity requirement for use in refrigeration and AC systems of new buildings.

4. Removed intermittent starting current of chillers.
5. Reduced capacity of fan coils and subsequent relevant motors due to the fluid's low temperature because of direct contact with ice, thus savings on initial costs as well as electricity consumption.
6. Diminished costs for electric power consumption.
7. Depreciation of additional investment related to purchase and installation of storage system in a relatively short period of time.
8. Increased range of spaces that can be ventilated without adding chillers to existing systems.
9. More-favorable conditions for correction of the power factor[f] at the place of consumption.
10. Possibility of using heat released during ice production to provide hot water and so on (heat pump system) [46].
11. Reduction of pollutants and greenhouse gas (GHG) emissions (particularly CO_2 and CFCs), lessening contribution to global warming [88–90].

7.5 The advantages of ice production via internal and external melt ice-on-coil

One of the most appropriate approaches for storing cold energy is to manufacture ice around tubes (or coils) submerged in a tank of water and then utilize the stored cold by melting the ice [91]. The methods of melt ice-on-coil, described in Section 7.2, can be either internal, through a refrigerant or secondary coolant (typically a solution of water and ethylene glycol[g]), or external, with the flow of water over and around the ice. Both approaches have advantages, as explained below.

– *The advantages of ice production using the method of ice-on-coil, internal melt.* This method has several advantages, such as [51, 52]:
   1. The chiller is utilized during off-peak hours, and the chiller and ice storage can be used at the same time during on-peak times.
   2. There is the possibility of creating a full ice block and increasing the amount of energy stored per unit volume of the tank.
   3. The tank can be manufactured in standard form according to the amount of cold energy storage required, and given the amount of

[f] In the terminology of electrical engineering, the "power factor" is defined as the ratio of the real (or working) power to apparent power [84–86]. In general, the power factor is utilized all over the world to measure and charge the active and reactive power of electrical systems [87].
[g] The most common mix is 25% ethylene glycol and 75% water [20].

cold storage required, a number of tanks can be utilized in parallel. They can also be produced with small dimensions and located in many suitable places, such as underground, which greatly reduces the space needed to house them.

4. The tank and the tubes (or coils) inside it are usually made of plastic (polyethylene), which has various advantages such as lightness, corrosion resistance, flexibility against ice expansion, and simplicity when used in the construction of spiral coils.
5. Spiral exchangers promote significantly heat transfer, because in the curvatures and bends, the flow is turbulent, and the heat transfer coefficient is higher.
6. A full ice block causes the tank to act as a very good insulator, so that only thin exterior insulation is sufficient for the tank.
7. Only one fluid is used for both the charging (ice producing) and discharging (ice melting) processes. Consequently, the cost of an additional pump is eliminated.
8. Control is relatively simple because the system is a closed loop [20].

– *The advantages of ice production using the method of ice-on-coil, external melt.* The advantages of this method include [51, 52]:
  1. The use of water as a discharge fluid,
  2. Forced heat transfer in the flow of water over the ice created around the tubes, and
  3. Higher total efficiency in comparison with other ice-based CTES systems because the chiller and related pumps are switched off at peak times and the only electrical energy input is from the ice water pump [20, 92].

7.6 Design and analysis of ice production and storage system for an office building

The performance of a cold storage system that produces ice and uses it to reduce the electricity required during on-peak consumption times was assessed for an office building in the city of Tehran, Iran. The three-story building has a total infrastructure of 900 m^2, and the height of each story is 3 m. The number of employees and clients on each floor during occupied times was considered to be 60 people [51, 52]. The working hours are assumed to be 12-h a day (from 7 am to 7 pm), 6 days a week (from Saturday to Thursday). Software was used to determine the heating and cooling loads (or demands) of the building at different hours of the day and night [93–95]. Due to factors such as the dimensions and physical characteristics of

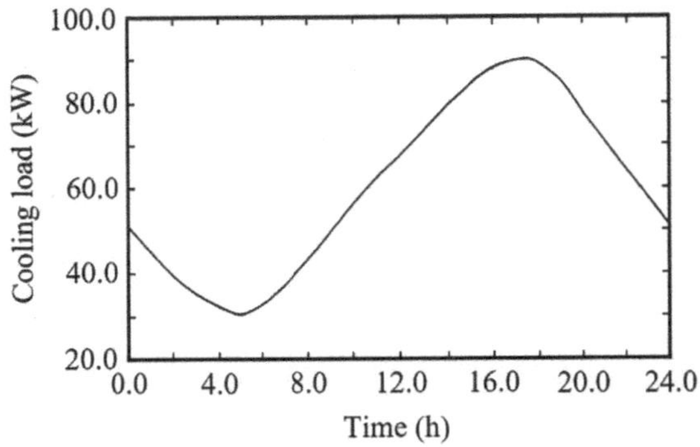

Fig. 7.5 Cooling demand of office building in Tehran City, Iran, at different hours of a normal working day.

different components of the building, its solar energy gain, heat exchange with the surrounding environment, the number of people present in the building at different hours, the heat generated within the building by the lighting system (for instance, light bulbs) and electrical appliances, the building's cooling needs at different hours of a normal working day in the month of August were determined. For this purpose, the maximum and minimum ambient air temperatures, respectively, 37.8°C and 22.8°C; the dew point of the ambient air, 19.5°C; and the sky clearness index, 0.65 (see Table 2.1) were selected and used to calculate the building's maximum and minimum cooling loads: 90 kW at 5 pm and 30 kW at 5 am, respectively. Fig. 7.5 shows the variations of the building's cooling load at different hours of a normal working day. The calculation results indicate that the cooling demand on a normal working day is 890 kWh or 3204 MJ per day.

7.6.1 Thermodynamic calculations for the system

As previously mentioned, for a daily (24-h) cycle, two cold storage systems, full storage and partial storage, can generally be designed to meet all or part of a building's cooling demand during on-peak hours. For the office building under consideration, both systems will be examined.

- *Partial storage system:*

  In this method, the refrigeration and AC machine (or chiller) works all 24-h of a day. When the building's cooling demand is lower than the capacity of the machine, the difference in capacity and demand is used for the production and storage of ice in the tank. In the hours when the building's cooling load is higher than the capacity of the machine, this difference is supplied through the melting of the ice produced and stored in the tank.

If the cooling need of the building in 12 h and the cooling capacity[h] (or cooling power) of the refrigeration and AC machine are indicated by Q_c and $\dot{q}$ (in kW), respectively, the cooling capacity of the machine can be defined by the following equation [51, 52]:

$$Q_c = \dot{q}\xi\,\Delta t \tag{7.1}$$

where ξ is the recovery efficiency of stored cold, and Δt is the duration of operating the selected refrigeration and AC machine. The parameter ξ equal to 85% and the duration of use of the machine (Δt) equal to 24 h were chosen. Thus, $\dot{q}$ is obtained equal to 43.63 kW from Eq. (7.1). In comparison with conventional refrigeration and AC system without cold storage, whose maximum capacity was 90 kW, the capacity for the system of interest was reduced by 51.5%.

- *Full storage system*:
  In this method, the selected refrigeration and AC machine is switched off at certain hours of the day when electrical energy consumption in the power grid is at its peak, and cooling demand is met by using the ice produced and stored in the tank during off-peak hours. If the duration of the shutdown of the refrigeration and AC machine is chosen to be 5 h, by means of Eq. (7.1), and 19 h for Δt and 85% for ξ are selected, the cooling capacity of the machine is determined to be 55.11 kW. Using this method, the capacity of the machine can be reduced by 38.77% in comparison with the conventional refrigeration and AC system without cold storage whose maximum capacity was 90 kW.

7.6.2 Determining the capacity of the ice storage tank

Herein, the capacity of the ice storage tank, for both systems of storage, is determined.

- *Partial storage system*:
  By drawing the curve of the cooling load changes over time (between 7 am and 7 pm), while the capacity ($\dot{q}$) of the refrigeration and AC machine with partial ice storage is constant and equal to 43.63 kW, it is observed that the cooling load of the considered building from 8 am to 7 pm is greater than the capacity of the selected refrigeration and AC machine, as shown in Fig. 7.6. In this figure, *SC* represents the storage capacity.

[h]The cooling capacity refers to the amount of heat withdrawal from a cooled space by the refrigeration machine (or system) over time [96, 97]. Its SI unit is Watt. It can also be quantified in British thermal unit (BTU) or, more commonly, tons.

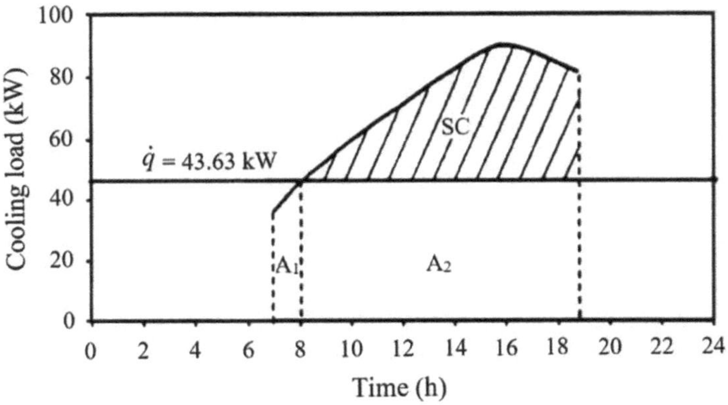

Fig. 7.6 Cooling need of the considered office building with a partial ice storage system.

According to Fig. 7.6, surfaces A_1 and A_2, combined, are determined to be equivalent to 520.25 kWh or 1873 MW, and the cold energy that needs to be stored in the form of ice is obtained using the following Equation [51, 52]:

$$SC = Q_c - (A_1 + A_2) = 3204 - 1873 = 1331\text{MJ} \tag{7.2}$$

The volume of ice that must be produced (V_i) is acquired from the equation below [51, 52]:

$$SC = m_i h_{if} = \rho_i V_i h_{if} \tag{7.3}$$

In this equation, m_i, ρ_i, and h_{if} represent, respectively, the ice mass, the ice density, and the enthalpy of the phase change of water to ice (or the enthalpy of the solidification of water). Referring to the thermodynamic tables of water, at 0°C, the parameters ρ_i and h_{if} are defined as 916.76 kg/m^3 and 333.42 kJ/kg, respectively [98]. By means of Eq. (7.3), the mass and volume of ice are determined to be 3992 kg and 4.354 m^3, respectively. The volume of the storage tank is selected to be 10% larger than the volume of ice obtained (4.35 m^3), namely 4.79 m^3.

- *Full storage system*:
  The curve of variations in the building's cooling demand from 7 am to 7 pm and the working method of the selected refrigeration and AC machine are shown in Fig. 7.7. It is assumed that the refrigeration and AC machine is switched off from 2 pm to 7 pm.

According to Fig. 7.7, surfaces A_1 and A_2, combined, are determined to be equivalent to 360 kWh or 1296 MW. Using Eq. (7.2), the parameter SC is obtained as follows:

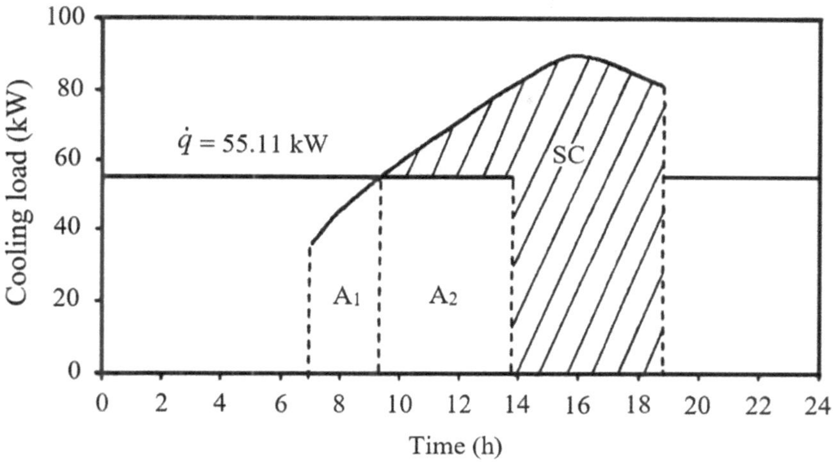

Fig. 7.7 Cooling load of the considered office building with a full ice storage system.

$$SC = 3204 - 1296 = 1908\ \text{MJ}$$

The mass and volume of ice can be specified by means of Eq. (7.3) and are, respectively, 5723 kg and 6.24 m^3. The volume of the storage tank is chosen to be 10% greater than the volume of ice obtained (6.24 m^3), meaning 6.86 m^3.

7.6.3 Determining the mass and volumetric flow rates of the interface fluid between evaporator and ice-making coil

In this subsection, the mass and volumetric flow rates of fluid flowing between the evaporator and ice-making coil are determined for both systems of storage.

– *Partial storage system:*

The mass and volumetric flow rates of the interface fluid for the partial storage system can be determined from the following equation [51, 52]:

$$\dot{q} = \dot{m}\, C \Delta T = \rho \dot{V} C \Delta T \tag{7.4}$$

In this equation, $\dot{m}$, $\dot{V}$, and C represent the mass flow rate, the volumetric flow rate, and the specific heat capacity of fluid, respectively. In addition, ΔT is the difference between the inlet and outlet temperatures of the fluid passing through the coil.

The selected fluid is a secondary coolant solution, a mixture of 30% ethylene glycol and 70% water, and its density and specific heat capacity, at an average temperature of $-5°C$, are, respectively, 1053.11 kg/m^3 and 3.574 kJ/kg K [99]. Choosing a temperature difference of 4°C between the inlet and outlet of the coil, and using Eq. (7.4), the mass and volumetric flow rates of the considered secondary coolant solution are found to be 3.052 kg/s and 2.898 L/s, respectively.

- *Full storage system*:
  By means of Eq. (7.4), and given that in this system the cooling capacity of the refrigeration and AC machine is 55.11 kW, the mass and volumetric flow rates of the interface fluid, respectively, are determined to be 3.856 kg/s and 3.661 L/s.

7.6.4 Analytical and numerical study of ice formation around tubes during internal melting

Considering a numerical model, the production and storage of ice-on-coil with internal melt has been investigated based on the quasi-stable analysis and the thermal resistance network methods. This model specifies the radius of the external freezing of water and the internal melting of ice at each time step and in each element of the coil. The cooling stage before the onset of freezing and the water heating stage after ice melting are also considered. In addition, this model determines the heat transfer of water around the coil during freezing (when the water temperature is higher than the melting point of the ice) and the heat transfer of the tank walls. In the model, the phenomenon of interference of ice and water layers at the end of the freezing and melting stages is considered. This model also specifies the pressure drop of the secondary coolant solution inside the coil [51, 52]. The modeled coil is divided into small elements. The law of conservation of energy has been written for each element and at each time step. The radius of ice in the freezing stage and also the radius of water in the melting stage are determined by trial and error. The outlet temperature of the interface fluid is obtained for each element and then utilized as the input temperature of the fluid in the next element. The time step in the program is considered to be 15 min and the number of coil elements to be 25 [51, 52]. According to the calculations, the time required for charging the tank, namely for producing 4.841 m^3 of ice in the storage tank, is 12 h, and the time for melting the same amount of ice is 5.5 h. The pressure drop in the ethylene glycol solution is 46.6 kPa.

7.6.5 Economic assessment and electricity consumption of the ice production and storage system in the office building

Economic analysis here means assessing the costs of ice production and storage systems and comparing them with those for conventional refrigeration and AC systems. The costs of the various parts of interest were obtained

Table 7.1 Comparison of annual electricity consumption of refrigeration and AC systems selected for the studied office building.

System type	Annual amount of electricity consumption GJ
Conventional refrigeration and AC system	52,755
Refrigeration and AC system with partial ice storage	52,770
Refrigeration and AC system with full ice storage	55,694

from the manufacturers. Table 7.1 shows the electrical energy consumption of the systems under study. Considering the prices obtained from the manufacturers of devices and equipment, the initial cost and electricity consumption of various refrigeration and AC systems, including those with partial ice storage, those with full ice storage, and a conventional system, are almost similar. Moreover, significant savings are not achieved by choosing a refrigeration and AC system that produces and stores ice.

If local and regional electric power companies impose special charges on electricity consumption during on-peak hours, or if the price of electrical energy during these hours is higher than in off-peak hours, the current cost of refrigeration and AC systems with cold storage will be less than that of conventional refrigeration and AC systems. In such cases, these systems are economically viable and even preferable. For example, with reference to a 2018 announcement by the Ontario Energy Board (OEB), Ontario, Canada [100], the TOU electricity rates for different hours of the day during the summer period have been calculated based on Table 7.2. As illustrated from this table, the TOU charge during on-peak hours (13.2 ¢/kWh) was more

Table 7.2 Time of Use (TOU) electricity rates for different hours of the day in summer in Ontario, Canada, effective May 1, 2018, to April 30, 2019 [100].

TOU price period	Summer period hours May 1 to October 31	TOU prices effective May 1, 2018, to April 30, 2019[a]	Change
Off-peak	Weekdays from 7 pm to 7 am, all-day weekends and holidays	6.5 ¢/kWh	No change
Mid-peak	Weekdays between 7 am–11 am and 5 pm–7 pm	9.4 ¢/kWh	Down 0.1 ¢/kWh
On-peak	Weekdays from 11 am to 5 pm	13.2 ¢/kWh	No change

[a] Prices are in Canadian dollars.

than double that of the TOU charge during off-peak hours (6.5 ¢/kWh). Therefore, the use of cold storage systems during the summer in Ontario is a cost-effective way of reducing electricity consumption and associated costs (especially in its southern cities such as Toronto and Ottawa where most residents live).

7.7 Conclusion

Energy and its associated issues such as energy sources, storage, conversion, and saving; energy management; increasing energy efficiency; and reducing energy costs are very important and hot topics because of the environmental and other challenges we face today [101–103]. For example, the storage of energy in different forms (e.g., mechanical, chemical, and thermal) plays a considerable role in most aspects of science and technology [67]. Thermal energy storage (heating or cooling) seems to be the most suitable, cost-effective, and reliable approach to balance, or even minimize, the mismatches between the production and demand of energy that sometimes occur [12, 67, 104]. For instance, cold storage, full or partial, can broadly be utilized for cooling and AC purposes for a variety of reasons, including load shifting; peak shaving; valley filling; flexibility in energy service provision; energy efficiency improvement; energy management and saving; and the reduction of expenses (such as capital, operating, and electrical energy consumption costs). In addition, CTES systems can help protect the environment through diminishing contaminant and GHG emissions, especially CO_2 and CFCs, resulting in lower global warming. Cold storage can in fact be an effective technology for achieving sustainable development goals in the modern world.

References

[1] E.M. Dobrescu, Reducing the polluting emissions, a source for diminishing the climate changes, Rev. Rom. Econ. 28 (2) (2009) 1–15.
[2] J.L. Dupont, P. Domanski, P. Lebrun, F. Ziegler, 38th Note on Refrigeration Technologies: The Role of Refrigeration in the Global Economy: Introduction, The International Institute of Refrigeration (IIR), Paris, France, June 2019.
[3] F. Birol, The Future of Cooling: Opportunities for Energy-Efficient Air Conditioning, Technology Report, International Energy Agency (IEA), Paris, France, May 2018, p. 92.
[4] N. Karali, N. Shah, W.Y. Park, N. Khanna, C. Ding, J. Lin, N. Zhou, Improving the energy efficiency of room air conditioners in China: costs and benefits, Appl. Energy 258 (2020) 114023.
[5] G.O. Watkins, A third factor in the variation of productivity: the load factor, Am. Econ. Rev. 5 (4) (1915) 753–786, American Economic Association.
[6] P. Chanhom, S. Nuilers, N. Hatti, A new V2G control strategy for load factor improvement using smoothing technique, Adv. Electr. Comp. Eng. 17 (3) (2017) 43–51.

[7] M.R. Aghaebrahimi, H. Taherian, Utilization of electric vehicles for improvement of daily load factor in the price-responsive environment of smart grids, Iran. J. Power Eng. 1 (1) (2016) 33–40.
[8] G. Li, Y. Hwang, R. Radermacher, Cold thermal energy storage materials and applications toward sustainability, in: X. Zhang, I. Dincer (Eds.), Energy Solutions to Combat Global Warming, Lecture Notes in Energy, vol. 33, Springer, Cham, 2017, pp. 67–117.
[9] C. Veerakumar, A. Sreekumar, Phase change material based cold thermal energy storage: materials, techniques and applications—a review, Int. J. Refrig. 67 (2016) 271–289.
[10] M.N. Bahadori, Reduction of peak electrical energy demands for cooling of industrial buildings, in: International Conference on Heat Transfer and Energy Conservation, 1988, pp. 153–156, Shenyang, China, Oct. 6–9.
[11] A. Arteconi, F. Polonara, Assessing the demand side management potential and the energy flexibility of heat pumps in buildings, Energies 11 (7) (2018) 1846.
[12] I. Dincer, M.A. Rosen, Thermal Energy Storage: Systems and Applications, second ed., John Wiley & Sons, Inc., Chichester, UK, 2011.
[13] I. Dincer, On thermal energy storage systems and applications in buildings, Energ. Buildings 34 (4) (2002) 377–388.
[14] N. Yu, R.Z. Wang, L.W. Wang, Sorption thermal storage for solar energy, Prog. Energy Combust. Sci. 39 (5) (2013) 489–514.
[15] D. Erdemir, Numerical investigation of thermal performance of geometrically modified spherical ice capsules during the discharging period, Int. J. Energy Res. 43 (9) (2019) 4554–4568.
[16] B. He, F. Setterwall, Technical grade paraffin waxes as phase change materials for cool thermal storage and cool storage systems capital cost estimation, Energ. Conver. Manage. 43 (13) (2002) 1709–1723.
[17] W.A. Qureshi, N.K.C. Nair, M.M. Farid, Impact of energy storage in buildings on electricity demand side management, Energ. Conver. Manage. 52 (5) (2011) 2110–2120.
[18] A.A. Dehghan, A.R. Dehghani, Experimental and theoretical investigation of thermal performance of underground cold-water reservoirs, Int. J. Therm. Sci. 50 (5) (2011) 816–824.
[19] S.Ø. Jensen, A. Marszal-Pomianowska, R. Lollini, W. Pasut, A. Knotzer, P. Engelmann, A. Stafford, G. Reynders, IEA EBC annex 67 energy flexible buildings, Energ. Buildings 155 (2017) 25–34.
[20] F.F. Kosi, D. Antonijevic, B. Zivkovic, M.A. Galil, M.S. Komatina, U.M. Milovancevic, Cold thermal energy storage, Chapter 20, in: P.D. Gaspar, P.D. da Silva (Eds.), Handbook of Research on Advances and Applications in Refrigeration Systems and Technologies, Engineering Science Reference, IGI Global, Hershey, PA, 2015.
[21] R.A. Parsons (Ed.), Thermal storage, Chapter 33, in: ASHRAE Handbook, HVAC Application (SI), American Society of Heating, Refrigeration, and Air-Conditioning Engineers (ASHRAE), Atlanta, GA, pp. 33.1–33.24, 1999.
[22] F.A. Wolak, An Experimental Comparison of Critical Peak and Hourly Pricing: The PowerCentsDC Program, Department of Economics Stanford University, Stanford, CA, March 2010.
[23] F.A. Wolak, Do residential customers respond to hourly prices? Evidence from a dynamic pricing experiment, Am. Econ. Rev. 101 (3) (2011) 83–87.
[24] S. Javaid, N. Javaid, M.S. Javaid, S. Javaid, U. Qasim, Z.A. Khan, Optimal scheduling in smart homes with energy storage using appliances' super-clustering, in: 2016 10th International Conference on Innovative Mobile and Internet Services in Ubiquitous Computing (IMIS), IEEE, Fukuoka, Japan, 2016, pp. 342–348, July 6–8.
[25] F. Chui, A. Elkamel, R. Surit, E. Croiset, P.L. Douglas, Long-term electricity demand forecasting for power system planning using economic, demographic and climatic variables, Eur. J. Ind. Eng. 3 (3) (2009) 277–304.
[26] S.M. Al-Alawi, S.M. Islam, Principles of electricity demand forecasting: I. Methodologies, Power Eng. J. 10 (3) (1996) 139–143.

[27] S.M. Al-Alawi, S.M. Islam, Principles of electricity demand forecasting: II. Applications, Power Eng. J. 11 (2) (1997) 91–95.
[28] C.E. Dorgan, J.S. Elleson, Design Guide for Cool Thermal Storage, American Society of Heating, Refrigeration and Air-Conditioning Engineers (ASHRAE), Atlanta, GA, 1993.
[29] B. Shen, G. Ghatikar, Z. Lei, J. Li, G. Wikler, P. Martin, The role of regulatory reforms, market changes, and technology development to make demand response a viable resource in meeting energy challenges, Appl. Energy 130 (2014) 814–823.
[30] A. Arteconi, F. Polonara, Demand side management in refrigeration applications, Int. J. Heat Technol. 35 (2017) S58–S63.
[31] E.M. Yahia (Ed.), Postharvest Biology and Technology of Tropical and Subtropical Fruits: Fundamental Issues, vol. 1, Woodhead Publishing Ltd., Cambridge, UK, 2011.
[32] C. Marella, K. Muthukumarappan, Food freezing technology, Chapter 13, in: M. Kutz (Ed.), Handbook of Farm, Dairy and Food Machinery Engineering, third ed., Academic Press, 2019, pp. 355–378.
[33] S. Reif-Acherman, The early ice making systems in the nineteenth century, Int. J. Refrig. 35 (2012) 1224–1252.
[34] S. Schiavon, K.H. Lee, F. Bauman, T. Webster, Simplified calculation method for design cooling loads in underfloor air distribution (UFAD) systems, Energ. Buildings 43 (2–3) (2011) 517–528.
[35] J.D. Spitler, Load Calculation Applications Manual, I-P edition, second ed., American Society of Heating, Refrigeration, and Air-Conditioning Engineers (ASHRAE), Atlanta, GA, 2014.
[36] C.D. MacCracken, Cooling storage: awakening giant, in: Advances in Energy Utilization Technology, World Energy Engineering Congress, Fairmont Press Inc., Atlanta, GA, Oct. 1981, pp. 217–220.
[37] F.J. Pearson, Ice Storage Can Reduce the Construction Cost of Office Buildings, vol. 95, American Society of Heating, Refrigeration and Air-Conditioning Engineers (ASHRAE) Transactions, 1989, pp. 1308–1316, Part 1.
[38] C.W. Sohn, J.J. Tomlinson, Diurnal Ice Storage Cooling Systems for Army Facilities, vol. 95, American Society of Heating, Refrigeration and Air-Conditioning Engineers (ASHRAE) Transactions, 1989, pp. 1079–1085, Part 1.
[39] N. Tran, J.F. Kreider, P. Brothers, Field Measurement of Chilled Water Storage Thermal Performance, vol. 95, American Society of Heating, Refrigeration and Air-Conditioning Engineers (ASHRAE) Transactions, 1989, pp. 1106–1112, Part 1.
[40] C.W. Sohn, Field Performance of an Ice Harvester Storage Cooling System, vol. 97, American Society of Heating, Refrigeration and Air-Conditioning Engineers (ASHRAE) Transactions, 1991, pp. 1187–1193, Part 2.
[41] S.C. Silver, A. Milibitz, J.W. Jones, J.L. Peterson, B.D. Hunn, Component Models for Computer Simulation of Ice Storage Systems, vol. 95, American Society of Heating, Refrigeration and Air-Conditioning Engineers (ASHRAE) Transactions, 1989, pp. 1214–1226, Part 1.
[42] J.W. Jones, G.S. Shiddapur, Evaluation of RP-459 Algorithms for Modeling External Melt, Ice-On-Pipe Thermal Storage System Components, vol. 101, American Society of Heating, Refrigeration and Air-Conditioning Engineers (ASHRAE) Transactions, 1995, pp. 1342–1351, Part 2.
[43] J.H.M. Neto, M. Krarti, Integral solutions for ice formation and melting outward of the external wall of a pipe with internal and external convection, in: Proceedings of the 30th Intersociety Energy Conversion Engineering Conference (IECEC), Orlando, FL, USA, vol. 2, 1995, pp. 243–248.
[44] T.B. Jekel, J.W. Mitchell, S.A. Klein, Modeling of Ice Storage Systems, vol. 99, American Society of Heating, Refrigeration and Air-Conditioning Engineers (ASHRAE) Transactions, 1993, pp. 1016–1023, Part 1.

[45] C.W. Carey, J.W. Mitchell, W.A. Beckman, The Control Storage Systems, vol. 101, American Society of Heating, Refrigeration and Air-Conditioning Engineers (ASHRAE) Transactions, 1995, pp. 1345–1352, Part 1.
[46] R.K. Strand, C.O. Pederson, G.N. Coleman, Development of Direct and Indirect Ice-Storage Models for Energy Analysis Calculations, vol. 100, American Society of Heating, Refrigeration and Air-Conditioning Engineers (ASHRAE) Transactions, 1994, pp. 1230–1244, Part 1.
[47] B. Vick, D.J. Nelson, X. Yu, Model of an Ice-On-Pipe Brine Thermal Storage Component, vol. 102, American Society of Heating, Refrigeration and Air-Conditioning Engineers (ASHRAE) Transactions, 1996, pp. 45–54, Part 1.
[48] J.H.M. Neto, M. Krarti, Deterministic Model for an Internal Melt Ice-On-Coil Thermal Storage Tank, vol. 103, American Society of Heating, Refrigeration and Air-Conditioning Engineers (ASHRAE) Transactions, 1997, pp. 113–124, Part 1.
[49] J.H.M. Neto, M. Krarti, Experimental Validation of a Numerical Model for an Internal Melt Ice-On-Coil Thermal Storage Tank, vol. 103, American Society of Heating, Refrigeration and Air-Conditioning Engineers (ASHRAE) Transactions, 1997, pp. 125–138, Part 1.
[50] A. Nasrollahi, Cold Storage System, Royan Mobaddel Co., Designer and Manufacturer of Heating and Cooling Equipment, Isfahan, Iran, 1997.
[51] H. Alifazli, Reducing the Maximum Electricity Load Required for Cooling Buildings with Daily Energy Storage, M.Sc. Thesis, School of Mechanical Engineering, Sharif University of Technology, Tehran, Iran, 1998 (in Farsi).
[52] M.N. Bahadori, H. Alifazli, Reduction of electricity required for refrigeration and air-conditioning systems of buildings during the peak hours of electricity consumption by producing and storing ice at off-peak hours, in: The 4th National Energy Congress, Tehran, Iran, May 2003 (in Farsi).
[53] A. Hajiah, M. Krarti, Optimal control of building storage systems using both ice storage and thermal mass—part I: simulation environment, Energ. Conver. Manage. 64 (2012) 499–508.
[54] A. Hajiah, M. Krarti, Optimal controls of building storage systems using both ice storage and thermal mass—part II: parametric analysis, Energ. Conver. Manage. 64 (2012) 509–515.
[55] F. Sehar, S. Rahman, M. Pipattanasomporn, Impacts of ice storage on electrical energy consumptions in office buildings, Energ. Buildings 51 (2012) 255–262.
[56] J.A. Candanedo, V.R. Dehkordi, M. Stylianou, Model-based predictive control of an ice storage device in a building cooling system, Appl. Energy 111 (2013) 1032–1045.
[57] J. Vetterli, M. Benz, Cost-optimal design of an ice-storage cooling system using mixed-integer linear programming techniques under various electricity tariff schemes, Energ. Buildings 49 (2012) 226–234.
[58] J. Jazaeri, T. Alpcan, R.L. Gordon, Model predictive control of residential demand in low voltage network using ice storage, in: 2018 Australian and New Zealand Control Conference (ANZCC), IEEE, Melbourne, Australia, 2018, pp. 51–55. Dec. 6–8.
[59] N. Luo, T. Hong, H. Li, R. Jia, W. Weng, Data analytics and optimization of an ice-based energy storage system for commercial buildings, Appl. Energy 204 (2017) 459–475.
[60] V. Kumar, K. Hewage, H. Haider, R. Sadiq, Techno-economic performance evaluation of building cooling systems: a study of snow storage and conventional chiller systems, Cold Reg. Sci. Technol. 130 (2016) 8–20.
[61] B. Habeebullah, R. Jassim, N. Turkmen, A. Bokhary, M. Alhazmy, Cost-effectiveness opportunities for thermal energy storage systems: a case study of school building in Saudi Arabia, Int. J. Sustain. Green Energy 5 (4) (2016) 59–70.
[62] B. Arcuri, C. Spataru, M. Barrett, Evaluation of ice thermal energy storage (ITES) for commercial buildings in cities in Brazil, Sustain. Cities Soc. 29 (2017) 178–192.

[63] A. Tam, D. Ziviani, J.E. Braun, N. Jain, Development and evaluation of a generalized rule-based control strategy for residential ice storage systems, Energ. Buildings 197 (2019) 99–111.
[64] F. Afsharpanah, S.S.M. Ajarostaghi, K. Sedighi, The influence of geometrical parameters on the ice formation enhancement in a shell and double coil ice storage system, SN Appl. Sci. 1 (2019) 1264.
[65] M. Elhelw, W.M. El-Maghlany, Thermodynamic analysis of two air conditioning systems with ice thermal storage in Egypt, J. Therm. Anal. Calorim. 140 (2020) 2563–2573.
[66] X. Song, L. Liu, T. Zhu, S. Chen, Z. Cao, Study of economic feasibility of a compound cool thermal storage system combining chilled water storage and ice storage, Appl. Therm. Eng. 133 (2018) 613–621.
[67] D. MacPhee, I. Dincer, Performance assessment of some ice TES systems, Int. J. Therm. Sci. 48 (12) (2009) 2288–2299.
[68] A.A. Altohamy, I.M. Elsemary, S. Abdo, M.A. Abdelrahman, A.A. Attia, R.Y. Sakr, Encapsulation surface roughness effect on the performance of cool storage systems, J. Energy Storage 28 (2020) 101279.
[69] Z. Sun, Y. Zhao, W. Xu, D. Wang, H. Li, X. Zhang, H. Li, Y. Li, Cold storage capacity for solar air-conditioning in office buildings in different climates, J. Therm. Sci. 28 (6) (2019) 1195–1204.
[70] G. Angrisani, E. Entchev, C. Roselli, M. Sasso, F. Tariello, W. Yaïci, Dynamic simulation of a solar heating and cooling system for an office building located in Southern Italy, Appl. Therm. Eng. 103 (2016) 377–390.
[71] P. Esparcieux, O. Baup, D. Mugnier, C. Weber, C. Marvillet, Development of a photovoltaic driven thermodynamic chiller, application to solar air conditioning and cooling storage, in: EuroSun 2018 Conference Proceedings, Rapperswil, Switzerland, Sep. 2018, pp. 1–12.
[72] F. Cascetta, L. Cirillo, A. Corte, S. Nardini, Comparison between different solar cooling thermally driven system solutions for an office building in Mediterranean Area, Int. J. Heat Technol. 35 (1) (March 2017) 130–138.
[73] X. Ma, W. Zhang, F. Li, S.B. Riffat, Solar ejector cooling technologies, in: X. Zhao, X. Ma (Eds.), Advanced Energy Efficiency Technologies for Solar Heating, Cooling and Power Generation, Springer, Cham, 2019, pp. 287–309.
[74] F. Reda, S. Paiho, R. Pasonen, M. Helm, F. Menhart, R. Schex, A. Laitinen, Comparison of solar assisted heat pump solutions for office building applications in Northern climate, Renew. Energy 147 (2020) 1392–1417.
[75] N.I. Ibrahim, M.M.A. Khan, I.M. Mahbubul, R. Saidur, F.A. Al-Sulaiman, Experimental testing of the performance of a solar absorption cooling system assisted with ice-storage for an office space, Energ. Conver. Manage. 148 (2017) 1399–1408.
[76] C. Yao, X. Kong, Y. Li, Y. Du, C. Qi, Numerical and experimental research of cold storage for a novel expanded perlite-based shape-stabilized phase change material wallboard used in building, Energ. Conver. Manage. 155 (2018) 20–31.
[77] M. Saffari, A. de Gracia, C. Fernández, M. Belusko, D. Boer, L.F. Cabeza, Optimized demand side management (DSM) of peak electricity demand by coupling low temperature thermal energy storage (TES) and solar PV, Appl. Energy 211 (2018) 604–616.
[78] A.M. Reda, A.H.H. Ali, M.G. Morsy, I.S. Taha, Design optimization of a residential scale solar driven adsorption cooling system in upper Egypt based, Energ. Buildings 130 (2016) 843–856.
[79] A. Boero, F. Agyenim, Modeling and simulation of a small-scale solar-powered absorption cooling system in three cities with a tropical climate, Int. J. Low Carbon Technol. 15 (1) (2020) 1–16.

[80] T.A. Reddy, J.F. Kreider, P.S. Curtiss, A. Rabl, Heating and Cooling of Buildings: Principles and Practice of Energy Efficient Design, third ed., CRC Press, Boca Raton, FL, 2016.
[81] M.A. Piette, E. Wyatt, J. Harries, Technology Assessment: Thermal Cool Storage in Commercial Buildings, Lawrence Berkeley Laboratory Report, LBL-25521, University of California, Berkeley, CA, Jan. 1988.
[82] S.M. Hasnain, Review on sustainable thermal energy storage technologies, Part 2: Cool thermal storage, Energ. Conver. Manage. 39 (1998) 1139–1153.
[83] H. Akbari, A. Mertol, Thermal energy storage for cooling of commercial buildings, in: B. Kılkış, S. Kakaç (Eds.), Energy Storage System, Kluwer Academic Publishers, Dordrecht, The Netherlands, 1989, pp. 315–347.
[84] S. Rustemli, M. Ates, Measurement and simulation of power factor using pic16f877, Electr. Rev. 88 (6) (2012) 290–294.
[85] D.K. Jang, J.H. Chang, Performance comparison of PM synchronous and PM vernier machines based on equal output power per unit volume, J. Electr. Eng. Technol. 11 (1) (2016) 150–156.
[86] T. Manglani, Y.S. Shishodia, Power factor improvement in radial distribution system using bionic random search algorithm, Int. J. Innov. Res. Stud. 2 (4) (2013) 377–387.
[87] C.D. Oancea, Power factor evaluation in data acquisition systems, in: 2017 9th IEEE International Conference on Intelligent Data Acquisition and Advanced Computing Systems: Technology and Applications (IDAACS), vol. 1, IEEE, Bucharest, Romania, 2017, pp. 481–484. Sep. 21–23.
[88] I. Dincer, S. Dost, X. Li, Performance analyses of sensible heat storage systems for thermal applications, Int. J. Energy Res. 21 (1997) 1157–1171.
[89] I. Dincer, M.A. Rosen, Energetic, environmental and economic aspects of thermal energy storage systems for cooling capacity, Appl. Therm. Eng. 21 (11) (2001) 1105–1117.
[90] M.A. Rosen, I. Dincer, N. Pedinelli, Thermodynamic performance of ice thermal energy storage systems, J. Energy Resour. Technol. 122 (4) (2000) 205–211.
[91] J.W. Mitchell, J.E. Braun, Principles of Heating, Ventilation, and Air Conditioning in Buildings, John Wiley & Sons, Inc., New York, NY, 2012.
[92] W. Yi, W. Dong, Modeling and simulation of discharging characteristics of external melt ice-on coil storage system, Int. J. Smart Home 9 (2) (2015) 179–192.
[93] A. Ahmadi-Nedoshan, The Potential for Reducing Cooling and Heating Energy in Residential Buildings in Tehran Using Ceiling Fans, M.Sc. Thesis, School of Mechanical Engineering, Sharif University of Technology, Tehran, Iran, 1998 (in Farsi).
[94] H. Safarzadeh, Thermal Performance Evaluation of Courtyards, PhD Thesis, School of Mechanical Engineering, Sharif University of Technology, Tehran, Iran, 2004 (in Farsi).
[95] H. Safarzadeh, M.N. Bahadori, Passive cooling effects of courtyards, Build. Environ. 40 (1) (2005) 89–104.
[96] Y.A. Cengel, M.A. Boles, Thermodynamics: An Engineering Approach, seventh ed., McGraw-Hill Education, New York, NY, 2011.
[97] A. Subiantoro, K.T. Ooi, A.Z. Junaidi, Performance and suitability comparisons of some R22 possible substitute refrigerants, in: 8th International Conference on Compressors and their Systems, Woodhead Publishing Ltd., Cambridge, UK, 2013, pp. 67–76. Sep. 9–10.
[98] R.E. Sonntag, C. Borgnakke, G.J. Van Wylen, Fundamentals of Thermodynamics, sixth ed., John Wiley & Sons, Inc., New York, NY, 2002.
[99] ASHRAE, 1997 ASHRAE Handbook: Fundamentals, SI ed., American Society of Heating, Refrigeration and Air-Conditioning Engineers (ASHRAE), Inc., Atlanta, GA, 1997.

[100] Ontario Energy Board (OEB), Electricity Prices Remain Virtually Unchanged Starting May 1, Ontario, Canada, 19 April, 2018. https://www.oeb.ca/newsroom/2018/electricity-prices-remain-virtually-unchanged-starting-may-1. (Accessed 21 July 2020).
[101] A.R. Dehghani-Sanij, E. Tharumalingam, M.B. Dusseault, R. Fraser, Study of energy storage systems and environmental challenges of batteries, Renew. Sustain. Energy Rev. 104 (2019) 192–208.
[102] D.G. Vlachos, S. Caratzoulas, The roles of catalysis and reaction engineering in overcoming the energy and the environment crisis, Chem. Eng. Sci. 65 (1) (2010) 18–29.
[103] A.R. Dehghani-Sanij, in: A. Sayigh (Ed.), Cisterns: Sustainable Development, Architecture and Energy, River Publishers, Denmark, 2016.
[104] P.T. Sardari, D. Gidding, M. Gillott, D. Grant, G. Walker, Effect of air channel geometry on the performance of a composite metal foam-PCM to air heat exchanger, in: Innovative Applied Energy (IAPE '19), Oxford, UK, 2019, March 14–15.

CHAPTER 8

Using various hybrid systems to supply thermal energy needs of buildings

8.1 Introduction

Global energy demand is significantly rising among all sectors (building, industry, transport, etc.) for a variety of reasons, the principal ones being fast-growing industrialization and populations worldwide, as described in Chapter 1. The building sector, which includes residential, office, commercial, and educational buildings, is responsible for around 40% of the world's total energy utilization [1–4], and about one-third of its greenhouse gas (GHG) emissions [5–7]. For example, in Europe, buildings account for ~40% of the total energy consumption and ~36% of the CO_2 emitted [8, 9]. In 2018, global primary energy use was $\sim 5.76 \times 10^{14}$ MJ [10], of which residential and commercial buildings had a substantial share (~29%) [11]. Fig. 8.1 shows the global consumption of primary energy by end-use sector, as well as the annual demand growth and the share of each sector to 2017 and a prediction to 2040. The consumption of energy in the building sector varies among countries [12, 13], influenced by factors such as weather and geographical conditions; building use, size, age, and design; the heating/cooling loads required; the accessibility and price of different types of energy; community welfare levels; and other less important factors [14–16]. For instance, buildings' energy use in developed countries ranges from approximately 35% to 40% of total energy use, of which around 50%–75% is from electricity [14]. As a specific example, the building sector in the United States consumes ~40% of the country's total energy produced [17–19], and ~75% of this comes from electrical energy [17]. According to Fig. 8.2, electricity plays a key role in meeting the energy needs of buildings [11], which is primarily produced by fossil fuel-based power plants.

Heating, ventilation, and air-conditioning (HVAC) systems are reported to account for over 60% of the energy consumed in buildings [1–3, 20]. For example, HVAC systems are responsible for ~65% of the total energy use in

Ice-Houses
https://doi.org/10.1016/B978-0-12-822275-1.00001-1
Copyright © 2021 Elsevier Inc.
All rights reserved.

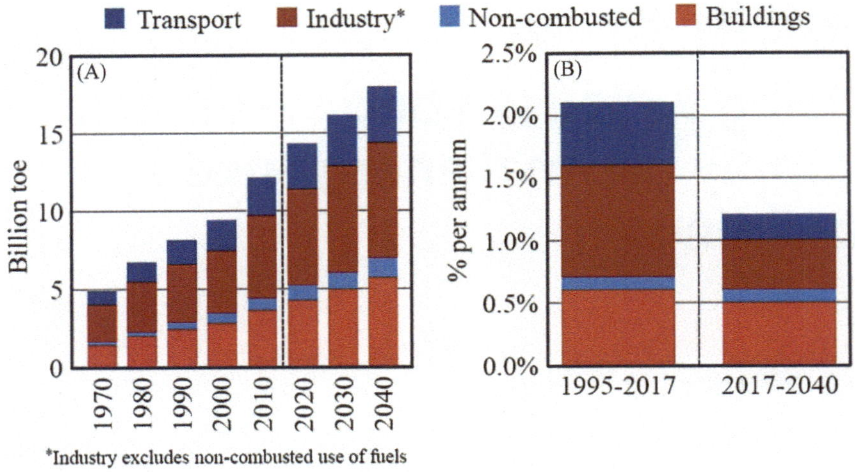

Fig. 8.1 (A) Worldwide demand for primary energy based on end-use sector, and (B) the annual demand growth and the share of each sector [11].

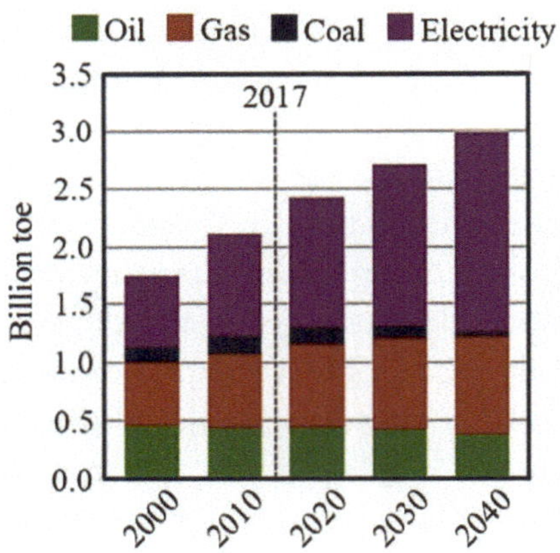

Fig. 8.2 Final consumption of energy in buildings related to energy source [11].

China's buildings [21, 22], and in Canada, around 63% of the total energy consumption in residential buildings is for heating/cooling needs [23]. The share of air-conditioning (AC) systems in the use of energy in buildings is around 70% in certain regions of the Middle East (e.g., in the Kingdom of Saudi Arabia (KSA)) during the hot months of the year [24, 25]. According to Pérez-Lombard et al. [15], Sohail and Qureshi [26] and Rashid et al. [27], HVAC systems account for a large share of energy use in commercial buildings: 55% in the UK, 52% in Spain, and 48% in the USA, for instance. It is expected that ongoing changes in technology and lifestyle choices, as well as increased economic prosperity, will dramatically escalate energy use in buildings for applications such as HVAC, lighting, appliances, refrigeration, and domestic hot water. Consequently, buildings can be expected to take a larger share of total energy consumption in the future [28, 29]. Based on the global HVAC systems' market report [30, 31], it is estimated that HVAC systems will have a compound growth rate of 5.5%–5.9% between 2015 and 2023, with space cooling taking the highest share: 37% of the total energy use in buildings by 2050 [32], as shown in Fig. 8.3.

Fossil fuels (most particularly, oil, coal, and natural gas) and electricity generated from fossil fuel-based power plants are currently the main sources of energy used to meet the needs of buildings (especially for space cooling and heating) and will remain so for at least the next few decades [33–37]. As an example, the energy consumption in the building sector of Iran, including residential, commercial, and public buildings, was ~489.1 million barrels of crude oil (equal to $\sim 3.0 \times 10^{12}$ MJ[a]) in 2016, which was around 41% of the total energy consumed among all sectors. The main energy carriers in this sector were natural gas (~75%), electricity (~15%), and oil products (~9%). In 2016, out of 76.2 GW of nominal power generated by various power plants in Iran, ~37% was produced by gas power plants, ~26% by combined cycle power plants, ~21% by steam power plants, ~15% by hydropower plants, ~1.3% by nuclear power plants, ~0.4% by diesel power plants, and ~0.1% by renewable power plants [38]. Although the efficiency of energy conversion in some of Iran's power plants (e.g., combined cycle and hydro) is relatively high, electricity is delivered to the consumer with reduced efficiency because of losses in the distribution and transmission networks [38]. For instance, in 2016, the losses in the electrical energy

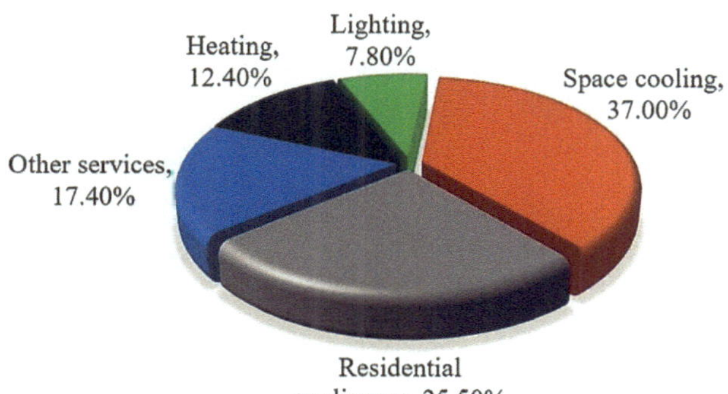

Fig. 8.3 Share of the growth of final electrical energy demand in buildings by 2050 [32].

[a] The energy equivalent of one barrel of crude oil is 6119 MJ [38].

distribution network and the transmission network were estimated to be ~11.5% and ~2.9%, respectively [38].

The high contribution of energy consumption in buildings suggests the need to adopt policies and long-term planning to reduce the energy demand by this sector, in particular, to reduce the use of fossil fuel-sourced energy in buildings. The motivation is not only to increase energy efficiency, but also to decrease air pollution and GHG emissions, which have well-defined environmental impacts. In order to reduce electrical energy dissipation in transmission lines and cut down or even eliminate the expense of power lines as much as possible, a sound alternative would be to generate the required energy at the place of consumption, using localized or dispersed production. If this localized energy generation utilizes thermal systems, according to the Second Law of Thermodynamics, any surplus heat produced can be employed to meet the thermal energy requirements of buildings or industry.

As illustrated in Fig. 8.4, various technologies/approaches, either as stand-alone or hybrid systems, can be used to meet energy demands in buildings. Renewable energy (such as solar, wind, geothermal, and biofuels) can be employed for producing both electrical and thermal (cooling or heating) energy locally. Energy storage systems can also play an important role in balancing electrical and thermal energy loads over time and enhancing energy efficiency in buildings. Passive cooling/heating and natural

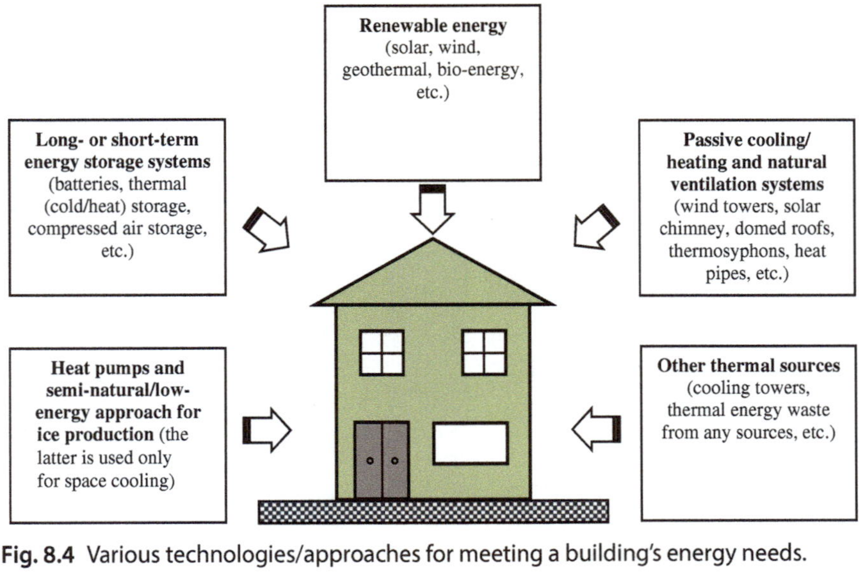

Fig. 8.4 Various technologies/approaches for meeting a building's energy needs.

ventilation technologies can help meet the cooling (e.g., by wind towers) and heating (e.g., by solar chimney) requirements of buildings in an environmentally friendly approach. Other thermal sources, for instance, cooling towers and thermal energy waste from any sources can be used for preheating or precooling purposes.

This chapter first provides an example of a hybrid HVAC system that integrates heat pumps, seasonal ice storage, and a solar energy system. The system is examined in detail, accompanied by an example demonstrating its performance efficiency. Thereafter, various methods of localized electrical energy production are briefly described, with a particular focus on solar energy. Some possibilities of using various active or passive technologies (individually or combined) to supply the building's thermal energy needs are then discussed.

8.2 Cooling/heating buildings using heat pumps, seasonal ice storage, and a solar energy system

In this section, a hybrid HVAC system is proposed for cooling/heating a building. It combines a heat pump, seasonal ice storage, and a solar energy system, as shown schematically in Fig. 8.5. In winter, if the building requires heat, the evaporator of the heat pump, using the electricity produced and stored in the solar energy system, receives heat from the water tanks (or ice storage tanks), freezes the water existing in these tanks, and thus meets the building's heat requirements through its condenser. In this case, according to Fig. 8.5, valves V_1 and V_2 are opened, and valves V_3 and V_4 are closed. In summer and autumn or whenever the building needs cooling, the heat pump system is switched off, valves V_1 and V_2 are closed, and valves V_3 and V_4 are opened. In this case, pumps P_1 and P_2 direct the secondary coolant fluid (typically a solution of water and ethylene glycol) into the circuit, and the melting of some of the ice produced and stored in the tanks cools this fluid, which can then be sent to the building's fan coil units by pump P_2 to meet its cooling demands.

If the ice stored in the tanks cannot supply the building's cooling requirements, which depend on the type and usage of the building and local climate conditions, a heat pump is needed to produce ice before the beginning of the hot season or even during the hours of the day in summer when the building does not need cooling, as described in Chapter 7. In this case, heat must be expelled by the condenser, and then either be dissipated in the surrounding environment or be used for different applications, for example, to meet the hot water demands of the building.

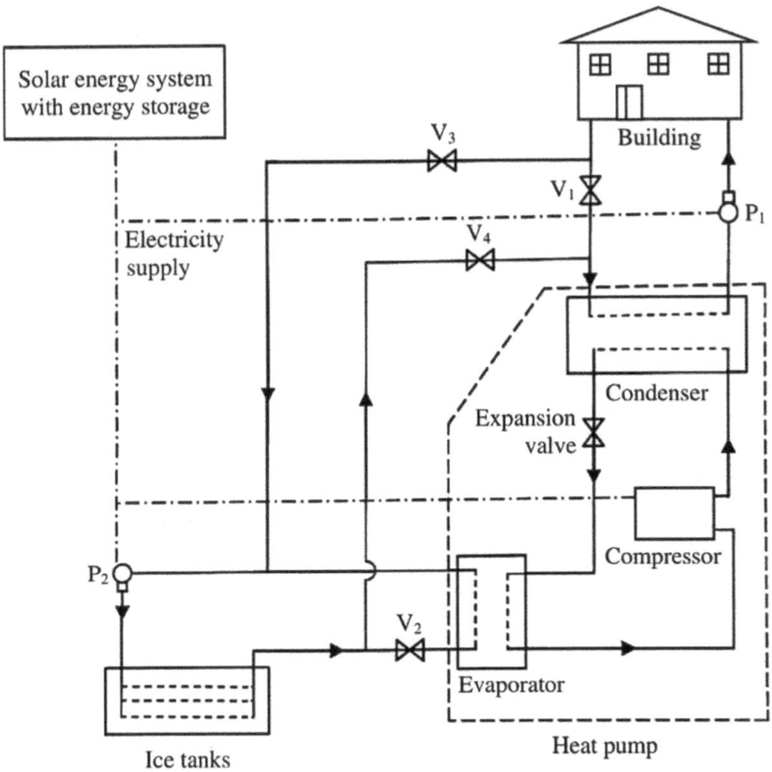

Fig. 8.5 An overall scheme for supplying thermal energy needs of a building through seasonal ice storage using a heat pump and a solar energy system.

8.3 Providing sanitary hot water for buildings through heat pumps, ambient air, and a solar energy system

The concept of providing sanitary hot water for a building is schematically illustrated in Fig. 8.6. In this scheme, the evaporator of the heat pump receives its required heat from the ambient air, and the heat dissipated in the condenser in this heat pump system is used to heat the sanitary water. To optimize energy use, the sanitary hot water consumed in the building is utilized to preheat the cold water entering a heat exchanger. The solar energy system supplies the electrical energy required to operate the compressor of the heat pump system and the ambient air fan. In buildings where cooling needs cannot be provided through the method shown in Fig. 8.5 alone, the building's sanitary hot water demand can be supplied if the heat required in the evaporator is provided by freezing the water in the tanks instead of using the ambient air; therefore, a part of the cooling needs of the building can be met by melting the ice stored in these tanks in summer.

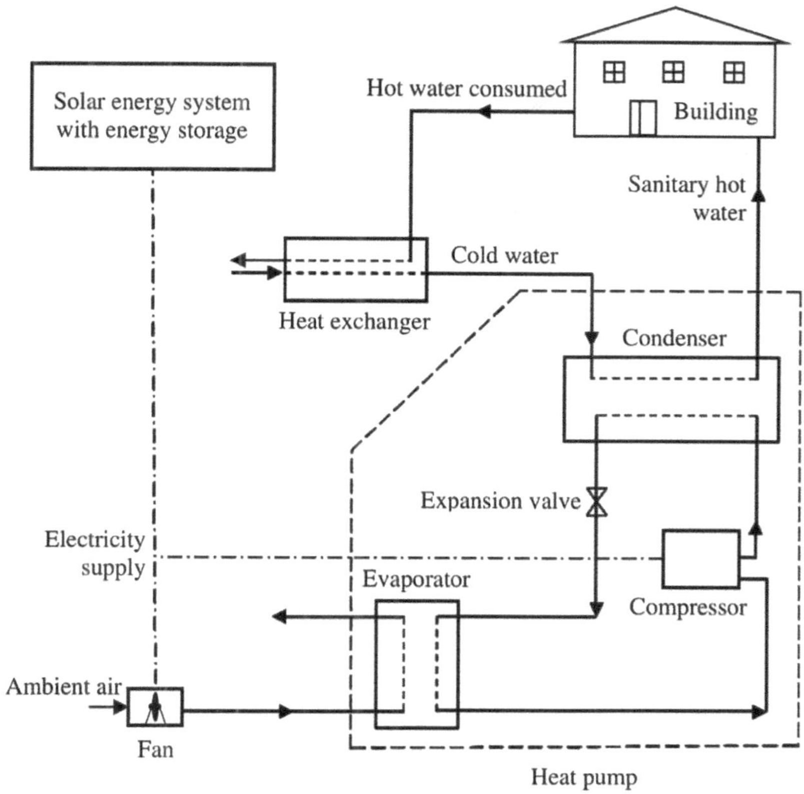

Fig. 8.6 An overall plan to provide sanitary hot water for a building using a heat pump, the ambient air, and a solar energy system.

8.4 Analyzing the performance of the proposed hybrid HVAC systems by an example

As an example, a one-floor small building with a basement and 120-m^2 infrastructure, located in the city of Tehran, Iran, is considered. Measures are taken to save energy, including insulation of walls and roof, double-glazed windows, shutters (exterior structures, closed on winter nights and summer days), and ceiling fans to create a gentle airflow in rooms during the summer. Occupants thus experience thermal comfort at 28°C instead of 25°C, and the heating and cooling loads of the building are consequently determined to be equal to 17.59 GJ/y and 16.1 GJ/y, respectively [39]. The refrigerant R-134a is used in the heat pump system. The evaporator and condenser temperatures are considered to be −15°C and 75°C, respectively. In addition, it is assumed that the processes in the evaporator and condenser are isobaric (or constant pressure), the compressor process is isentropic (or constant entropy), and the

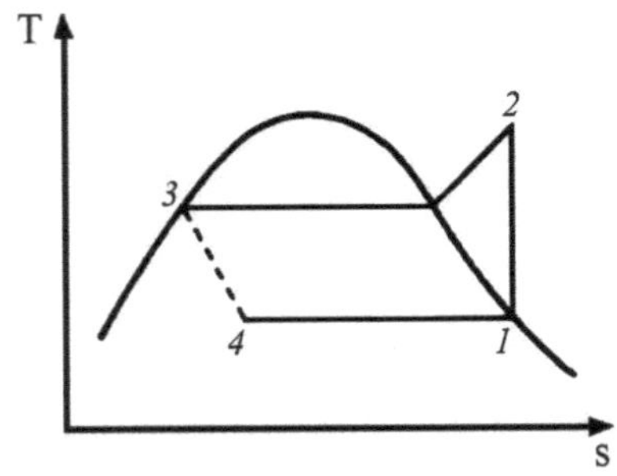

Fig. 8.7 *T-s* diagram of the considered heat pump cycle.

expansion valve process is isenthalpic (or constant enthalpy). Fig. 8.7 shows a *T-s* diagram of the heat pump process.

For the refrigerant R-134a, the following values can be obtained from thermodynamics tables [40]:

In temperature of (T_1) −15°C: $P_1 = 165\,\text{kPa}$, $h_1 = 389.2\,\text{kJ/kg}$, and $s_1 = 1.735\,\text{kJ/kgK}$.

In temperature of (T_3) 75°C: $P_3 = 2364.4\,\text{kPa}$ and $h_3 = 313.51\,\text{kJ/kg}$.

For the diagram shown in Fig. 8.7, we have:

$$s_2 = s_1,\ h_3 = h_4,\ P_1 = P_4 \text{ and } P_2 = P_3$$
$$h_2 = 440 \text{ kJ / kg and } T_2 = 90°\text{C}$$

Therefore, we obtain:

$$q_L = (h_1 - h_4) = 75.69\text{kJ / kg}$$
$$q_H = (h_2 - h_3) = 126.49\text{kJ / kg}$$
$$w = q_H - q_L = 126.49 - 75.69 = 50.80\text{kJ / kg}$$
$$(\text{COP})_{hp} = q_H / w = 126.49 / 50.80 = 2.49$$

Given the assumptions and for the conditions selected, we obtain a coefficient of performance (COP) for the heat pump equivalent to 2.49. For the heat pump system considered in Fig. 8.5, a COP of 2.4 is selected. With this COP value, the energy required for the compressor is 7.33 GJ/y, and the heat that must be provided to the evaporator of the system is equal to 10.3 GJ/y. Assuming the initial water temperature of 10°C in the ice production and storage tanks and considering the enthalpy of melting of ice equal to 333.42 kJ/kg [40], the amount of water mass that must be frozen is obtained to be 27,300 kg. Taking into account 20% losses or an 80% recovery coefficient, melting this amount of ice can supply the equivalent of 7.3

GJ/y of the building's cooling requirements, which is less than the amount of cold energy required, namely 16.1 GJ/y, for the building in summer.

To supply 16.1 GJ/y of the building's cooling load and assuming an 80% recovery coefficient, the amount of 60,435 kg of water needs to be frozen. Thus, in addition to working the heat pump to meet the building's heating load during the winter, it must be kept operating for ice production at other times of the year when the building does not need heating. At these times, heat rejected by the condenser can either be dissipated to the surrounding environment or utilized for other purposes. To provide the sanitary hot water required (Fig. 8.6), water and its solidification can be used instead of air to transfer heat to the evaporator (as shown in Fig. 8.5). In this case, to meet a part of the building's cooling needs after the ice has melted completely (according to Fig. 8.5), the flow of the solution of water and ethylene glycol must be directed to the water and ice tanks of the sanitary hot water system. The amount of water that needs to be cooled and frozen for this application depends on the annual hot water consumption.

8.5 Methods of localized electricity production

Electrical energy can be generated locally by different renewable or conventional energy systems in the specified power capacity ranges. This approach to electricity production has a number of advantages, such as [41–43]: (1) diminishing the load on main networks, (2) decreasing electrical energy prices, (3) minimizing transmission losses, (4) helping to save electricity, (5) enhancing grid flexibility and reliability, (6) creating local jobs and economic benefits, (7) lessening the adverse environmental effects, (8) improving public safety, and (9) lessening vulnerability to terrorism. Some of these systems (renewable or conventional) are as follows:

- Systems based on renewable energy, such as
  1. *Photovoltaic (PV) systems*: The power capacity of PV systems ranges between ~10W (for a small solar cell) and megawatt-scale [44]—greater than 100MW for very large-scale enterprises [45]. For example, one of the world's biggest PV plants, situated in the Tengger Desert Solar Park in Zhongwei, Ningxia, China, has a total capacity of around 1500MW [46].
  2. *Wind turbines*: Different types and sizes of turbines range in power capacity from around 100W (in what is known as a stand-alone microsystem) to more than 3MW (in grid-connected systems) [47, 48]. In addition, a small number to several hundred wind turbines that

are designed in arrays, called "wind farms," are used to generate electricity in both onshore and offshore sites. Wind energy is one of the fastest-growing sources of electrical energy production worldwide [49]. As an example, Alta Wind Energy Center, located in Kern County, California, USA, is one of the largest onshore wind farms in the world, with a current capacity of around 1500 MW [50].

3. *Hydrogen fuel cells*: Their power capacity ranges typically from about 10 W to a megawatt-scale [44]. The power capacity of current hydrogen fuel cells is higher than 20 MW [51, 52].
4. *Geothermal power systems*: These power plants have a power capacity ranging from 4 to 10 MW (small-scale) [53, 54] to over 200 MW [55]. The Geysers Geothermal Complex, situated in Sonoma and Lake Counties, California, USA, with a current capacity of about 1500 MW is now the biggest geothermal power plant in the world [56].

Other types of renewables such as hydropower [57], bioenergy (e.g., biogas [58], biomass, and municipal waste [59]), as well as tides and sea or ocean waves [60], can produce electricity locally in different ranges based on their availability and accessibility, and on local geographical conditions.

- Systems based on conventional energy or fossil fuels, such as
  1. *Diesel generators*: Worldwide, because of their low capital expense, they are most widely utilized as small electrical power producing units in places with no connection to a power network, or as an emergency power supply in case of network failure [61, 62]. Diesel generators in different sizes can provide electricity in various ranges, for instance, from 8 kW to 30 kW for homes, small shops, and offices, as well as from several hundred kilowatts up to 2000 kW utilized for large office complexes, factories, etc. [63, 64].
  2. *Microturbines*: These small-scale gas turbines can generate electricity with a power capacity normally from around 25 kW to 500 kW (commercial-sized systems) [65–67] and are mostly employed in large buildings, for instance, hotels, offices, and schools [66];
  3. *Internal combustion engines (ICEs)*: Their power generation capacity as off-grid systems is in the range of 5 kW to 7 MW [44];
  4. *Natural gas fuel cells*: Their power generation capacity ranges from about 1 kW to 250 MW [44].

The electricity production systems must usually be accompanied by an energy storage system (mainly batteries) in order to balance, or even minimize, mismatches between the production and consumption of electrical energy, and to create a more flexible and reliable network system, for example, through storing excess energy generated for peak consumption times.

8.6 Using solar energy for electricity generation

Among all the renewable energy sources, solar energy has been selected for the following detailed description of producing electricity locally, because of its maturity, high global acceptance, and availability most of the year worldwide. Solar energy, considered to be green, sustainable, and renewable, is a suitable means of generating electricity, particularly in sunny regions. For example, Iran, with over 7 h of sunshine on more than 300 days, over more than two-thirds of its land area [68–70], for a total of 2800 h of sunshine on average [71, 72] during a year, as well as about 4.5–5.5 kWh/m^2/day average solar irradiation [73–75] (~2200 kWh/m^2/yr [69, 70]), has one of the world's highest potentials to use solar energy to generate a significant proportion of its electrical energy. Fig. 8.8 indicates possible regions for

Fig. 8.8 Daily average solar radiation in Iran [76].

solar energy generation in the country. According to Bahadori [77], it is estimated that the daily solar energy gain in Iran is, on average, 18 MJ/day for a one-square-meter horizontal plate, and 28 MJ/day for a plate that is always perpendicular to the sun's rays. In sunny regions, the solar radiation would be even higher.

Solar energy can be converted to electricity directly by means of PV generators, or through thermal conversion. In thermal conversion, solar energy is the heat source for a thermal power plant, but a sufficient power level is needed to operate the plant. PV conversion is more applicable for low power needs (up to 100 kW).

The main components of PV systems are the solar cells, which work using semi-conductor materials. Much research is now focused on boosting cell efficiency above the current level of about 15% while maintaining low production costs. PV systems need storage because of sunlight's intermittent nature, and most PV systems use batteries for this purpose, which increases the cost of solar energy systems.

A PV system has the following components:

1. *Arrays of solar cells.* The array forms the main part of the PV system (Fig. 8.9). Several solar cells that lie next to each other form a module, and several modules aligned together form a panel. Finally, several panels that fit together form an array.
2. *Charge controller.* This device controls the current flow from the array to the battery; thus, its main function is to prevent overcharging of the battery. The charge controller also blocks battery current from leaking back into the PV array at night or on cloudy days, which would drain the battery's charge. If a consumer wants to use more than battery power, it stops the current, thereby increasing battery life.
3. *Converter.* This device is used to convert direct current (DC) to alternating current (AC) and vice versa, at which point it becomes an inverter. The inverter converts the DC produced by the solar array into AC to use in consumption (or electrical) loads.
4. *Battery.* This device provides electricity to use at night or other times when sufficient sun is not available, for example, in cloudy or rainy conditions. The batteries used for solar systems are usually lead-acid (Pb-A) type, and ~80% of their capacity is commonly used. The choice among different types of batteries depends on weather conditions, the type of chemicals used to store the electricity, and cost.

Auxiliary components consist of piles for inserting arrays at a specific slope, combiner boxes, switches, cables, and other circuit components [78].

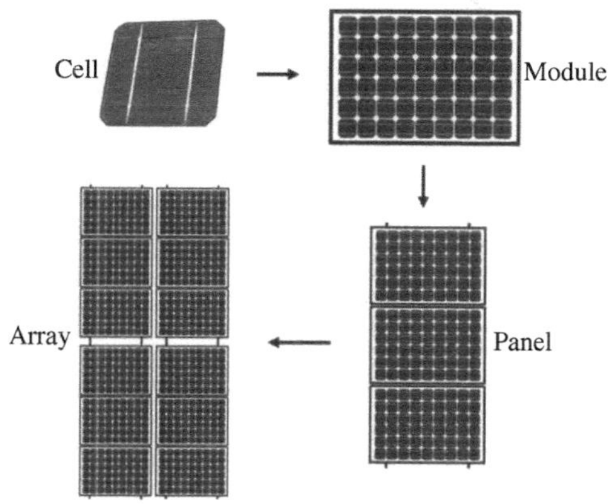

Fig. 8.9 A photovoltaic (PV) array.

One type of storage involves the integration of a solar power system with a fuel cell. The generated electricity is utilized for electrolysis and the production and storage of hydrogen. When needed, hydrogen can be used to generate electricity in a fuel cell. Fig. 8.10 indicates a schematic diagram of a solar energy system and a fuel cell designed to produce electrical energy.

Some of the advantages of PV systems are as follows [79–84]:

- Silent because of the lack of mechanically moving parts, and visually unobtrusive;
- Almost negligible operating and maintenance costs in comparison with the expenses of other renewable energy systems;
- Easy maintenance and operation due to lack of mechanically moving parts;
- Needs no water to generate electricity (this issue in areas that face water shortages is very important);
- Can be designed for any capacity based on energy requirements;
- Slow rate of depreciation, and long life;
- Little need for spare parts inventory;
- Energy is free, abundant, and supplied naturally;
- No GHG emissions.

Several disadvantages of PV systems are listed in the following [83, 84]:

- High cost of initial investment;
- Intermittency problems arising from the sun not shining at night or on cloudy, rainy, or snowy days;
- Some toxic chemicals such as cadmium and arsenic used in PV production have environmental impacts;

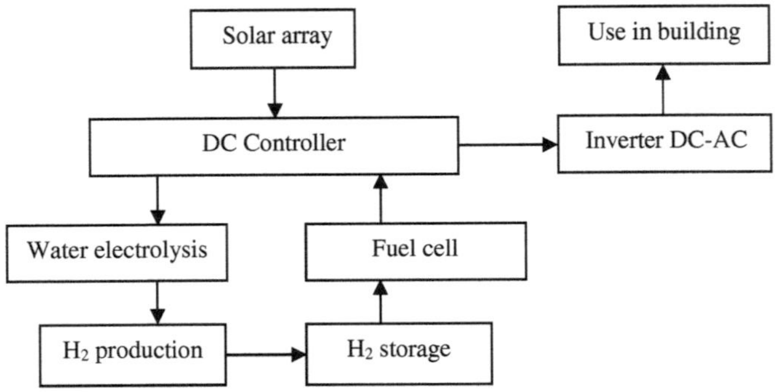

Fig. 8.10 Combination of a solar energy system and a fuel cell to generate electricity.

- Relatively low efficiency levels compared to other renewable energy systems;
- High costs of energy storage systems (for instance, a large battery capacity is needed);
- Requires sufficient land for deployment.

8.7 Integrating various technologies to meet buildings' thermal energy needs

Nowadays, providing a safe, healthy, productive, and comfortable environment for a building's occupants is very important, because humans spend almost 80%–90% of their time inside buildings or confined spaces [85, 86]. Various HVAC technologies/approaches, either individually or integrated with each other, can improve thermal comfort and provide good indoor environmental quality (IEQ) for buildings. Therefore, HVAC systems play a key role in the provision of a suitable and pleasant environment for inhabitants. This section explores the use of active or passive technologies, as stand-alone systems or integrated as hybrid systems, to supply buildings' thermal energy needs. The focus here is primarily on space cooling and AC.

Some passive cooling/heating and natural ventilation systems such as wind towers, solar chimney, thermosiphons, and also renewable energy technologies (for instance, solar and geothermal systems) can alone supply all or part of buildings' thermal energy (cooling or heating) requirements, although local climate and geographical conditions play an important role in this provision. These systems supply a building's required thermal energy in an environmentally friendly way.

Wind towers (or wind-catchers), as examples of passive and natural AC systems, can provide thermal comfort for a building's inhabitants throughout the warm months in hot/warm regions. Traditional wind towers, in different forms and sizes, have been employed in a number of countries located in the Middle East (particularly in Iran, the origin of this structure) and Egypt for many years (over one thousand years in Persia [1, 2, 87]) to ventilate buildings naturally [1, 87–92]. Traditional wind towers can be categorized into four overall types [1, 87]: (1) one-sided; (2) two-sided; (3) four-, six-, and eight-sided; and (4) cylindrical forms. Fig. 8.11 indicates widespread utilization of traditional wind-catchers in the old part of Yazd City, Iran, as an example. In regions with specified and consistent wind direction, one-sided wind-catchers have been utilized, but regions, where the wind direction varies, are capriciously relied on other forms [2, 87].

The efficiency of traditional wind-catchers is low when the ambient wind speed is zero or very low. In addition, they have a number of limitations, such as [1, 2, 87]:

- The head of wind-catchers is fixed and cannot spin to catch the maximum ambient airflow;
- They allow small birds, animals, and dust to come into the building;
- These exposed structures gradually erode due to natural phenomena (for instance, precipitation, wind, and sun);

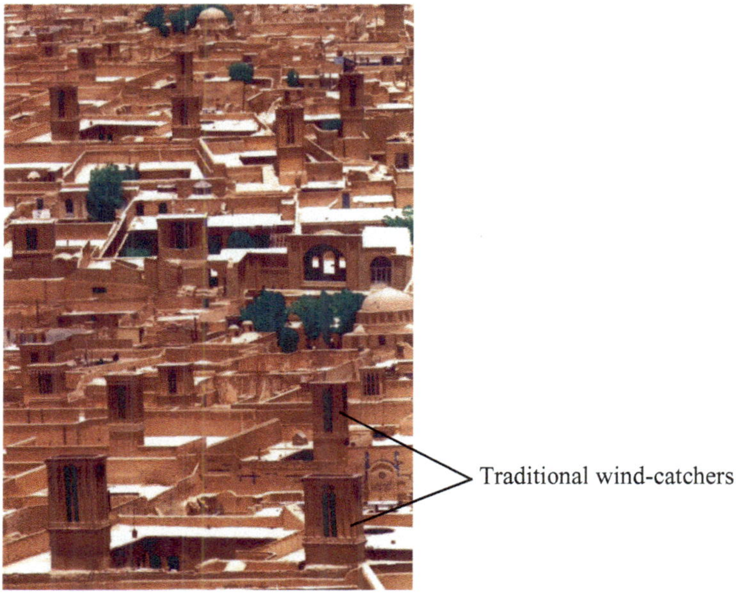

Fig. 8.11 Traditional wind-catchers located in the old part of Yazd City, Yazd, Iran [2, 87].

- A portion of inlet airflow via wind-catchers (excluding one-sided type) exits with no circulation in the building;
- The quantity of coolness that can be generated by traditional wind-catchers is normally restricted and may not be sufficient to meet cooling needs on hot days; and
- In very still weather, circulation is purely due to density differences in the air, which can prevent adequate flow.

To reduce or even eliminate the limitations of traditional wind-catchers, as well as to enhance their performance and improve the conditions of thermal comfort in buildings, several new wind tower designs have been proposed [1, 2, 87–89, 93–105], each with its own advantages and restrictions. Some researchers have also suggested the integration of a wind tower with a solar chimney (or an air heater) to increase the rate of air ventilation passively through dwellings [1, 87, 106–108]. A solar chimney is a passive heating system that utilizes solar radiation to enhance the air temperature and provide upward movement to inflowing air through buoyancy effect, meaning the air density difference between the inlet and outlet parts of the chimney, converting the thermal energy into kinetic energy [109–112].

A notable point associated with the use of conventional AC systems is potential pathogen transmission (for instance, of COVID-19, seasonal flu, and Legionnaire's disease) via these systems in buildings and enclosed spaces. They do not provide enough fresh air turnaround because cooled air is recycled to conserve energy, sometimes with filtering or disinfection. Conventional room-sized and building-sized AC systems need to be modified or even redesigned to meet these concerns. New-generation air conditioners must be designed as stand-alone systems and/or as hybrid, integrated systems combined with other means of ventilation and cooling.

A combination of new wind tower designs with cold storage and shallow geothermal systems can address these concerns while sustaining thermal comfort levels for the building's inhabitants, including heating. It is also possible to use different cold sources, depending on their availability, local weather, and geographical conditions. However, the goal of design for these systems is not to achieve an extremely cool output at all times, but to enhance the comfort of inhabitants to a reasonable level. For example, in a system designed for a hot and dry climate, temperatures lower than 28°C, but above 20°C, are targeted (see Fig. 3.4), with continuous fresh air turnover. This range remains in the warm comfort zone that people can easily cope with, and the system does not have to meet stringent and energy-intensive standards such as 22 ± 2°C with extreme forced-air circulation.

Two proposed hybrid AC technologies are presented below for two different climate conditions: (1) hot and humid and (2) hot and dry.

8.7.1 Hot and humid climates

Fig. 8.12 shows the integration of a new wind tower design with a cold storage facility and a shallow geothermal system combined to form a hybrid AC system for cooling confined spaces in hot and humid climates. Via a heat exchanger, fresh air entering through a modern wind tower is cooled by chilled water. A cold storage system, described in Chapter 7, and a shallow geothermal heat pump cooling system provide the chilled water. Driven by a small pump, chilled water flows through pipes installed within the wind tower's column, perhaps enhanced by a low-power fan, cooling air to provide AC and achieving a high rate of air turnover. Water condensing on the exterior surfaces of the cooling pipes collects in a small pool at the base of the wind tower's column.

The wind tower's head rotates to orient itself in the maximum ambient wind speed direction. For this purpose, a vane above the wind entrance

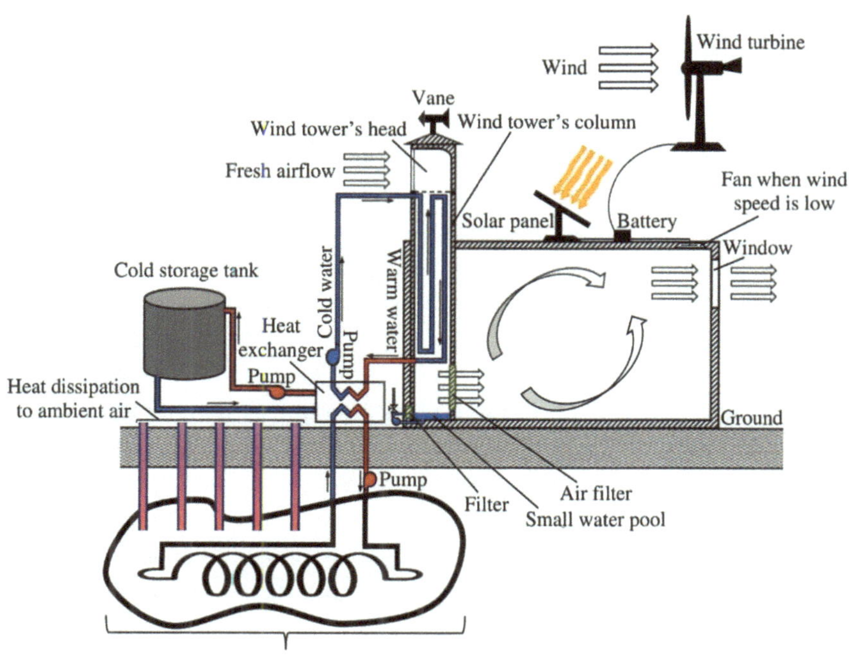

Fig. 8.12 Configuration of a hybrid AC system for space ventilation and cooling in hot and humid regions.

detects the maximum wind direction to adjust the air opening perpendicular to this direction manually or electronically. When there is no need for cooling and ventilation, the air opening is closed. A screen on the air opening prevents the entry of small birds and animals, and transparent material panes can be used in the tower and in the building to help introduce diffuse natural light to the living or working space. The materials forming the interior of the wind tower are moisture-proof (against the effects of condensation, some rain entry), mold resistant, and easily cleaned through access panels. The cross-section of the modern wind tower can be in the form of a rectangle, square, or circle.

To address additional issues related to particulate matter or pathogens, intake air can be treated by two approaches:

1. The small basal water pool will collect some particulate matter (also called particle pollution). In humid climates, condensation resupplies the water pool, and local water systems can supplement this supply so that the water can be circulated through a suitable filter for local use.
2. An air filter can further purify the airflow before it enters the living or working space. This filter is passive, and a low-power fan helps sustain adequate airflow, giving steady flow in the wind tower during calm conditions.

Exhaust airflow, after distribution and circulation inside to cool the space, is still reasonably cool. It can exit directly, or pass through a passive heat exchanger with intake air at the upper part of the wind tower, precooling the airflow. Alternatively, cool exhaust air can also help chill the geothermal system or cold storage tank to help conserve energy. The efficiency of this proposed hybrid AC system can be improved through optimal management of the three cold sources, namely the storage tank, the ground, and the exhaust air.

When wind speed is insufficient, low-power fans can give some "forced air" capability to sustain the system. Low-power fans may be used within the wind tower, just in front of a large-area filter, or even in the location of air exhaust, after the heat exchanger has extracted the cooling potential from the exhaust air. Small solar or wind energy systems with simple battery storage can supply electricity to run fans and other electrical equipment (e.g., small circulating pumps). Proper natural airflow inside a building is promoted by means of various passive approaches, such as strategic placing of inter-room grates and exhaust systems, especially in buildings with several floors. Building inhabitants control other aspects themselves by the management of windows, doors, grate openings, and local fans.

If a shallow geothermal system or a ground source heat pump (GSHP) is used, warmed water can be pumped into the ground through a closed cycle to dissipate its heat and return cooled water to the heat exchanger. As shown in Fig. 8.12, passive thermosiphons also can dissipate excess ground heat into the ambient air, keeping the ground as cool as possible. The thermosiphon principle is based on a vertical condensing fluid passive heat transfer system that is self-activated only when the ambient temperature falls below a set level. For example, when the air temperature falls, for instance, below 24°C, the system starts to transfer heat from the ground to be dissipated atmospherically. As cooler temperatures occur mostly at night in hot and humid climates, this period can be viewed as "recharging" the cooling potential of the geothermal repository.

Rejected heat may, in some circumstances and configurations, particularly with some heat pump deployments, be used to sustain a tank of warm water for personal and domestic use. The closed geothermal loop system can be either horizontal or vertical in form, sized appropriately [36]. A pond (through evaporative cooling) or even a subsurface aquifer can be used as a cold storage source instead of the ground if local and regional conditions and regulations allow. It should be noted that the geothermal repository (also called a "thermal battery") needs to be charged when the system does not work, so it does not lose its effectiveness during operation either for cooling or heating purposes. This can be done through different approaches.

The cold thermal energy storage (CTES) system can use chilled water, ice, or other phase change materials (PCMs) as a storage medium, as described in Chapter 7, and it can be placed underground during construction or as a retrofit. The dimensions and capacity of the storage tank depend on the load required to cool the water flowing inside the pipes installed in the wind tower's column.

Other designs of the proposed hybrid AC system are feasible, such as using a horizontal wind channel. During colder months, especially if a geothermal GSHP is used, warm water is directed to the pipes inside the wind tower to heat the incoming airflow, or to preheat air for use in heating systems, leading to energy savings. The heat needed can be provided through geothermal systems (shallow, deep, or a combination of both) for larger applications.

8.7.2 Hot and dry climates

In hot and dry climates, a new and different approach is used to cool the airflow entering a building. Moist curtains or porous clay conduits installed in the wind tower's column [87–90, 104, 105], or porous pads installed in the wind tower air opening [1, 2], are used to take the advantage of evaporative cooling to reduce the temperature of hot and dry air entering the building. As shown in Fig. 8.13, several thick curtains (made of suitable materials such as sack or rough textile), as sprayed moistened surfaces, are mounted inside the column and kept moist with chilled water. The hot, dry air passes through these curtains for evaporative cooling, reducing its temperature (by up to 8°C in extreme conditions) and increasing its humidity (important for dusty climates). Additional water is collected, filtered, and reused by recirculating to the heat exchanger and subsequently back to the wind tower. Eventually, the airflow passes through an air filter to remove particles as it enters the living or working space, passes on to other rooms, and leaves the building through windows, designed gratings, or heat exchangers.

When the wind speed in the ambient air is zero or very low, this hybrid AC system can also operate passively. The air around moistened curtains inside the wind tower's column can exchange heat with the wetted surfaces, leading to a rise in the density of air along the column. Thus, due to the difference in air density between the inside and outside (ambient air)

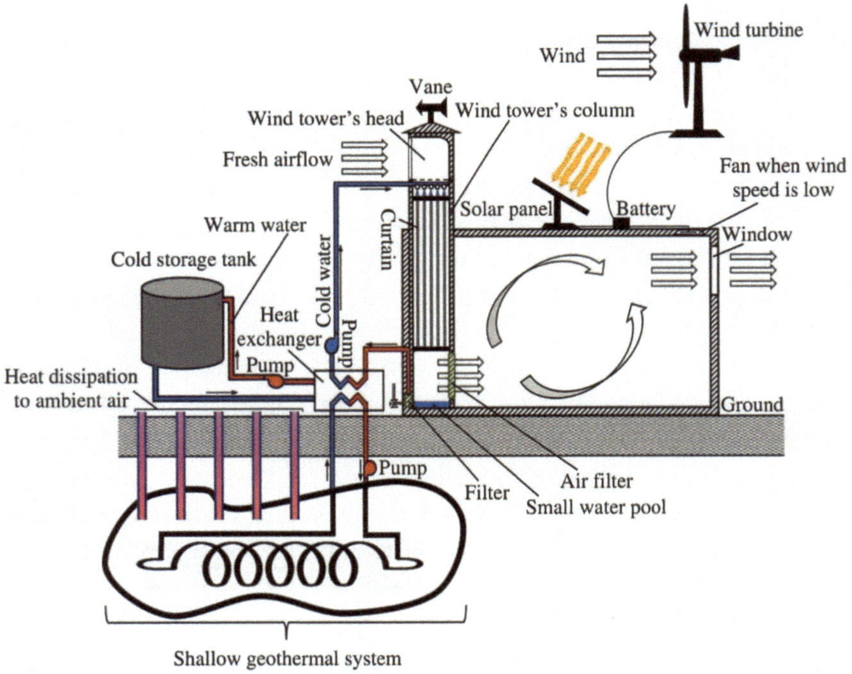

Fig. 8.13 Configuration of a hybrid AC system for space ventilation and cooling in hot and dry areas.

the column, a density-driven advective flow pattern is generated, and the downward airflow is increased because of buoyancy force. Low-power fans enhance this airflow as needed, and the system provides some heat in winter months, especially if the system includes a GSHP system. Alternately, for a very simple design, chilled water supplied to the upper edge of the wind tower's head by a small pump wets a pad (or straw mat) installed at the air opening; the fresh air passing through the moistened pad is evaporatively cooled and enters the wind tower's column [1, 2].

In areas where sufficient wind velocity is available most of the time in warm weather, if the ambient air temperature and humidity are within acceptable ranges, there is no need to use air cooling systems (either evaporative or refrigerated) because only steady air movement and exchange in the building are needed. Modern wind towers, or several of them, capture the wind and direct fresh airflow into the building, and the airflow in the rooms sustains personal thermal comfort and provides rapid air turnover, resulting in saving energy.

A central (or district) thermal system with the modern wind towers and geothermal approaches can help meet the cooling needs of an array of buildings, high-rise towers, and even a small community during the warmer months, while providing heat in the cold months. This system can enhance buildings' energy efficiency, optimize overall energy utilization, and manage energy loads and related expenses (e.g., capital, operating, and energy consumption costs).

At a large scale, the system design includes energy storage systems such as compressed air storage, pumped hydrostorage, and hydrogen-based storage. Batteries are less desirable for large-scale energy storage due to environmental burdens from raw material provision, storage hazards, recycling, and waste disposal [34].

To improve ventilation and thermal performance, other technologies, and natural ventilation and passive systems (for instance, solar chimneys, thermosiphons, and domed roof protections) can also be integrated, depending on local climate and geographical conditions; building use, size, age, and design; and required thermal energy loads. Solar chimneys and passive thermosiphons can passively warm inner spaces during the cold months when wind towers are not operating. New green building architecture can factor these integrated and hybrid systems into efficient and sustainable design.

A smart control system needs to be utilized for the operation and management of the proposed hybrid AC systems (or other integrated HVAC scenarios) to achieve the desired goals, including increasing inhabitants' thermal comfort, enhancing buildings' energy efficiency, reducing the consumption of fossil fuel-based energy, diminishing energy costs, decreasing public health risks, while reducing environmental challenges, and threats (such as air pollution and climate impacts).

8.8 Conclusion

Nowadays, heating, ventilation, and air-conditioning (HVAC) are among the main energy-consuming processes in different types of buildings. Nevertheless, they are essential in providing the conditions of thermal comfort and good indoor environmental quality (IEQ) for buildings' inhabitants. Different strategies are being developed to reduce thermal (and also electrical) energy consumption in buildings, alongside supplying their energy demands. In today's world, efforts are being made to design new green buildings, which can be called "net zero energy buildings (NZEB)." In addition to using various solutions in existing buildings such as enhancing their energy efficiency, managing energy loads over time, and optimizing overall energy use—depending on various factors, the primary ones being buildings' specifications (e.g., size, age, usage, and design) and local weather and geographical conditions—different strategies/technologies, either as stand-alone or hybrid systems, can be employed to meet the buildings thermal energy (cooling or heating) needs. For example, integrating renewable energy systems (for generating both electrical and thermal energy locally) with either passive and natural ventilation technologies, heat pumps, semi-natural/low-energy approach for ice production (only for space cooling), or other thermal sources (e.g., cooling towers and thermal energy waste from any sources) accompanied by energy storage systems, can be an efficient, ecologically friendly, economical, and practical solution not only for supplying the thermal comfort requirements of residents, but also for decreasing local and global environmental problems (e.g., global warming and climate change) and achieving a degree of sustainability.

References

[1] A.R. Dehghani-Sanij, M. Soltani, K. Raahemifar, A new design of wind tower for passive ventilation in buildings to reduce energy consumption in windy regions, Renew. Sustain. Energy Rev. 42 (2015) 182–195.

[2] M. Soltani, A.R. Dehghani-Sanij, A. Sayadnia, F.M. Kashkooli, K. Gharali, S.B. Mahbaz, M.B. Dusseault, Investigation of airflow patterns in a new design of wind tower with a wetted surface, Energies 11 (2018) 1–23.

[3] O. Saadatian, H.L. Chin, K. Sopian, M.Y. Sulaiman, Review of wind catcher technologies, Renew. Sustain. Energy Rev. 12 (2012) 1477–1495.
[4] O.T. Masoso, L.J. Grobler, The dark side of occupants' behavior on building energy use, Energ. Buildings 42 (2010) 173–177.
[5] A. Robert, M. Kummert, Designing net-zero energy buildings for the future climate, not for the past, Build. Environ. 55 (2012) 150–158.
[6] M.M. Stojiljković, M.G. Ignjatović, G.D. Vucković, Greenhouse gases emission assessment in residential sector through buildings simulations and operation optimization, Energy 92 (3) (2015) 420–434.
[7] Y. Xing, N. Hewitt, P. Griffiths, Zero carbon buildings refurbishment—a hierarchical pathway, Renew. Sustain. Energy Rev. 15 (6) (2011) 3229–3236.
[8] S. Buffa, M. Cozzini, M. D'Antoni, M. Baratieri, R. Fedrizzi, 5th generation district heating and cooling systems: a review of existing cases in Europe, Renew. Sustain. Energy Rev. 104 (2019) 504–522.
[9] H. Nami, A. Anvari-Moghaddam, A. Arabkoohsar, Application of CCHPs in a centralized domestic heating, cooling and power network—thermodynamic and economic implications, Sustain. Cities Soc. 60 (2020) 102151.
[10] B. Dudley, BP Statistical Review of World Energy 2019, sixty-eighth ed., BP p.l.c., London, UK, June 2019, p. 61.
[11] B. Dudley, BP Energy Outlook: 2019 edition, BP p.l.c., London, UK, 2019, p. 141.
[12] D.A. Asimakopoulos, M. Santamouris, I. Farrou, M. Laskari, M. Saliari, G. Zanis, G. Giannakidis, K. Tigas, J. Kapsomenakis, C. Douvis, S.C. Zerefos, T. Antonakaki, C. Giannakopoulos, Modelling the energy demand projection of the building sector in Greece in the 21st century, Energ. Buildings 49 (2012) 488–498.
[13] C.H. Pout, F. MacKenzie, R. Bettle, Carbon Dioxide Emissions from Non-domestic Buildings: 2000 and Beyond, BRE Energy Technology Centre, Watford, UK, 2002.
[14] A. Mardiana, S.B. Riffat, Building energy consumption and carbon dioxide emissions: threat to climate change, J. Earth Sci. Clim. Chang. S3 (2015) 001.
[15] L. Pérez-Lombard, J. Ortiz, C. Pout, A review on buildings energy consumption information, Energ. Buildings 40 (2008) 394–398.
[16] B. Noh, J. Son, H. Park, S. Chang, In-depth analysis of energy efficiency related factors in commercial buildings using data cube and association rule mining, Sustainability 9 (11) (2017) 2119.
[17] R. Hart, S. Selkowitz, C. Curcija, Thermal performance and potential annual energy impact of retrofit thin-glass triple-pane glazing in US residential buildings, Build. Simul. 12 (2019) 79–86.
[18] T.A. Nguyen, M. Aiello, Energy intelligent buildings based on user activity: a survey, Energ. Buildings 56 (2013) 244–257.
[19] S. Roumi, F. Razi Astaraei, R. Ghasempour, H. Yousefi, M. Hamlehdar, S.M. Razavi, New insulation replacement in buildings' walls and its impact on air pollution reduction in Tehran, Intell. Build. Int. 11 (2) (2019) 65–74.
[20] H.Y. Chan, S.B. Riffat, J. Zhu, Review of passive solar heating and cooling technologies, Renew. Sustain. Energy Rev. 14 (2010) 781–789.
[21] Y. Wu, Chinese building energy conservation: existing situation, problems and policy, in: Proceedings of the International Conference on Sustainable Development in Building and Environment, Chongqing, China, Oct. 24–27, 2003.
[22] X. Liu, G.J.D. Hewings, S. Wang, Evaluation on the impacts of the implementation of civil building energy efficiency standards on Chinese economic system and environment, Energ. Buildings 41 (2009) 1084–1090.
[23] Energy Fact Book 2016-2017, Natural Resources Canada (NRCan), Ottawa, Ontario, Canada, 2017.

[24] N. Enteria, K. Mizutani, The role of thermally activated desiccant cooling technologies in the issue of energy and environment, Renew. Sustain. Energy Rev. 15 (2011) 2095–2122.
[25] A.M. Reda, A.H.H. Ali, M.G. Morsy, I.S. Taha, Design optimization of a residential scale solar driven adsorption cooling system in upper Egypt based, Energ. Buildings 130 (2016) 843–856.
[26] M. Sohail, M.U.D. Qureshi, Energy—efficient buildings in Pakistan, Sci. Vis. 46 (17) (2011) 27–38.
[27] S.A. Rashid, Z. Haider, S.M.C. Hossain, K. Memon, F. Panhwar, M.K. Mbogba, P. Hu, G. Zhao, Retrofitting low-cost heating ventilation and air-conditioning systems for energy management in buildings, Appl. Energy 236 (2019) 648–661.
[28] G. Gong, W. Zeng, L. Wang, C. Wu, A new heat recovery technique for air-conditioning/heat-pump system, Appl. Therm. Eng. 28 (2008) 2360–2370.
[29] S. Wang, C. Fang, X. Guan, B. Pang, H. Ma, Urbanisation, energy consumption, and carbon dioxide emissions in China: a panel data analysis of China's provinces, Appl. Energy 136 (2014) 738–749.
[30] ReportLinker, HVAC Systems Market: Global Industry Analysis, Trends, Market Size & Forecasts to 2023, May 2019. https://www.reportlinker.com/p05026385. (Accessed 21 July 2020).
[31] L. Cattani, A. Magrini, P. Cattani, Water extraction from air by refrigeration—experimental results from an integrated system application, Appl. Sci. 8 (11) (2018) 2262.
[32] IEA, The Future of Cooling: Opportunities for Energy-Efficient Air Conditioning, Technology Report, Key Findings from the Future of Cooling, International Energy Agency (IEA), May 2018. https://www.iea.org/futureofcooling/?utm_campaign=IEA%20newsletters&utm_source=SendGrid&utm_medium=Email. (Accessed 21 July 2020).
[33] A.R. Kazemi, S.B. Mahbaz, A.R. Dehghani-Sanij, M.B. Dusseault, R. Fraser, Performance evaluation of an enhanced geothermal system in the Western Canada Sedimentary Basin, Renew. Sustain. Energy Rev. 113 (2019) 109278.
[34] A.R. Dehghani-Sanij, E. Tharumalingam, M.B. Dusseault, R. Fraser, Study of energy storage systems and environmental challenges of batteries, Renew. Sustain. Energy Rev. 104 (2019) 192–208.
[35] A.R. Dehghani-Sanij, S.R. Dehghani, G.F. Naterer, Y.S. Muzychka, Marine icing phenomena on vessels and offshore structures: prediction and analysis, Ocean Eng. 143 (2017) 1–23.
[36] M. Soltani, F.M. Kashkooli, A.R. Dehghani-Sanij, A.R. Kazemi, N. Bordbar, M.J. Farshchi Heydari, M. Elmi, K. Gharali, M.B. Dusseault, A comprehensive study of geothermal heating and cooling systems, Sustain. Cities Soc. 44 (Jan. 2019) 793–818.
[37] M. Soltani, F.M. Kashkooli, A.R. Dehghani-Sanij, A. Nokhosteen, A. Ahmadi-Joughi, K. Gharali, S.B. Mahbaz, M.B. Dusseault, A comprehensive review of geothermal energy evolution and development, Int. J. Green Energy 16 (13) (2019) 971–1009.
[38] Energy Balance-Sheet in 2016, Office of planning and macroeconomics of Electricity and Energy, Deputy Minister of Electricity and Energy Affairs, Ministry of Energy of Iran, 2018 (in Farsi).
[39] H. Safarzadeh, Thermal Performance Evaluation of Courtyards, PhD Thesis, School of Mechanical Engineering, Sharif University of Technology, Tehran, Iran, 2004 (in Farsi).
[40] R.E. Sonntag, C. Borgnakke, G.J. Van Wylen, Fundamentals of Thermodynamics, sixth ed., John Wiley and Sons, Inc., New York, NY, 2002.
[41] R.S. Karki, S. Chanana, Simulation of energy management system for local energy market in microgrids, in: 2016 IEEE Students' Conference on Electrical, Electronics and Computer Science (SCEECS), IEEE, Bhopal, India, 2016, pp. 1–6. March 5–6.

[42] R. Khalid, N. Javaid, A. Almogren, M.U. Javed, S. Javaid, M. Zuair, A blockchain-based load balancing in decentralized hybrid P2P energy trading market in smart grid, IEEE Access 8 (March 2020) 47047–47062.
[43] W. Sweet, Get-rich-quick scheme [innovative electricity conservation], IEEE Spectr. 43 (1) (Jan. 2006) 10–13.
[44] M. Aliehyaei, Determination of Optimum State of Localized Electricity, Heating and Cooling Systems in Different Climate Conditions of Iran, PhD Thesis, School of Mechanical Engineering, Sharif University of Technology, Tehran, Iran, 2007 (in Farsi).
[45] A. Cabrera-Tobar, E. Bullich-Massagué, M. Aragüés-Peñalba, O. Gomis-Bellmunt, Topologies for large scale photovoltaic power plants, Renew. Sustain. Energy Rev. 59 (2016) 309–319.
[46] E. Rakhshani, K. Rouzbehi, A.J. Sánchez, A.C. Tobar, E. Pouresmaeil, Integration of large scale PV-based generation into power systems: a survey, Energies 12 (2019) 1425.
[47] K. Kaygusuz, Wind energy: progress and potential, Energy Source. 26 (2) (2004) 95–105.
[48] A.T. de Almeida, P.S. Moura, C.W. Gellings, K.E. Parmenter, Distributed generation and demand-side management, in: J. Kreider, Y. Goswami (Eds.), The Handbook of Conservation and Renewable Energy, CRC Press, pp. 5-1–5-53, 2007.
[49] O.P. Mahela, A.G. Shaik, Comprehensive overview of grid interfaced wind energy generation systems, Renew. Sustain. Energy Rev. 57 (2016) 260–281.
[50] M.S. Nazir, F. Alturise, S. Alshmrany, H. Nazir, M. Bilal, A.N. Abdalla, P. Sanjeevikumar, Z.M. Ali, Wind generation forecasting methods and proliferation of artificial neural network: a review of five years research trend, Sustainability 12 (9) (2020) 3778.
[51] T.M. Masaud, K. Lee, P.K. Sen, An overview of energy storage technologies in electric power systems: what is the future? in: North American Power Symposium, IEEE, Arlington, TX, 2010, pp. 1–6. Sep. 26–28.
[52] R. Elliman, C. Gould, M. Al-Tai, Review of current and future electrical energy storage devices, in: 2015 50th International Universities Power Engineering Conference (UPEC), IEEE, Stoke on Trent, UK, 2015, pp. 1–5. Sep. 1–4.
[53] R. DiPippo, Small geothermal power plants: design, performance and economics, in: Geo-Heat Center Quarterly Bulletin, A Quarterly Progress and Development Report on the Direct Utilization of Geothermal Resources, vol. 20, June 1999, pp. 1–8. No. 2.
[54] J.W. Lund, Small geothermal power projects, in: Geo-Heat Center Quarterly Bulletin, A Quarterly Progress and Development Report on the Direct Utilization of Geothermal Resources, vol. 20, June 1999, pp. 9–26. No. 2.
[55] R. Bertani, Geothermal power generation in the world 2010–2014 update report, Geothermics 60 (2016) 31–43.
[56] Statista, Ranking of largest geothermal plants worldwide as of March 2020 (in megawatts), 2020. https://www.statista.com/statistics/525206/geothermal-complexes-worldwide-by-size/. (Accessed 30 July 2020).
[57] W.E. Mabee, J. Mannion, T. Carpenter, Comparing the feed-in tariff incentives for renewable electricity in Ontario and Germany, Energy Policy 40 (2012) 480–489.
[58] J.Y. Lim, B.S. How, G. Rhee, S. Hwangbo, C.K. Yoo, Transitioning of localized renewable energy system towards sustainable hydrogen development planning: P-graph approach, Appl. Energy 263 (2020) 114635.
[59] Y. Uemura, T. Kai, R. Natori, T. Takahashi, Y. Hatate, M. Yoshida, Potential of renewable energy sources and its applications in Yakushima Island, Renew. Energy 29 (4) (2004) 581–591.
[60] M.J. Lewis, S.P. Neill, M.R. Hashemi, M. Reza, Realistic wave conditions and their influence on quantifying the tidal stream energy resource, Appl. Energy 136 (2014) 495–508.
[61] F. Brikké, M. Bredero (Eds.), Linking Technology Choice with Operation and Maintenance in the Context of Community Water Supply and Sanitation, a Reference

Document for Planners and Project Staff, World Health Organization (WHO) and IRC Water and Sanitation Centre, Geneva, Switzerland, p. 136, 2003.
[62] A.Q. Jakhrani, A.R.H. Rigit, A.K. Othman, S.R. Samo, S.A. Kamboh, Estimation of carbon footprints from diesel generator emissions, in: 2012 International Conference on Green and Ubiquitous Technology, IEEE, Jakarta, Indonesia, 2012, pp. 78–81. July 7–8.
[63] J. Valencia, V. Henríquez, A. Ramos, E.F. Cabrera, D. Echeverria, Characterization of surge phenomenon by the temperature tracking in power plants turbochargers, Electr. Eng. Electron. J. 1 (3) (May 2016) 68–72.
[64] M. Khaled, M. Ramadan, M.G. El-Rab, Application of waste heat recovery concepts to generators—thermal modeling and analysis, in: International Conference on Renewable Energies for Developing Countries 2014, IEEE, Beirut, Lebanon, 2014, pp. 82–88. Nov. 26–27.
[65] P. Breeze, Microturbines, Chapter 8, in: P. Breeze (Ed.), Gas-Turbine Power Generation, Academic Press, 2016, pp. 77–82.
[66] M. Mokhtari, G.B. Gharehpetian, S.M. Mousavi Agah, Distributed energy resources, Chapter 1, in: G.B. Gharehpetian, S.M.M. Agah (Eds.), Distributed Generation Systems: Design, Operation and Grid Integration, Butterworth-Heinemann, Oxford, UK, 2017, pp. 1–19.
[67] C. Soares, Microturbines: Applications for Distributed Energy Systems, Butterworth-Heinemann, Oxford, UK, 2007.
[68] H. Ahmadi, F. Ahmadi, Evaluation of sunshine duration and temporal-spatial distribution based on geostatistical methods in Iran, Int. J. Environ. Sci. Technol. 16 (3) (March 2019) 1589–1602.
[69] F. Fazelpour, N. Soltani, M.A. Rosen, Feasibility of satisfying electrical energy needs with hybrid systems for a medium-size hotel on Kish Island, Iran, Energy 73 (2014) 856–865.
[70] M. Baneshi, F. Hadianfard, Techno-economic feasibility of hybrid diesel/PV/wind/battery electricity generation systems for non-residential large electricity consumers under southern Iran climate conditions, Energ. Conver. Manage. 127 (2016) 233–244.
[71] S. Nadizadeh Shorabeh, M. Karimi Firozjaei, O. Nematollahi, H. Karimi Firozjaei, M. Jelokhani-Niaraki, A risk-based multi-criteria spatial decision analysis for solar power plant site selection in different climates: a case study in Iran, Renew. Energy 143 (2019) 958–973.
[72] H. Kazemi, A. Zahedi, Wind and Solar Energy Developments in Iran, The Australasian Universities Power Engineering Conference (AUPEC), Melbourne, Australia, 2002.
[73] A. Razmjoo, M. Qolipour, R. Shirmohammadi, S.M. Heibati, I. Faraji, Techno-economic evaluation of standalone hybrid solar-wind systems for small residential districts in the central desert of Iran, Environ. Prog. Sustain. Energy 36 (4) (2017) 1194–1207.
[74] E. Noorollahi, D. Fadai, M. Akbarpour Shirazi, S.H. Ghodsipour, Land suitability analysis for solar farms exploitation using GIS and fuzzy analytic hierarchy process (FAHP)—a case study of Iran, Energies 9 (8) (2016) 643.
[75] G. Najafi, B. Ghobadian, R. Mamat, T. Yusaf, W.H. Azmi, Solar energy in Iran: current state and outlook, Renew. Sustain. Energy Rev. 49 (2015) 931–942.
[76] A. Shahsavari, F.T. Yazdi, H.T. Yazdi, Potential of solar energy in Iran for carbon dioxide mitigation, Int. J. Environ. Sci. Technol. 16 (1) (2019) 507–524.
[77] M.N. Bahadori, The position of solar energy in Iran, J. Mech. Eng. 1 (1) (1992) 8–20 (in Farsi).
[78] B. Farahmandpour, Designing a Solar Power System to Supply Electricity for an Office Building in Tehran, M.Sc. Thesis, School of Mechanical Engineering, Sharif University of Technology, Tehran, Iran, Nov. 2005 (in Farsi).

[79] M.A. Mohamed, A.M. Eltamaly, H.M. Farh, A.I. Alolah, Energy management and renewable energy integration in smart grid system, in: 2015 IEEE International Conference on Smart Energy Grid Engineering (SEGE), Oshawa, ON, Canada, 2015, pp. 1–6. Aug. 17–19.
[80] M.A. Mohamed, A.A.Z. Diab, H. Rezk, Partial shading mitigation of PV systems via different meta-heuristic techniques, Renew. Energy 130 (2019) 1159–1175.
[81] A.A. Babatunde, S. Abbasoglu, M. Senol, Analysis of the impact of dust, tilt angle and orientation on performance of PV plants, Renew. Sustain. Energy Rev. 90 (2018) 1017–1026.
[82] P. Chen, B. Yan, C. Liu, S. Wang, Y. Liu, A comparative study on MPPT for photovoltaic generation systems, in: 2015 IEEE 2nd International Future Energy Electronics Conference (IFEEC), Taipei, Taiwan, Nov. 1–4, 2015, pp. 1–6.
[83] I. Guney, N. Onat, G. Kocyigit, Software design for life cycle analysis of a stand-alone PV system in Turkey, in: Proceedings of the 3rd WSEAS International Conference on Renewable Energy Sources, Canary Islands, Spain, 2009, pp. 347–352. July 1–3.
[84] N.S. Jayalakshmi, Study of hybrid photovoltaic/fuel cell system for stand-alone applications, Chapter 9, in: A. Basile, A. Cassano, A. Figoli (Eds.), Current Trends and Future Developments on (Bio-) Membranes, Renewable Energy Integrated with Membrane Operations, Elsevier, 2019, pp. 213–234.
[85] ASHRAE, Guideline 10P, Interactions Affecting the Achievement of Acceptable Indoor Environments, Second Public Review, The American Society of Heating, Refrigerating and Air-Conditioning Engineers (ASHRAE), Inc., Atlanta, GA, 2010.
[86] M. Salvador, S. Grieu, Methodology for the design of energy production and storage systems in buildings: minimization of the energy impact on the electricity grid, Energ. Buildings 47 (2012) 659–673.
[87] M.N. Bahadori, A.R. Dehghani-Sanij, in: A. Sayigh (Ed.), Wind Towers: Architecture, Climate and Sustainability, Springer International Publishing, New York, NY, 2014.
[88] S. M.R.Khani, M.N. Bahadori, A.R. Dehghani-Sanij, A. Nourbakhsh, Performance evaluation of a modular design of wind tower with wetted surfaces, Energies 10 (845) (2017) 1–22.
[89] S. M.R.Khani, M.N. Bahadori, A.R. Dehghani-Sanij, Experimental investigation of a modular wind tower in hot and dry regions, Energy Sustain. Dev. 39 (2017) 21–28.
[90] M.N. Bahadori, M. Mazidi, A.R. Dehghani, Experimental investigation of new designs of wind towers, Renew. Energy 33 (2008) 2273–2281.
[91] M. Mazidi, A.R. Dehghani, C. Aghanajafi, Wind towers' role in natural air conditioning and passive cooling of buildings in hot, arid regions, WSEAS Trans. Fluid Mech. 1 (11) (2006) 959–966.
[92] A.R. Dehghani, C. Aghanajafi, Experimental study of performance of two new wind tower designs and their comparison with traditional wind towers, Iran. J. Energy 9 (21) (2005) 14–26.
[93] N. Goudarzi, W.D. Zhu, H. Bahari, Wind energy conversion: the potential of a novel ducted turbine, for residential and commercial applications, in: Proceedings of the ASME International Mechanical Engineering Congress & Exposition IMECE, San Diego, CA, USA, 2013.
[94] A.A.M. Elzaidabi, Low Energy, Wind Catcher Assisted Indirect—Evaporative Cooling System for Building Applications, PhD Thesis, University of Nottingham, Nottingham, UK, Sep. 2008.
[95] D. Pearlmutter, E. Erell, Y. Etzion, A multi-stage down-draft evaporative cool tower for semi-enclosed spaces: experiments with a water spraying system, Sol. Energy 82 (2008) 430–440.

[96] D. Pearlmutter, E. Erell, Y. Etzion, I.A. Meir, H. Di, Refining the use of evaporation in an experimental down-draft cool tower, Energ. Buildings 23 (1996) 191–197.
[97] E. Erell, D. Pearlmutter, Y. Etzion, A multi-stage down-draft evaporative cool tower for semi-enclosed spaces: aerodynamic performance, Sol. Energy 82 (2008) 420–429.
[98] S. Soutullo, R. Olmedo, M.N. Sanchez, M.R. Heras, Thermal conditioning for urban outdoor spaces through the use of evaporative wind towers, Build. Environ. 46 (2011) 2520–2528.
[99] S. Soutullo, C. Sanjuan, M.R. Heras, Energy performance evaluation of an evaporative wind tower, Sol. Energy 86 (2012) 1396–1410.
[100] S. Soutullo, M.N. Sanchez, R. Olmedo, M.R. Heras, Theoretical model to estimate the thermal performance of an evaporative wind tower placed in an open space, Renew. Energy 36 (2011) 3023–3030.
[101] R.J. Issa, B. Chang, Performance prediction of a multi-stage wind tower for indoor cooling, J. Therm. Sci. 21 (4) (2012) 327–335.
[102] Z. Ji, Y. Su, N. Khan, Performance evaluation and energy saving potential of wind catcher natural ventilation systems in China, Int. J. Archit. Eng. Constr. 1 (2) (June 2012) 84–95.
[103] A.H. Javid, Modernizing the Traditional Ventilation Principles of Yazd to Minimize the Energy Consumption for Hot Climate, M.Sc. Thesis, Department of Architecture, University of Sheffield, Sheffield, UK, Sep. 2007.
[104] A.R. Pakzad, Analysis of Two New Designs of Wind Towers, M.Sc. Thesis, School of Mechanical Engineering, Sharif University of Technology, Tehran, Iran, Sep. 1997 (in Farsi).
[105] A.R. Dehghani, An Experimental Evaluation of Two New Designs of Wind Towers and a Comparison with a Conventional Wind Tower, B.Sc. Thesis, Yazd University, Yazd, Iran, 2001 (in Farsi).
[106] F. Tavakolinia, Wind-Chimney: Integrating the Principles of a Wind-Catcher and a Solar-Chimney to Provide Natural Ventilation, M.Sc. Thesis, California Polytechnic State University, San Luis Obispo, CA, Dec. 2011.
[107] M.N. Bahadori, A passive cooling/heating system for hot, arid regions, in: Proceedings of the American Solar Energy Society Conference, MIT, Cambridge, MA, 1988, pp. 364–367. June 20–24.
[108] L. Moosavi, M. Zandi, M. Bidi, E. Behroozizade, I. Kazemi, New design for solar chimney with integrated windcatcher for space cooling and ventilation, Build. Environ. 181 (2020) 106785.
[109] A. Koonsrisuk, S. Lorente, A. Bejan, Constructal solar chimney configuration, Int. J. Heat Mass Transf. 53 (1–3) (2010) 327–333.
[110] N. Pasumarthi, S.A. Sherif, Experimental and theoretical performance of a demonstration solar chimney model—part I: mathematical model development, Int. J. Energy Res. 22 (3) (1998) 277–288.
[111] N. Pasumarthi, S.A. Sherif, Experimental and theoretical performance of a demonstration solar chimney model—part II: experimental and theoretical results and economic analysis, Int. J. Energy Res. 22 (5) (1998) 443–461.
[112] A.S. Bachrun, C. Vidiyanti, L.H. Ismail, I.A. Wahab, House's solar chimney a numerical analysis on the thermal performance in Jakarta, Dermatol. Sin. 24 (3) (2020) 245–252.

CHAPTER 9

Pictures of traditional ice-houses

In this chapter, the selected pictures show traditional ice-houses (known as "Yakhchals" or "Yakhdans") located in different villages or cities of Iran, and also in Merv, Turkmenistan. These pictures, along with the time period of some of them, are as follows:

- Iran

  Kerman Province

  Fig. 9.1. Moayedi ice-house, Kerman City (late Safavid period).

  Fig. 9.2. Zarisef ice-house from two different views, before and after repair, Kerman City (Qajar era).

  Fig. 9.3. Rig Abad ice-house, Kerman City (Qajar era).

  Fig. 9.4. Haji Agha Ali ice-house, located in Abbas Abad Haji village, Rafsanjan City (Qajar era).

  Fig. 9.5. Kabutarkhan ice-house from two different views, located in Kabutarkhan village, Rafsanjan City (Qajar era).

  Fig. 9.6. A view of an ice-house in Ravar City (late Qajar period).

  Fig. 9.7. Haj Rashid ice-houses (or "twin Yakhdans" in local language) in Sirjan City (Qajar era).

  Fig. 9.8. Another view of the Haj Rashid ice-houses ("twin Yakhdans") in Sirjan City.

  Fig. 9.9. Arg-e Bam ice-house, before an earthquake in 2003, in Bam City (Safavid period).

  Fig. 9.10. A view of the Langar ice-house in Langar village, Mahan City (Qajar era).

  Fig. 9.11. A view of an ice-house in Anar City (Qajar era).

  Yazd Province

  Fig. 9.12. A view of an ice-house's shadowing adobe walls and shallow pond in Yazd City.

  Fig. 9.13. A view of a shadowing adobe wall and shallow pond of another ice-house in Yazd City.

  Fig. 9.14. An old picture of an ice-house located near Yazd City.

Ice-Houses
https://doi.org/10.1016/B978-0-12-822275-1.00006-0
Copyright © 2021 Elsevier Inc.
All rights reserved.

Fig. 9.15. An old picture of an ice-house beside Bagh-e Khan mansion, near Yazd City.
Fig. 9.16. A view of the Aghazadeh ice-house's dome in Abarkouh City (Qajar era).
Fig. 9.17. A top view of the Aghazadeh ice-house in Abarkouh City.
Fig. 9.18. A view of the Barzan ice-house in Abarkouh City (Qajar era).
Fig. 9.19. A top view of the Barzan ice-house in Abarkouh City.
Fig. 9.20. An internal view of an ice-house's dome in Abarkouh City.
Fig. 9.21. A view of the Meybod ice-house, Meybod City (Safavid era).
Fig. 9.22. A view of the Meybod ice-house's dome, Meybod City.
Fig. 9.23. An old picture of an ice-house in Yakhdan Neiborhood, Meybod City.

Isfahan Province
Fig. 9.24. A view of Jalali ice-houses in Kashan City (Safavid era).
Fig. 9.25. Another view of one of the Jalali ice-houses in Kashan City.
Fig. 9.26. An internal view of one of the Jalali ice-houses in Kashan City.
Fig. 9.27. A view of an ice-house located in Josheghan, Kashan City (late Safavid period).
Fig. 9.28. An ice-house in Zavareh district, Ardestan City (late Safavid period).
Fig. 9.29. An ice-house in Mahabad village, Ardestan City.
Fig. 9.30. A view of the much-deteriorated ice-house's dome in Mahabad village, Ardestan City.
Fig. 9.31. Bafran ice-house from two different views, located in Bafran village, Naein City (Safavid era).
Fig. 9.32. A view of the Mohammadieh ice-house, before and after repair, in Naein City.

Razavi Khorasan Province
Fig. 9.33. A view of the ice-houses, known as "the triplets," in Sabzevar City (Qajar era).
Fig. 9.34. A view of two of three ice-houses in Sabzevar City.
Fig. 9.35. An old picture (pre-restoration) of several ice-houses in Sabzevar City.
Fig. 9.36. A view of an ice-house in Kusar Bidokht village, Gonabad City (late Qajar period).
Fig. 9.37. Another view of the ice-house in Kusar Bidokht village, Gonabad City.

Fig. 9.38. An ice-house in Khaf City (Qajar era).
Fig. 9.39. A view of the Namadmal ice-house in Quchan City (late Qajar period).
Fig. 9.40. An internal view of the Namadmal ice-house's dome in Quchan City.

South Khorasan Province
Fig. 9.41. A view of an ice-house in Tabas City.
Fig. 9.42. Another view of the ice-house in Tabas City.
Fig. 9.43. An internal view of the ice-house's dome in Tabas City.
Fig. 9.44. A view of an ice-house in Bahlgerd village, Birjand City (Qajar era).
Fig. 9.45. Another view of the ice-house in Bahlgerd village, Birjand City.
Fig. 9.46. A view of an ice-house in Shokat Abad village, Birjand City (Qajar era).
Fig. 9.47. Twin ice-houses from two different views, located in Kalateh Nayeb village, Birjand City (Qajar era).

Qom Province
Fig. 9.48. Taghrud ice-house's shadowing adobe walls and underground ice reservoir in Jafarabad village, Qom City (Safavid era).

Tehran Province
Fig. 9.49. A view of an ice-house's shadowing adobe walls and shallow pond in the south-east of Tehran City.
Fig. 9.50. A view of the Tarshanbeh ice-house in Islamshahr City (Safavid era).
Fig. 9.51. An overview of the Tarshanbeh ice-house in Islamshahr City.
Fig. 9.52. An ice-house from two different views, located in Koleyn village, Shahr-e-Rey City (Safavid era).

Semnan Province
Fig. 9.53. A view of an ice-house in Shahroud City, Semnan.
Fig. 9.54. An ice-house in Mazaj village, Shahroud City, Semnan (Qajar era).
Fig. 9.55. A view of an ice-house in Deh-Namak, Aradan City (Safavid era).
Fig. 9.56. A view of an ice-house in Garmsar City.
Fig. 9.57. A view of another ice-house in Garmsar City.

West Azerbaijan Province
Fig. 9.58. An internal view of the Noh Pelleh ice-house (or ice chamber) in Urmia City (Qajar era).
Fig. 9.59. A view of the Alagh Beyk ice-house in Takab City (Qajar era).

East Azerbaijan Province
Fig. 9.60. An internal view of the Sadeghieh ice-house (during repair) in Tabriz City (Qajar era).
Fig. 9.61. Internal and external views of the Moradian ice-house in Maragheh City (Qajar era).

Markazi Province
Fig. 9.62. A view of the Mahdi Abad Mamounieh ice-house, Zarandieh City (Qajar era).
Fig. 9.63. A view of the Nimehvar ice-house, Mahalat City (Qajar era).

Hamadan Province
Fig. 9.64. Mirfattah ice-house from two different views, located in Malayer City (Qajar era).

Fars Province
Fig. 9.65. Haj Mohammad Rahim ice-house from two different views, Estahban City (Qajar era).

- Turkmenistan
  Merv[a]
  Fig. 9.66. A view of an ice-house located south of Sultan Kala.
  Fig. 9.67. Internal and external views of an ice-house.

[a] Three ice-houses were built in Merv (a major oasis-city on the "Silk Road"), estimated to be of the Timurid period (1370–1507) [14]. Due to the existence of only three ice-houses in the city of Merv and the absence of this structure in other Central Asia cities, as well as the similarity of these ice-houses to old Iranian ice-houses (especially in eastern parts of Iran such as the cities of Sabzevar, Shahroud, and Garmsar), it seems most likely that they were inspired by Persian ice-houses and that their estimated construction date is not correct.

9.1 Selected pictures of traditional ice-houses

Fig. 9.1 Moayedi ice-house, Kerman City, Kerman, Iran.

Fig. 9.2 Zarisef ice-house from two different views, before and after repair, Kerman City, Kerman, Iran [1]. Local people called this ice-house "Yakhdan-e Ganj."

Fig. 9.3 Rig Abad ice-house, Kerman City, Kerman, Iran [1].

Fig. 9.4 Haji Agha Ali ice-house, located in Abbas Abad Haji village, Rafsanjan City, Kerman, Iran.

Fig. 9.5 Kabutarkhan ice-house from two different views, located in Kabutarkhan village, Rafsanjan City, Kerman, Iran.

Fig. 9.6 A view of an ice-house in Ravar City, Kerman, Iran.

Fig. 9.7 Haj Rashid ice-houses (or "twin Yakhdans" in local language) in Sirjan City, Kerman, Iran.

Fig. 9.8 Another view of the Haj Rashid ice-houses ("twin Yakhdans") in Sirjan City, Kerman, Iran.

Fig. 9.9 Arg-e Bam ice-house, before an earthquake in 2003, in Bam City, Kerman, Iran.

Fig. 9.10 A view of Langar ice-house in Langar village, Mahan City, Kerman, Iran.

Fig. 9.11 A view of an ice-house in Anar City, Kerman, Iran.

Fig. 9.12 A view of an ice-house's shadowing adobe walls and shallow pond in Yazd City, Yazd, Iran.

Fig. 9.13 A view of a shadowing adobe wall and shallow pond of another ice-house in Yazd City, Yazd, Iran.

Fig. 9.14 An old picture of an ice-house located near Yazd City, Yazd, Iran [2]. This ice-house has since been completely destroyed.

Fig. 9.15 An old picture of an ice-house beside Bagh-e Khan mansion, near Yazd City, Yazd, Iran [2]. This ice-house has since been completely destroyed.

Fig. 9.16 A view of the Aghazadeh ice-house's dome in Abarkouh City, Yazd, Iran.

Fig. 9.17 A top view of Aghazadeh ice-house in Abarkouh City, Yazd, Iran [3, 4].

Fig. 9.18 A view of Barzan ice-house in Abarkouh City, Yazd, Iran.

Fig. 9.19 A top view of Barzan ice-house in Abarkouh City, Yazd, Iran.

Fig. 9.20 An internal view of an ice-house's dome in Abarkouh City, Yazd, Iran [5].

Fig. 9.21 A view of Meybod ice-house, Meybod City, Yazd, Iran.

Fig. 9.22 A view of the Meybod ice-house's dome, Meybod City, Yazd, Iran.

Fig. 9.23 An old picture of an ice-house in Yakhdan Neiborhood, Meybod City, Yazd, Iran [6]. This ice-house has since been completely destroyed.

Fig. 9.24 A view of Jalali ice-houses in Kashan City, Isfahan, Iran.

Fig. 9.25 Another view of one of the Jalali ice-houses in Kashan City, Isfahan, Iran.

Fig. 9.26 An internal view of one of the Jalali ice-houses in Kashan City, Isfahan, Iran.

Fig. 9.27 A view of an ice-house located in Josheghan, Kashan City, Isfahan, Iran.

Fig. 9.28 An ice-house in Zavareh district, Ardestan City, Isfahan, Iran.

Fig. 9.29 An ice-house in Mahabad village, Ardestan City, Isfahan, Iran.

Fig. 9.30 A view of the much-deteriorated ice-house's dome in Mahabad village, Ardestan City, Isfahan, Iran.

Fig. 9.31 Bafran ice-house from two different views, located in Bafran village, Naein City, Isfahan, Iran [7].

Fig. 9.32 A view of the Mohammadieh ice-house, before and after repair, in Naein City, Isfahan, Iran.

Fig. 9.33 A view of the ice-houses, known as "the triplets," in Sabzevar City, Razavi Khorasan, Iran.

Fig. 9.34 A view of two of three ice-houses in Sabzevar City, Razavi Khorasan, Iran.

Fig. 9.35 An old picture (pre-restoration) of several ice-houses in Sabzevar City, Razavi Khorasan, Iran.

Fig. 9.36 A view of an ice-house in Kusar Bidokht village, Gonabad City, Razavi Khorasan, Iran.

Fig. 9.37 Another view of the ice-house in Kusar Bidokht village, Gonabad City, Razavi Khorasan, Iran [8].

Fig. 9.38 An ice-house in Khaf City, Razavi Khorasan, Iran.

Fig. 9.39 A view of the Namadmal ice-house in Quchan City, Razavi Khorasan, Iran.

Fig. 9.40 An internal view of the Namadmal ice-house's dome in Quchan City, Razavi Khorasan, Iran.

Fig. 9.41 A view of an ice-house in Tabas City, South Khorasan, Iran [9].

Fig. 9.42 Another view of the ice-house in Tabas City, South Khorasan, Iran [9].

Fig. 9.43 An internal view of the ice-house's dome in Tabas City, South Khorasan, Iran [9].

Fig. 9.44 A view of an ice-house in Bahlgerd village, Birjand City, South Khorasan, Iran.

Fig. 9.45 Another view of the ice-house in Bahlgerd village, Birjand City, South Khorasan, Iran.

Fig. 9.46 A view of an ice-house in Shokat Abad village, Birjand City, South Khorasan, Iran.

Fig. 9.47 Twin ice-houses from two different views, located in Kalateh Nayeb village, Birjand City, South Khorasan, Iran [10].

Fig. 9.48 Taghrud ice-house's shadowing adobe walls and underground ice reservoir in Jafarabad village, Qom City, Qom, Iran.

Fig. 9.49 A view of an ice-house's shadowing adobe walls and shallow pond in the south-east of Tehran City, Tehran, Iran.

Fig. 9.50 A view of the Tarshanbeh ice-house in Islamshahr City, Tehran, Iran.

Fig. 9.51 An overview of the Tarshanbeh ice-house in Islamshahr City, Tehran, Iran.

Fig. 9.52 An ice-house from two different views in Koleyn village, Shahr-e-Rey City, Tehran, Iran.

Fig. 9.53 A view of an ice-house in Shahroud City, Semnan, Iran.

Fig. 9.54 An ice-house in Mazaj village, Shahroud City, Semnan, Iran.

Fig. 9.55 A view of an ice-house in Deh-Namak, Aradan City, Semnan, Iran.

Fig. 9.56 A view of an ice-house in Garmsar City, Semnan, Iran.

Fig. 9.57 A view of another ice-house in Garmsar City, Semnan, Iran.

Fig. 9.58 An internal view of the Noh Pelleh ice-house (or ice chamber) in Urmia City, West Azerbaijan, Iran.

Fig. 9.59 A view of the Alagh Beyk ice-house in Takab City, West Azerbaijan, Iran.

Fig. 9.60 An internal view of the Sadeghieh ice-house (during repair) in Tabriz City, East Azerbaijan, Iran.

Fig. 9.61 Internal and external views of the Moradian ice-house in Maragheh City, East Azerbaijan, Iran.

Fig. 9.62 A view of the Mahdi Abad Mamounieh ice-house, Zarandieh City, Markazi, Iran.

Fig. 9.63 A view of the Nimehvar ice-house, Mahalat City, Markazi, Iran [11].

Fig. 9.64 Mirfattah ice-house from two different views, Malayer City, Hamadan, Iran [12].

Fig. 9.65 Haj Mohammad Rahim ice-house from two difference views, Estahban City, Fars, Iran [13].

Fig. 9.66 A view of an ice-house located south of Sultan Kala, Merv, Turkmenistan [15].

Fig. 9.67 Internal and external views of an ice-house in Merv, Turkmenistan [16, 17].

9.2 Conclusion

The pictures shown in this chapter confirm that the Iranian architects and/or engineers who built traditional ice-houses—in different forms, styles, and sizes and of various materials—were highly knowledgeable about the principles of heat transfer, thermodynamics, refrigeration, statics, the strengths of materials, aerodynamics, etc., and that they correctly applied these principles in their designs and structures according to the geographic and climatic conditions involved. The fact that some of these structures have lasted for more than two or three centuries, despite little upkeep, is a testament to the quality of the art of the builders.

References

[1] J. Ghazbanpour, New Life—Old Structure: A Selection of Valuable Historical Structures, second ed., Ministry of Roads and Urban Development in collaboration with Cultural Heritage, Handicrafts and Tourism Organization of Iran, Tehran, Iran, Aug. 1995 (in Farsi).

[2] E. Beazley, M. Harverson, Living with the Desert: Working Buildings on the Iranian Plateau, Orchid Press Publishing Ltd., Bangkok, Thailand, 2006.

[3] S. Golriz Khatami, A. Akrami, The ice-house in entrance of Abarkouh City, Tourism Center of Abarkouh Azad University, 22 May, 2010. http://tourismabarkouh.blogfa.com/cat-8.aspx. (Accessed 21 July 2020).

[4] A. Ghasemi, Abarkouh mud-brick ice-house, Tishineh, 2020. https://www.tishineh.com/post/2431. (Accessed 30 September 2020).

[5] P. Javaheri, M. Javaheri, Challenge for Water in Ancient Fars, vol. 2, Iranian National Committee on Irrigation & Drainage (IRNCID) and Water National Museum of Iran (WNMI), Tehran, Iran, 2000 (in Farsi).

[6] I. Afshar, Yadgarha-ye Yazd (Remembrances of Yazd), vol. 2, Association of Cultural Works and Honors, Tehran, Iran, 1969 (in Farsi).

[7] M. Soltani Mohammadi, M. Boluri Bonab, Study of physical structure of ice-houses (Yakhdans) of Naein City, Asar J. 73 (Summer 2016) 43–58 (in Farsi).

[8] R.A. Labbaf Khaniki, Gonabad: The Origin of the Hidden Epics, Cultural Heritage, Handicrafts and Tourism Organization of Iran, Tehran, Iran, 2014 (in Farsi).

[9] Y. Daneshdoost, Tabas, the City that Was, Soroush Publications, Tehran, Iran, 1990 (in Farsi).

[10] H. Jorgensen, Ice Houses of Iran: Where, How, Why, Mazda Publishers, Costa Mesa, CA, 2012.

[11] M. Mollahosseini, Acquaintance with old ice-house of Nimehvar–Markazi, Hamshahri online, 22 Oct., 2012. http://www.hamshahrionline.ir/details/189516. (Accessed 21 July 2020).

[12] M. Mollahosseini, Acquaintance with old ice-house of Mirfattah–Hamadan, in: Hamshahri online, 22 Oct., 2012. http://www.hamshahrionline.ir/details/189970. (Accessed 21 July 2020).

[13] Anjirestan, Estahban Ice-house, 21 April, 2015. https://www.anjirestan.ir/. (Accessed 27 September 2020).

[14] P. Brummell, Turkmenistan: The Bradt Travel Guide, Bradt Travel Guides Ltd., Bucks, UK, 2005.

[15] Photo of Icehouse 1, Ancient Merv in Mary, Turkmenistan, Flickr, 28 Aug 2009. http://www.flickr.com/photos/cyark/3866399208/lightbox/. (Accessed 21 July 2020).
[16] P. Partensky, Ice House, Merv, Flickr, 4 Sep. 2009. https://commons.wikimedia.org/wiki/File:Ice_House_Merv_(3891851927).jpg. (Accessed 21 July 2020).
[17] P. Partensky, Ice House, Merv, Flickr, 4 Sep. 2009. https://www.flickr.com/photos/ifl/3891851361/in/photostream. (Accessed 21 July 2020).

Nomenclature

A	surface area, m^2
C	specific heat capacity, kJ/kg K
D_{ab}	diffusivity of air-water vapor, m^2/s
$e_s(T)$	saturated vapor pressure, Pa
F	shape factor, –
F_r	coolness recovery factor, –
G_{bh}	intensity of the solar radiation on a horizontal surface, W/m^2
G_{bn}	intensity of the solar radiation on a surface normal to the direction of the sun, W/m^2
G_{bt}	intensity of the solar radiation on an inclined surface, W/m^2
G_{sc}	solar constant, W/m^2
h	enthalpy, kJ/kg, and height, m
h_c	convective heat transfer coefficient, W/m^2K
h_{if}	enthalpy of the solidification of water, kJ/kg
I_{bh}	hourly direct solar radiation on the horizontal surface, W/m^2
I_{dh}	hourly diffuse solar radiation on the horizontal surface, W/m^2
I_h	hourly global solar radiation on the horizontal surface, W/m^2
I_N	direct normal solar radiation on horizontal surfaces, W/m^2
I_{oh}	hourly extraterrestrial radiation, W/m^2
I_T	total radiation on a tilted surface, W/m^2
K_T	daily clearness index, –
k	thermal conductivity, W/m K
k_T	hourly clearness index, –
L	characteristic length, m
L_{loc}	local longitude at the position of interest, °
L_{std}	local longitude of the standard time meridian, °
l_v	latent heat of vaporization, J/kg
m	mass, kg
$\dot{m}$	mass flow rate, kg/s
N	length of the day, h
Nu	Nusselt number, –
n	number of the day in the year, –
P	pressure, Pa
Pr	Prandtl number, –
P_{vs}	saturation pressure, Pa
Q	heat transfer, J
Q_c	cooling load, W
Q_i	total coolness stored in the ice mass, J
Q_L	annual cooling demand, J
$\dot{Q}$	heat transfer rate, W
$\dot{Q}_{co}$	heat flow at the cold end, W
$\dot{Q}_H$	heat transferred to a source at higher temperature, W
$\dot{Q}_{ho}$	heat flow at the hot end, W

$\dot{Q}_L$	heat absorbed from a source at low temperature, W
$\dot{q}$	cooling capacity, kW
R	gas constant, J/kg K
Re	Reynolds number, –
r	adiabatic coefficient, –
S	thickness, m
s	entropy, kJ/kg K
$\dot{S}_i$	rate of ice production, mm/h
SC	storage capacity, J
Sc	Schmidt number, –
T	temperature, °C
T_H	high junction temperature, °C
T_L	low junction temperature, °C
T_{wi}	initial temperature, °C
t	time, s
U	wind velocity, m/s
u	internal energy, kJ/kg
V	volume, m^3
$\dot{V}$	volumetric flow rate, m^3/s
$\dot{W}$	work required by the considered system, W

Greek and Latin Symbols

α	thermal diffusivity, m^2/s, and solar altitude angle, °
β	tilt angle, °, and coefficient of performance, –
γ	surface azimuthal angle, °
δ	declination angle, °
ε	emissivity, –
$\in$	ratio of the molecular weight of water vapor/dry air, –
η	efficiency, –
θ	solar incidence angle, °
θ_z	solar zenith angle, °
μ	dynamic viscosity, Pa s
μ_J	Joule–Thomson coefficient, K/Pa
υ	kinematic viscosity, m^2/s
ξ	recovery efficiency, –
ρ	density, kg/m^3
ρ_g	reflection coefficient of the ground, –
σ	Stefan-Boltzmann constant, $W/m^2 K^4$
ϕ	relative humidity, %
φ	latitude of location, °
ω	solar hour angle, °
ω_s	hour angle, °

Subscripts

a	air
c	convection
dp	dew point
e	evaporation
f	final
g	ground
hp	heat pump
i	ice
k	conduction
m	midnight
mis	miscellaneous
n	minimum
p	pressure
r	radiation
ref	refrigeration
s	solar radiation
sky	sky
w	water
x	maximum

Abbreviations

AC	air-conditioning; or alternating current
AD	anno domini
AMR	active magnetic regenerative
API	American Petroleum Institute
ASHRAE	American Society of Heating, Refrigerating and Air-Conditioning Engineers
ASRE	American Society of Refrigerating Engineers
AST	apparent solar time
BC	before christ
BES	battery energy storage
BOE	barrel of oil equivalent
BTU	British thermal unit
CAES	compressed air energy storage
CESSs	chemical energy storage systems
CFCs	chlorofluorocarbons
COP	coefficient of performance
CSP	concentrating solar power
CTES	cold thermal energy storage
DC	direct current
EESSs	electrochemical energy storage systems
EESSs*	electrical energy storage systems
EoT	equation of time
EPA	Environmental Protection Agency
ESSs	energy storage systems
EVs	electric vehicles
EU	European Union
FES	flywheel energy storage
GAN	gaseous nitrogen
GDP	gross domestic product
GHG	greenhouse gas
GJ	gigajoule
GMT	Greenwich mean time
GW	gigawatt
HCFCs	hydrochlorofluorocarbons
HDI	human development index
HRE	hybrid renewable energy

HVAC	heating, ventilation, and air-conditioning
IEA	International Energy Agency
IEEE	Institute of Electrical and Electronics Engineers
IEQ	indoor environmental quality
IIR	International Institute of Refrigeration
ICE	internal combustion engine
KSA	Kingdom of Saudi Arabia
Li-ion	lithium-ion
LST	local standard time
MCE	magneto-caloric effect
MESSs	mechanical energy storage systems
MHDI	Municipal Human Development Index
MJ	megajoule
Na-S	sodium-sulfur
Ni-Cd	nickel-cadmium
NREAP	National Renewable Energy Action Plan
NZEB	net zero energy building
OEB	Ontario Energy Board
OECD	Organization for Economic Cooperation and Development
PCMs	phase change materials
PHES	pumped hydro energy storage
Pb-A	lead-acid
PV	photovoltaic
SDGs	sustainable development goals
SI	International System of Units
SMES	superconducting magnetic energy storage
SSSF	steady-state steady-flow
SRWE	statistical review of world energy
SWOT	strengths, weaknesses, opportunities, and threats
RTP	real time pricing
TES	thermal energy storage
TOE (or toe)	ton of oil equivalent
TOU	time of use
UAE	United Arab Emirates
UK	United Kingdom
UN	United Nations
UNESCO	United Nations Educational, Scientific and Cultural Organization
USA	United States of America
WCED	World Commission on Environment and Development

Index

Note: Page numbers followed by *f* indicate figures, *t* indicate tables, and *np* indicate footnotes.

M

N

P

Q

R

U

V

W

Y

Z

FSC
www.fsc.org
MIX
Papier | Fördert gute Waldnutzung
FSC® C083411

Kolibri 360 GmbH
Ferdinand-Jühlke-Straße 7
99095 Erfurt, Deutschland
produktsicherheit@kolibri360.de